Lecture Notes in Physics

New Series m: Monographs

The Editorial Policy for Monographs

The series Lecture Notes in Physics reports new developments in physical research and teaching - quickly, informally, and at a high level. The type of material considered for publication in the New Series m includes monographs presenting original research or new angles in a classical field. The timeliness of a manuscript is more important than its form, which may be preliminary or tentative. Manuscripts should be reasonably self-contained. They will often present not only results of the author(s) but also related work by other people and will provide sufficient motivation, examples, and applications.
The manuscripts or a detailed description thereof should be submitted either to one of the series editors or to the managing editor. The proposal is then carefully refereed. A final decision concerning publication can often only be made on the basis of the complete manuscript, but otherwise the editors will try to make a preliminary decision as definite as they can on the basis of the available information.
Manuscripts should be no less than 100 and preferably no more than 400 pages in length. Final manuscripts should preferably be in English, or possibly in French or German. They should include a table of contents and an informative introduction accessible also to readers not particularly familiar with the topic treated. Authors are free to use the material in other publications. However, if extensive use is made elsewhere, the publisher should be informed. Authors receive jointly 50 complimentary copies of their book. They are entitled to purchase further copies of their book at a reduced rate. As a rule no reprints of individual contributions can be supplied. No royalty is paid on Lecture Notes in Physics volumes. Commitment to publish is made by letter of interest rather than by signing a formal contract. Springer-Verlag secures the copyright for each volume.

The Production Process

The books are hardbound, and quality paper appropriate to the needs of the author(s) is used. Publication time is about ten weeks. More than twenty years of experience guarantee authors the best possible service. To reach the goal of rapid publication at a low price the technique of photographic reproduction from a camera-ready manuscript was chosen. This process shifts the main responsibility for the technical quality considerably from the publisher to the author. We therefore urge all authors to observe very carefully our guidelines for the preparation of camera-ready manuscripts, which we will supply on request. This applies especially to the quality of figures and halftones submitted for publication. Figures should be submitted as originals or glossy prints, as very often Xerox copies are not suitable for reproduction. In addition, it might be useful to look at some of the volumes already published or, especially if some atypical text is planned, to write to the Physics Editorial Department of Springer-Verlag direct. This avoids mistakes and time-consuming correspondence during the production period.
As a special service, we offer free of charge LaTeX and TeX macro packages to format the text according to Springer-Verlag's quality requirements. We strongly recommend authors to make use of this offer, as the result will be a book of considerably improved technical quality. The typescript will be reduced in size (75% of the original). Therefore, for example, any writing within figures should not be smaller than 2.5 mm.
Manuscripts not meeting the technical standard of the series will have to be returned for improvement.
For further information please contact Springer-Verlag, Physics Editorial Department V, Tiergartenstrasse 17, D-69121 Heidelberg, FRG.

Giampiero Esposito

Quantum Gravity, Quantum Cosmology and Lorentzian Geometries

Second Corrected and Enlarged Edition

Springer-Verlag Berlin Heidelberg GmbH

Author

Giampiero Esposito
International Centre for Theoretical Physics, P. O. Box 586
Strada Costiera 11, I-34014 Trieste, Italy

Scuola Internazionale Superiore di Studi Avanzati
Via Beirut 2-4, I-34013 Trieste, Italy

Istituto Nazionale di Fisica Nucleare
Gruppo IV, Sezione di Napoli
Mostra d'Oltremare Padiglione 20, I-80125 Napoli, Italy

Dipartimento di Scienze Fisiche
Mostra d'Oltremare Padiglione 19, I-80125 Napoli, Italy

Dipartimento di Fisica Teorica
Università degli Studi di Salerno
I-84081 Baronissi, Italy

ISBN 978-3-662-14933-1 ISBN 978-3-540-47295-7 (eBook)
DOI 10.1007/978-3-540-47295-7

Originally published by Springer-Verlag Berlin Heidelberg New York in 1994
Softcover reprint of the hardcover 2nd edition 1994

2158/3140-543210 - Printed on acid-free paper

a Maria Gabriella

PREFACE

This book is aimed at theoretical and mathematical physicists and mathematicians interested in modern gravitational physics. I have thus tried to use language familiar to readers working on classical and quantum gravity, paying attention both to difficult calculations and to existence theorems, and discussing in detail the current literature.

The first aim of the book is to describe recent work on the problem of boundary conditions in one-loop quantum cosmology. The motivation of this research was to understand whether supersymmetric theories are one-loop finite in the presence of boundaries, with application to the boundary-value problems occurring in quantum cosmology. Indeed, higher-loop calculations in the absence of boundaries are already available in the literature, showing that supergravity is not finite. I believe, however, that one-loop calculations in the presence of boundaries are more fundamental, in that they provide a more direct check of the inconsistency of supersymmetric quantum cosmology from the perturbative point of view. It therefore appears that higher-order calculations are not strictly needed, if the one-loop test already yields negative results. Even though the question is not yet settled, this research has led to many interesting, new applications of areas of theoretical and mathematical physics such as twistor theory in flat space, self-adjointness theory, the generalized Riemann zeta-function, and the theory of boundary counterterms in supergravity. I have also compared in detail my work with results by other authors, explaining, whenever possible, the origin of different results, the limits of my work and the unsolved problems.

The second aim of the book is to present a recent study of the singularity problem for space-times with torsion. Indeed, the singularity problem in cosmology and theories of gravity with torsion play a fundamental role in motivating quantum cosmology and supergravity respectively. The reader can thus find the nonperturbative continuation of the first two parts of my work within the framework of classical theories of gravitation. It was my intention to write a treatise not too specialized in a single topic, but dealing with some fundamental problems of both classical and quantum gravity. I hope this will stimulate further interaction between these two branches of theoretical and mathematical physics. In fact, research workers interested in classical gravity are not always aware of the conceptual and technical problems of quantum gravity, whereas those working on quantum cosmology do not frequently use the elegant, powerful and rigorous global techniques of general relativity.

The book is based on my J.T. Knight Prize Essay and Ph.D. thesis at Cambridge University, a paper published in Fortschritte der Physik, a series of papers written with Dr. Peter D'Eath, other papers of mine on classical gravity and my post-doctoral work in Napoli and in Trieste. The chapters here contain more details than the papers, and the presentation of the arguments is different. Much work appearing in this book, however, has not been previously published. Where appropriate, sections on background material appear because I tried to write the book in self-contained form. I have chosen to present my results in the order they were derived. In many cases, problems have been initially formulated in the simplest possible way, and finally presented and solved at increasing levels of complexity. Quantum cosmology is not an isolated field of research but lies at the very heart of fundamental theoretical physics. I hope the reader will appreciate this after reading the book.

I am especially indebted to Dr. Peter D'Eath for encouraging, correcting and supervising my work on quantum cosmology over many years, to Professors W. Beiglböck, J. Ehlers and J. Wess of the Springer-Verlag Editorial Board for several suggestions which led to a substantial improvement of the original manuscript, and to Professor Stephen Hawking for inspiring all my work on classical and quantum cosmology.

Special thanks are also due to Professors John Beem, James Hartle, Friedrich Hehl, Chris Isham, Giuseppe Marmo, Renato Musto, Cesare Reina, Abdus Salam and Dennis Sciama; Drs. Stewart Dowker, Gary Gibbons, Domenico Giulini, Jonathan Halliwell, David Hughes, Bernard Kay, Gerard Kennedy, Jorma Louko, Hugh Luckock, Ian Moss, Richard Pinch, Stephen Poletti and Kristin Schleich; graduate students Giuseppe Bimonte, Eduardo Ciardiello, Paola Diener and Hugo Morales Técotl, and undergraduate student Gabriele Gionti, for enlightening conversations. Last but not least, I gratefully acknowledge lots of help from Professors Ruggiero de Ritis and Giovanni Platania, and Drs. Paolo

Scudellaro, John Stewart and Cosimo Stornaiolo in understanding theories with torsion and related literature, and I thank Drs. Marcos Bordin, Mark Manning and Paolo Lo Re for solving computer problems while I typed the manuscript.

Over the past five years my research has been financially supported by St. John's College, the Istituto Nazionale di Fisica Nucleare and the International Atomic Energy Agency.

Giampiero Esposito

Trieste

June 1992

PREFACE TO THE SECOND EDITION

Since there has been a favourable response to the first edition of **Quantum Gravity, Quantum Cosmology and Lorentzian Geometries**, it has been decided to publish a second edition. The main changes are as follows.

In section 2.3, I have corrected the statements following Eq. (2.3.22). Thus, Eqs. (2.3.23)-(2.3.33) do not appear, since that review part is no longer necessary. In section 3.3, I have corrected the misprints occurring in Eqs. (3.3.6) and (3.3.14)-(3.3.15). In section 6.5, I have modified or omitted a few lines, bearing in mind the addendum to chapter six inserted thereafter. In section 7.2, I have chosen a convention I find more appropriate to the real Riemannian case studied therein. This leads to some sign changes in Eqs. (7.2.18), (7.2.20), (7.2.22)-(7.2.25), (7.2.32). New and important references have been added.

The addendum to chapter six has been included since it describes unpublished work on a longstanding problem in quantum field theory, i.e. the mode-by-mode analysis of the ghost-field contribution to the Faddeev-Popov one-loop amplitudes of Euclidean Maxwell theory in the presence of a three-sphere boundary. This detailed analysis may lead to a better understanding of the quantization of gauge fields and gravitation in the presence of boundaries. Since the addendum studies a research topic, it is likely to become superseded in due course. I very much hope that graduate students and research workers will find it helpful in their attempt to solve the problem.

Notation and style have also been improved, and I have tried to clarify some statements appearing in the review parts of the first edition. When a difficult scientific problem has (or has not) been completely solved, what survives is our effort to understand the physical world. I think the elliptic boundary-value problems studied in Part II of this monograph provide a relevant example of this fascinating property of theoretical physics.

I am indebted to Professor W. Beiglböck for suggesting that a second edition should be published. I am also very grateful to Bruce Allen and Andrei Barvinsky for enlightening conversations on gauge-averaging functionals in quantum field theory, and I enjoyed correspondence with Alexander Kamenshchik on one-loop quantum cosmology. This second edition has been prepared while I was a post-doctoral research assistant at the Department of Theoretical Physics of the University of Salerno.

Giampiero Esposito

Napoli and Salerno
September 1993

HOW TO READ THIS BOOK

Readers interested in a general overview of classical and quantum gravity should study the first three chapters, all sections from 6.1 to 6.4, all sections from 10.1 to 10.6. They can thus find the detailed description of the motivations for studying quantum gravity and quantum cosmology, with achievements and unsolved problems; Dirac's theory of constrained Hamiltonian systems and its application to the quantization of general relativity; the background-field method and the one-loop approximation in perturbative quantum gravity; the Batalin-Fradkin-Vilkovisky and Faddeev-Popov methods of quantizing gauge theories; the mathematical foundations of classical general relativity, i.e. Lorentzian geometry, spinor structure, causal structure, asymptotic structure and Hamiltonian structure of space-time.

Note that the reader may well study Part III before Part I, or at least before chapter 2. However, I found it more satisfactory to write a continuous sequence of chapters on quantum gravity before Part III. The first three chapters enable one to become familiar with the basic tools of canonical and perturbative quantum gravity (nonexperts may limit themselves to these chapters and to Part III). It is then possible to understand the one-loop calculations presented from chapter four to chapter nine. The asymptotic heat kernel for manifolds with a boundary is first studied in the case of the Dirac operator subject to global boundary conditions. This is the completion of previous work by theoretical physicists on the role of fermionic fields in quantum cosmology, and is motivated by the mathematical study of spectral asymmetry and Riemannian geometry.

However, only local boundary conditions respect supersymmetry. One possible set of supersymmetric local boundary conditions involves field strengths for spins 1, $\frac{3}{2}$ and 2, the undifferentiated spin-$\frac{1}{2}$ field, and a mixture of Dirichlet and Neumann conditions for scalar fields. The corresponding one-loop properties, and the relation of these boundary conditions to twistor theory in flat space, are derived in chapters five, seven and eight. A detailed proof of self-adjointness of the boundary-value problem for the Dirac operator with these local boundary conditions is also given in chapter five.

An alternative set of boundary conditions can be motivated by studying transformation properties under local supersymmetry, as in chapter nine; these are in general mixed, and involve in particular Dirichlet conditions for the perturbed three-metric of pure gravity, Dirichlet conditions for the transverse modes of the vector potential of electromagnetism, a mixture of Dirichlet and Neumann conditions for scalar fields, and local boundary conditions for the spin-$\frac{1}{2}$ field and the spin-$\frac{3}{2}$ potential. Remarkably, the one-loop results for fermionic fields are equal to the ones obtained using nonlocal boundary conditions (chapters eight and nine). Moreover, no exact cancellation of one-loop divergences is found, in the presence of boundaries, for simple supergravity and extended supergravity theories (chapter nine). All one-loop calculations are performed in great detail, so as to enable graduate students and research workers to learn these techniques.

However, evidence exists that restriction of gauge theories to a set of physical degrees of freedom leads to different one-loop results with respect to the quantization of the full theory in Becchi-Rouet-Stora-Tyutin-invariant fashion. This problem is also investigated in chapter six (section 6.5), in the case of electromagnetism.

New results on the singularity problem for space-times with torsion are finally derived and discussed in sections 10.7-8, whereas the research results obtained in the whole book are summarized in chapter eleven.

At the end of Part IV, I have proposed a series of problems for the reader. I encourage all readers to work very hard on these problems, since this is the best way to make sure they have learned the techniques and the ideas described in the book.

The only prerequisites are the knowledge of the basic differential geometry described in chapter two of Hawking and Ellis 1973, and of the path-integral formalism for quantum field theory at the level of many introductory textbooks in the current literature. I have omitted the treatment of these topics since I believe they are very well described by other authors. I have instead focused on quantum gravity, quantum cosmology and Lorentzian geometries, since there are not many textbooks which study all of them. I hope the resulting monograph will be useful to a very large audience.

CONTENTS

PART I:

QUANTUM GRAVITY

CHAPTER ONE

QUANTUM GRAVITY, QUANTUM COSMOLOGY AND CLASSICAL GRAVITY

1.1 Quantum Gravity: Approaches, Achievements and Unsolved Problems

The first nine chapters of this monograph deal mainly with quantum cosmology, where some mathematical techniques used in quantum gravity are applied so as to get a better understanding of the early universe. It is therefore very important to discuss and clarify some basic points about the problem of quantum gravity at the beginning of our work. In our opinion, the two main motivations for studying quantum gravity are the following :

(a) The singularity theorems of Penrose, Hawking and Geroch show that Einstein's general relativity leads to the occurrence of singularities in cosmology in a rather generic way (Geroch 1966-1967, Hawking 1966a-b, Hawking 1967, Hawking and Penrose 1970, Hawking and Ellis 1973). One might define the quantum gravity era as the one when all physics is confined to a region whose linear size is of the order of 10^{-33} cm. In other words we are asking the questions : is there a theory which describes gravitational interactions at these length scales ? Does this theory avoid singularities in a generic way ?

(b) The electroweak and strong interactions are described by renormalizable quantum field theories (Warr 1988). However, Einstein's general relativity cannot be renormalized (Duff 1982). Some authors (De Witt 1964) tried to rearrange and sum infinite subclasses of Feynman graphs, but in so doing the effective propagators may be shown to pick up new poles which destroy unitarity (Warner 1982).

In order to study these problems, many efforts have been produced so far. The main approaches seem to be : (1) covariant (De Witt 1964, 1967b-c); (2) canonical (De Witt 1967a, Isham and Kakas 1984a-b, Ashtekar 1988); (3) path integral (Hawking 1979a-b); (4)

asymptotic quantization (Ashtekar 1987); (5) quantization of supergravity (van Nieuwenhuizen 1981, Warner 1982, D'Eath 1984); (6) higher-derivative theories (Stelle 1977, Barth and Christensen 1983, Boulware 1984); (7) lattice theories (Menotti and Pelissetto 1987); (8) application of Regge calculus (Rocek and Williams 1982, Warner 1982); (9) null-strut calculus (Kheyfets et al. 1989, 1990a-b); (10) string theories (Green et al. 1987); (11) twistor theory (Penrose 1975-1987); (12) topological quantization (Isham 1989, Isham et al. 1990). A complete description of all these approaches would require by itself a book, but a few important comments on some ideas can be made here.

(I) Using the Arnowitt-Deser-Misner formalism for general relativity (Misner et al. 1973, MacCallum 1975, Hanson et al. 1976), one makes a $3+1$ split of the space-time metric, which may be cast in the form :

$$g = -\left(N^2 - N_i N^i\right) dt \otimes dt + N_i \left(dx^i \otimes dt + dt \otimes dx^i\right) + h_{ij} dx^i \otimes dx^j \quad . \tag{1.1.1}$$

Using the Gauss-Codazzi equations (Lightman et al. 1975, Hanson et al. 1976), one finds that the action for general relativity (York 1972-1986, Gibbons and Hawking 1977)

$$16\pi G I_g = \int_M R\sqrt{-g}\, d^4x + 2\int_{\partial M} \chi\sqrt{h}\, d^3x \tag{1.1.2}$$

gives rise to the Lagrangian :

$$L = \int_{\partial M} N\sqrt{h}\left(\chi_{ij}\chi^{ij} - \chi^2 + {}^{(3)}R\right) d^3x \quad . \tag{1.1.3}$$

One thus finds the primary constraints (De Witt 1967a) : $\pi \equiv \frac{\delta L}{\delta N} \approx 0$, $\pi^i \equiv \frac{\delta L}{\delta N_i} \approx 0$. Requiring the preservation in time of these constraints (Dirac 1964), one finds the secondary constraints :

$$\mathcal{H} \equiv \sqrt{h}\left(\chi_{ij}\chi^{ij} - \chi^2 - {}^{(3)}R\right) \approx 0 \quad , \tag{1.1.4}$$

$$\mathcal{H}^i \equiv -2\pi^{ij}{}_{,j} - h^{il}\left(2h_{jl,k} - h_{jk,l}\right)\pi^{jk} \approx 0 \quad , \tag{1.1.5}$$

where $\pi^{ij} \equiv \frac{\delta L}{\delta h_{ij}} = -\sqrt{h}\,(\chi^{ij} - h^{ij}\chi)$. The constraint (1.1.4) is called Hamiltonian constraint. On quantization, Poisson brackets become commutators, and first-class constraint equations (section 2.4) become conditions on the state vector ψ (De Witt 1967a) :

$$\pi\psi = 0 \quad , \quad \pi^i\psi = 0 \quad , \tag{1.1.6}$$

$$\mathcal{H}\psi = 0 \quad , \quad \mathcal{H}^i\psi = 0 \quad . \tag{1.1.7}$$

It is indeed true that the $3+1$ split of the metric may seem contrary to the whole spirit of relativity (Hawking 1979b, Ashtekar 1987). It is also true that the underlying manifold structure has been assumed to be $R \times \Sigma$ (where Σ is a three-manifold) and usually kept fixed (see, however, Ashtekar 1987), whereas one would expect quantum gravity to allow also for those topologies which are not a product (Hawking 1979b). But the main problem is due to the difficulty in solving the quantum constraints. In fact, the equation $\mathcal{H}\psi = 0$ (the Wheeler-De Witt equation) is an equation on a space, called superspace (Fisher 1970, Francaviglia 1975), whose points are equivalence classes of metrics related by the action of the diffeomorphism group of a compact spacelike three-surface. More precisely, the superspace $S(\mathcal{M})$ is defined as $S(\mathcal{M}) \equiv Riem(\mathcal{M})/Diff(\mathcal{M})$. With this notation, $\mathcal{M}$ is a compact, connected, orientable, Hausdorff, C^∞ three-manifold without boundary. $Riem(\mathcal{M})$ is the space of C^∞ Riemannian metrics on $\mathcal{M}$, and $Diff(\mathcal{M})$ is the group of C^∞ orientation-preserving diffeomorphisms of $\mathcal{M}$. Thus one has to deal with an infinite number of degrees of freedom, and in addition operator-ordering problems are found to arise, because the Hamiltonian $\mathcal{H}$ is a quadratic function of the momenta π^{ij}.

Later on, we shall see how supergravity can be cast in Hamiltonian form, and how this formalism can be used so as to study quantum cosmological problems. However, it should be recalled that progress has been made in the last few years in our understanding of the canonical structure of general relativity owing to Ashtekar (Ashtekar 1988). In his new formalism, the space-time metric is a secondary object, while the new configuration variable is the restriction to a three-manifold of a $SL(2,C)$ spin-connection. The momentum conjugate to this variable is a $SU(2)$ soldering form which turns internal $SU(2)$ indices into $SU(2)$ spinor indices. Therefore we are now able to use in quantum gravity some techniques which were already very useful for other gauge theories, and there is now hope to solve the quantum constraints in a nonperturbative way. It is remarkable that, in terms of Ashtekar's variables, the constraints are polynomial.

(II) The basic postulate of the path-integral approach to quantum gravity (Hawking 1979a-b) is that the probability amplitude A of going from a three-metric h and a matter-field configuration ϕ on a spacelike surface Σ to a three-metric h' and a field configuration ϕ' on a spacelike surface Σ' is formally given by :

$$A\Big(h', \phi' \mid h, \phi\Big) = \int_{C'} D\,[g]\, D\,[\Phi]\, e^{iI[g,\Phi]} \quad , \tag{1.1.8}$$

where C' is the class of all four-metrics inducing h on Σ and h' on Σ', and of all field configurations matching ϕ on Σ and ϕ' on Σ'. In computing the amplitude (1.1.8), we have to fix the gauge, and it is worth discussing this problem. In so doing we closely follow Teitelboim 1983. The Hamiltonian form of the action for the gravitational field in a closed universe (here studied for simplicity) is :

$$I = \int_{t_1}^{t_2} dt \int d^3x \left(\pi^{ij}\dot{h}_{ij} - N^{\perp}H_{\perp} - N^{i}H_{i}\right) \quad , \tag{1.1.9}$$

where $N^{\perp}$ is equal to $g^{-\frac{1}{2}}$ times the usual lapse N. Denoting by $\{\ ,\ \}$ the Poisson brackets, we have that the action (1.1.9) is invariant under the gauge transformation :

$$\delta h_{ij} = \{h_{ij}, H[\epsilon^{\mu}]\} \quad , \tag{1.1.10}$$

$$\delta \pi^{ij} = \{\pi^{ij}, H[\epsilon^{\mu}]\} \quad , \tag{1.1.11}$$

plus a more complicated relation involving $\delta N^{\perp}$ and δN^{i}, whose form is not strictly needed here. In (1.1.10-11), one has :

$$H[\epsilon^{\mu}] = \int d^3x \Big(\epsilon^{\perp}H_{\perp} + \epsilon^{i}H_{i}\Big) \quad , \tag{1.1.12}$$

with the boundary condition :

$$\epsilon^{\perp}(x, t_2) = \epsilon^{\perp}(x, t_1) = 0 \quad . \tag{1.1.13}$$

Moreover, since we are fixing the three-geometries at the end points, we also have to require that :

$$\epsilon^i(x,t_2) = \epsilon^i(x,t_1) = 0 \quad . \tag{1.1.14}$$

Let us now consider the intervals $I_1 =]t_1, t'[$ and $I_2 =]t', t_2[$, and let us require that :

$$\dot{N}^{\perp} = 0 \qquad N^i = 0 \qquad \forall\, t \in I_1 \quad , \tag{1.1.15}$$

$$\dot{N}^i = 0 \qquad N^{\perp} = 0 \qquad \forall\, t \in I_2 \quad . \tag{1.1.16}$$

These conditions imply that $N^{\perp} = 0$, $\forall t > t'$. However, in so doing we allow for an adjustable change of spatial coordinates during I_2. Namely, the dependence of N^i on x is not fixed during I_2, which in turn allows to set any given coordinate system on the final surface Σ'. Another way of fixing the gauge, together with a detailed study of the ghost action and of the path integral can be found in Teitelboim 1983 as well. The application of the ghost formalism to quantum cosmology can be found in Halliwell 1988, Laflamme 1988, Moss and Poletti 1990a, and is also studied in our chapter six. Following Hawking 1979a-b, we now assume that an analytic continuation to the Euclidean regime (where $\tau = it$) is sometimes possible, so as to deal with probability amplitudes of the form :

$$\widetilde{A}\Big(h', \phi' \mid h, \phi\Big) = \int_C D[g]\, D[\Phi]\, e^{-I_E[g,\Phi]} \quad , \tag{1.1.17}$$

where I_E is the action integral for gravitation and matter fields in the Euclidean regime, and where the metrics belonging to the class C are now required to be positive-definite, so that we deal with Riemannian geometry. Alternatively, (1.1.17) has been taken as the starting point of the path-integral formulation of quantum gravity, working from the beginning in the Euclidean regime. Indeed, there is not general agreement on this approach because it has several drawbacks. In fact :

(i) The measure in (1.1.17) has not yet been given a rigorous mathematical meaning (Allen 1983a, Glimm and Jaffe 1987). In particular, translational invariance in Feynman-type integrals may be violated (see Tarski 1980 and literature given therein). However, a rigorous approach to path integration (though not yet in quantum gravity) has become available in recent years using the concept of prodistribution and stochastic calculus (De Witt-Morette 1987 and references therein);

(ii) There are topological obstructions to a Wick rotation if the Euler number is not zero (Warner 1982). Therefore, it is very difficult to perform a meaningful analytic continuation from the Lorentzian to the Euclidean regime, or viceversa (Hawking 1984a);

(iii) In formally defining (1.1.17), we are also summing over all possible topologies. This idea may seem very appealing, but there is no proof that a space-time foam (Hawking 1979a-b) exists. Indeed, some authors (De Witt and Anderson 1986) disagree with Hawking about the idea that the space-time topology might change;

(iv) The Euclidean action of the gravitational field is not positive-definite (Hawking 1979b). Thus, the integrand on the right-hand side of (1.1.17) is not exponentially damped, but it may blow up exponentially (see, however : Gibbons et al. 1978, Schleich 1987).

Despite all these problems, at least two very important results have been obtained using path integrals :

(a) The thermodynamic properties of black holes have been derived in a very powerful way (Gibbons and Hawking 1977). It should be emphasized that black holes provide the first example in physics where entropy is related to a purely geometric quantity such as the area of the event horizon (Birrell and Davies 1982);

(b) Gravitational instantons are complete, nonsingular, positive-definite solutions of the Euclidean Einstein field equations $R_{\mu\nu} = \Lambda g_{\mu\nu}$ (Gibbons and Hawking 1991). They play an important role in quantum gravity because they are assumed to give the dominant contribution to the calculation of the right-hand side of (1.1.17). Compact, asymptotically flat and asymptotically Euclidean instantons were studied in detail (Hawking 1979b, Pope 1980-1981). The fixed points of the Euclidean action are of two types : isolated points called nuts and two-surfaces called bolts. Gibbons and Hawking were also able to relate numbers and types of nuts and bolts to the topological invariants of the instantons. Moreover, they derived a formula for the gravitational action of the instantons in terms of the areas of the bolts and certain nut charges and potentials (Gibbons and Hawking 1979).

The path-integral approach is also the most used in studying cosmological problems. This will be discussed in the next section. Here we just mention that a very remarkable approach to quantum gravity is still the one based on twistors. In twistor theory one does not take space-time (M, g) as the background for physical processes. Instead, the space L of light rays is proposed as the support of quantum fields. The transformations from M to L and viceversa are studied in the theory of the Penrose transform, for which we refer the reader to Manin 1988 and Ward and Wells 1990. Twistors in flat space are defined in some detail in section 5.6. A lot of progress has been made in twistor theory over the past fifteen years (Penrose 1975, Huggett and Tod 1985, Penrose 1986, Bailey and Baston 1990, Ward and Wells 1990), but unfortunately it is not yet clear how to apply twistor theory in getting a consistent theory of quantum gravity. In fact, for example, it is not clear how to find a twistor-space description of general vacuum space-times.

1.2 Quantum Cosmology: Motivations and Some Recent Developments

The aim of quantum cosmology is to study the early universe and its evolution so as to shed new light on the following problems : (1) boundary conditions in cosmology (Hawking 1982); (2) singularity problem (Hawking 1984a); (3) other problems of the hot big-bang model (Hawking 1984a, Linde 1984, Brandenberger 1985); (4) origin of structure in the universe (Halliwell and Hawking 1985); (5) large-scale properties of the observed universe (Hawking and Luttrell 1984a-b, Amsterdamski 1985); (6) arrow of time in cosmology (Hawking 1985, Page 1985, Hawking 1987, Laflamme 1988).

A review of all problems studied so far cannot be attempted here, but the basic ideas and techniques should be briefly recalled. Let us assume for simplicity that only scalar fields are present (see section 11.1 about this problem). The basic object in quantum cosmology is a wave functional ψ. In a closed universe, this functional can only depend on the three-geometry and on the matter field configuration on a compact spacelike three-surface (De Witt 1967a, Hartle and Hawking 1983). It is given by the Euclidean path integral :

$$\psi\left[h, \phi_0\right] = \int_C D\left[g\right] D\left[\Phi\right] e^{-I_E\left[g,\Phi\right]} \quad . \tag{1.2.1}$$

It has been claimed (Hawking 1984a) that the right-hand side of (1.2.1) satisfies the Wheeler-De Witt equation :

$$\left[-G_{ijkl}\frac{\delta^2}{\delta h_{ij}\delta h_{kl}}+\sqrt{h}\left(-{}^{(3)}R+16\pi T_{00}\left(\frac{\delta}{\delta\phi},\phi\right)\right)\right]\psi\left[h,\phi\right]=0\ , \tag{1.2.2}$$

and the momentum constraints :

$$\left(\frac{\delta\psi}{\delta h_{ij}}\right)_{|i}=8\pi T^{0j}\psi\quad . \tag{1.2.3}$$

More precisely, also this one is a delicate point. In fact a partial proof of the derivation of the Wheeler-De Witt equation from the path integral has been only recently given in Halliwell 1988 for models with a finite number of degrees of freedom, called minisuperspace models. More recently, a complete but formal proof has been obtained in three different ways in Halliwell and Hartle 1991. They look at sums over histories defined with an invariant action, invariant measure and an invariant class of paths summed over. Wave functions constructed from these invariant sums over histories are then shown to be annihilated by the constraints. This result is then used to derive the Wheeler-De Witt equation either via BRST techniques, or from the gauge-fixed Hamiltonian path integral, or relying on the work in Isham and Kuchar 1985a-b. In Halliwell and Hartle 1991 it is emphasized that the symmetry which actually leads to the Wheeler-De Witt equation is the canonical symmetry of the Hamiltonian form of the action of general relativity. However, their proofs remain formal because they do not solve operator-ordering problems, the last of their five assumptions has not been verified a posteriori (although it is valid for time-slicing implementation of path integrals), and they frequently use Dirac's delta.

In (1.2.2), $G_{ijkl}=\frac{1}{2\sqrt{h}}\left(h_{ik}h_{jl}+h_{il}h_{jk}-h_{ij}h_{kl}\right)$ (see Pekonen 1987 for a modern study of G_{ijkl}), and (1.2.3) holds since ψ is invariant under the action of the diffeomorphism group (Higgs 1958-59). We may think that (1.2.1) is obtained from (1.1.17) by letting the initial three-surface shrink to zero (Horowitz 1985). Thus ψ would be the amplitude for the universe to appear from *nothing* (Hartle and Hawking 1983). However, the interpretation of ψ is an open question, and is still receiving careful consideration (Halliwell 1987). Nobody knows how to exactly solve the Wheeler-De Witt equation on superspace, but in the last few years progress has been made in choosing the boundary conditions and computing ψ in approximate models.

It is indeed clear that, in studying (1.2.1), we must specify the class C of four-metrics and matter fields involved in the path integral. For this purpose, in Hartle and Hawking 1983 the following proposal was made (see page 2965 therein). The path integral (1.2.1) for the ground-state wave function should be only taken over all compact Riemannian four-metrics which induce h on Σ, and all matter field configurations which are regular on the compact four-manifolds having Σ as their only boundary. It should be emphasized that also alternative proposals have been made (e.g. Vilenkin 1986-1988). Moreover, an usual criticism of quantum field theorists is that regular matter field configurations should not dominate in the path integral, because on the contrary discontinuous fields play the central role in Feynman's approach to quantum field theory (Schulman 1981, pp 319-320). With reference to this problem, we think it is worth mentioning the point of view on page 236 of Schulman 1981. There, the author points out that for gravitation the irregular fields might not be a serious problem. In fact, the amplitude is $e^{\left(\frac{iI}{L}\right)}$, where I is the action and $L = 1.6 \times 10^{-33}$ cm is the Planck length. Thus, any submicroscopic pathologies allowed by the quantization might have been smoothed and rendered harmless on a scale of 10^{-13} cm. Needless to say, this is not a rigorous argument, and the question remains open.

However, the work in this monograph is relevant for the Hartle-Hawking proposal, and it is thus worth describing the original argument of these authors (Esposito 1988). One is assuming the dominant contribution to (1.2.1) is given by solutions of the Euclidean Einstein field equations : $R_{\mu\nu} = \Lambda g_{\mu\nu}$. When Λ is > 0, the four-metric is compact. In fact, a theorem due to Myers (Milnor 1962) states that in an n-dimensional Riemannian manifold (M, g_R) whose Ricci curvature is $\geq \frac{(n-1)}{r}$ (r being a positive constant) everywhere, every geodesic whose length is $> \pi\sqrt{r}$ contains conjugate points and hence is not minimal. Moreover, if M is assumed to be geodesically complete, the Hopf-Rinow theorem states that any two points of M can be joined by a minimal geodesic. We also know that in geodesically complete manifolds any closed bounded set is compact. Thus, in light of these theorems, a Riemannian, geodesically complete, four-dimensional manifold (M, g_R) whose

Ricci curvature is $\geq \frac{3}{r}$ everywhere, is compact with a diameter $\leq \pi\sqrt{r}$. If Λ is ≤ 0, there are noncompact solutions of $R_{\mu\nu} = \Lambda g_{\mu\nu}$; the most symmetric are flat Euclidean space and Euclidean anti-de Sitter space. The latter is the maximally symmetric space $SO(4,1)/SO(4)$, i.e. a four-hyperboloid of constant negative curvature, with R^4 topology. Thus its curvature tensor is : $R_{abcd} = \lambda(g_{ac}g_{bd} - g_{ad}g_{bc})$; moreover : $R_{ab} = 3\lambda g_{ab}$, $R = 12\lambda$, where λ is < 0. In this space, if we identify τ and $\tau + \beta$, quantum fields can be held at any temperature $\frac{1}{\beta}$ (Allen 1983). One can show (Hawking 1984b) that connected asymptotically Euclidean (i.e. flat Euclidean space out of a compact region) or anti-de Sitter metrics which have an inner boundary at Σ do not give the dominant contribution to (1.2.1), because there is a scale transformation which violates a certain inequality which ought to hold. One is thus left with disconnected metrics given by a compact part with boundary at Σ and an asymptotically Euclidean or anti-de Sitter metric without inner boundary. But in Hawking 1984a-b it is argued that these metrics are suitable for scattering problems, which is not what happens in cosmology. Some authors have also used the choice of compact metrics so as to set boundary conditions for the numerical solutions of the Wheeler-De Witt equation in minisuperspace models (e.g. Laflamme and Shellard 1987). However, there is not yet general agreement about these numerical techniques.

Another important point is how the Hartle-Hawking (hereafter referred to as HH) Euclidean path integral enables one to describe classical Lorentzian space-times. The basic idea is that a semiclassical approximation of ψ is possible, so that :

$$\psi \sim \sum_k A_k e^{-\frac{(I_E)_k}{\hbar}} + \text{higher} - \text{order terms} \quad , \tag{1.2.4}$$

where $(I_E)_k$ is the Euclidean action computed along a compact solution of the Euclidean field equations inducing h on Σ. If $(I_E)_k$ is complex and $Re\ (I_E)_k$ is slowly varying compared with $Im\ (I_E)_k$ (Louko 1988a), each component of the sum in (1.2.4) behaves as Ce^{iS}, where, setting : $S = S(q_1, ..., q_N)$, one has : $\frac{\partial S}{\partial q_i} \gg \frac{\partial(\log C)}{\partial q_i}$, $\forall i = 1, ..., N$. For example, in a minisuperspace model with a massive or massless scalar field ϕ (Hawking 1984a-b, Laflamme and Shellard 1987, Esposito and Platania 1988) and a single scale factor a, a wave function of the form Ce^{iS} is peaked about the first integral : $p_a = \frac{\partial S}{\partial a}$, $p_\phi = \frac{\partial S}{\partial \phi}$, where S is the analytic continuation of the action for compact four-metrics and regular matter fields. This first integral consists of two first-order ordinary differential equations, so that the solution involves just two arbitrary constants. Therefore the HH wave function is peaked about a set of solutions which are a two-parameter subset of the general solution of the field equations, which involves three arbitrary parameters. It has been shown that this is a subset of inflationary solutions (see again : Hawking 1984a-b, Laflamme and Shellard 1987, Esposito and Platania 1988). Nevertheless, Vilenkin disagrees about this point. Another alternative approach based on the inclusion of all possible boundaries in the path integral for quantum cosmology can be found in Suen and Young 1989.

A brief mention of the singularity problem in the HH approach can be found in section 11.1. Let us now recall the basic results about the canonical formalism for supergravity.

1.3 Introduction to Supergravity

In chapters five and nine we will need the Hamiltonian formulation of simple supergravity in four dimensions. This is why we here summarize the main ideas of this theory, following D'Eath 1984.

Simple supergravity is a supersymmetric gauge theory of gravitation. Thus the graviton, a spin-2 excitation corresponding to the gravitational field, acquires a supersymmetric partner represented by a spin-$\frac{3}{2}$ particle, called gravitino. From now on, we omit the word *simple* for convenience of notation. The basic quantities in the action integral of supergravity are the tetrad, given by the Hermitian spinor-valued one-form $e^{AA'}_{\ \ \ \mu}$, the spin-$\frac{3}{2}$ field, given by the spinor-valued one-form $\psi^{A}_{\ \ \mu}$, and its Hermitian conjugate $\overline{\psi}^{A'}_{\ \ \mu}$. The $e^{AA'}_{\ \ \ \mu}$ commute with all other variables, whereas $\psi^{A}_{\ \ \mu}$ and $\overline{\psi}^{A'}_{\ \ \mu}$ anticommute among themselves. In terms of the connection forms : $\omega^{AA'BB'}_{\ \ \ \ \ \ \ \mu} = \omega^{AB}_{\ \ \ \mu}\epsilon^{A'B'} + \overline{\omega}^{A'B'}_{\ \ \ \ \mu}\epsilon^{AB}$, the derivative operator D_μ acting on spinor-valued forms is defined by : $D_\mu e^{AA'}_{\ \ \ \nu} \equiv$

$\partial_\mu e^{AA'}{}_\nu + \omega^A{}_{B\mu} e^{BA'}{}_\nu + \overline{\omega}^{A'}{}_{B'\mu} e^{AB'}{}_\nu$, $D_\mu \psi^A{}_\nu \equiv \partial_\mu \psi^A{}_\nu + \omega^A{}_{B\mu} \psi^B{}_\nu$. Torsion is thus given by the spinor-valued two-form :

$$S^{AA'}{}_{\mu\nu} = D_{[\mu} e^{AA'}{}_{\nu]} = -4\pi i \, \overline{\psi}^{A'}_{[\mu} \psi^A{}_{\nu]} \quad . \tag{1.3.1}$$

Moreover, $R^{AB}{}_{\mu\nu} \equiv 2\left(\partial_{[\mu} \omega^{AB}{}_{\nu]} + \omega^A{}_{C[\mu} \omega^{CB}{}_{\nu]}\right)$ is the definition of the spinor-valued curvature two-forms, so that the curvature scalar is : $R = e_{AA'}{}^\mu e_B{}^{A'\nu} R^{AB}{}_{\mu\nu} + H.C.$ The action integral of supergravity is :

$$I = \int d^4x \left[\frac{(\det e)R}{16\pi} + \frac{1}{2}\epsilon^{\mu\nu\rho\sigma}\left(\overline{\psi}^{A'}{}_\mu e_{AA'\nu} D_\rho \psi^A{}_\sigma + H.C.\right)\right] \quad . \tag{1.3.2}$$

The action (1.3.2) should be supplemented by boundary terms at spatial infinity and on all boundary surfaces which might be present. The auxiliary fields do not appear in (1.3.2) because they vanish classically and are set to zero in the Hamiltonian formalism. If boundary surfaces are absent, (1.3.2) is invariant under the following transformations:

(a) local Lorentz transformations :

$$\delta e^{AA'}{}_\mu = N^A{}_B e^{BA'}{}_\mu + \overline{N}^{A'}{}_{B'} e^{AB'}{}_\mu \quad , \tag{1.3.3}$$

$$\delta \psi^A{}_\mu = N^A{}_B \psi^B{}_\mu \quad , \quad \delta \overline{\psi}^{A'}{}_\mu = \overline{N}^{A'}{}_{B'} \overline{\psi}^{B'}{}_\mu \quad , \tag{1.3.4}$$

where $N^{AB} = N^{(AB)}$.

(b) coordinate transformations :

$$\delta e^{AA'}{}_\mu = \xi^\nu \partial_\nu e^{AA'}{}_\mu + e^{AA'}{}_\nu \partial_\mu \xi^\nu \quad , \tag{1.3.5}$$

$$\delta \psi^A{}_\mu = \xi^\nu \partial_\nu \psi^A{}_\mu + \psi^A{}_\nu \partial_\mu \xi^\nu \quad , \tag{1.3.6}$$

(c) local supersymmetry transformations :

$$\delta e^{AA'}{}_\mu = -2i\sqrt{2\pi}\left(\epsilon^A \overline{\psi}^{A'}{}_\mu + \overline{\epsilon}^{A'} \psi^A{}_\mu\right) \quad , \tag{1.3.7}$$

$$\delta \psi^A{}_\mu = \frac{1}{4\sqrt{2\pi}} D_\mu \epsilon^A \quad , \quad \delta \overline{\psi}^{A'}{}_\mu = \frac{1}{4\sqrt{2\pi}} D_\mu \overline{\epsilon}^{A'} \quad . \tag{1.3.8}$$

In (1.3.7-8), ϵ^A and $\bar{\epsilon}^{A'}$ are anticommuting fields which depend on the space-time position. Defining $p_{AA'}{}^i \equiv \frac{\delta I}{\delta \dot{e}^{AA'}{}_i}$, the basic dynamical variables of the theory are : $e^{AA'}{}_i$, $p_{AA'}{}^i$, $\psi^A{}_i$ and $\overline{\psi}^{A'}{}_i$. The remaining variables N, N^i (see (A.8) of appendix A), $\psi^A{}_0$ and $\overline{\psi}_0^{A'}$ can be freely specified. Defining :

$$\pi_A{}^i \equiv -\frac{1}{2}\epsilon^{ijk}\,\overline{\psi}^{A'}{}_j e_{AA'k} \quad , \tag{1.3.9}$$

$$J_{AB} \equiv e_{(A}{}^{A'i} p_{B)A'i} + \psi_{(A}{}^i \pi_{B)i} \quad , \tag{1.3.10}$$

one finds the primary constraints :

$$J_{AB} \approx 0 \quad , \quad \overline{J}_{A'B'} \approx 0 \quad , \tag{1.3.11}$$

in view of the invariance of (1.3.2) under (1.3.3-4). The Hamiltonian H is found to be :

$$H = \int d^3x \left[N\left({}_1H_\perp\right) + N^i\left({}_1H_i\right) + \psi^A_0\left({}_1S_A\right) + \left({}_1\overline{S}_{A'}\right)\overline{\psi}^{A'}{}_0 + M_{AB}J^{AB} + \overline{M}_{A'B'}\overline{J}^{A'B'} \right].$$

In this formula, $\psi^A{}_0$, $\overline{\psi}^{A'}{}_0$, M_{AB} and $\overline{M}_{A'B'}$ act as Lagrange multipliers. The quantities $({}_1H_\perp)$, $({}_1H_i)$ and $({}_1S_A)$ only depend on the basic variables $e^{AA'}{}_i$, $p_{AA'}{}^i$, $\psi^A{}_i$ and $\overline{\psi}^{A'}{}_i$. Their complicated expression may be found in D'Eath 1984. There are also primary constraints in view of the vanishing of the momenta conjugate to N, N^i, $\psi^A{}_0$ and $\overline{\psi}^{A'}{}_0$. Requiring the preservation in time of these primary constraints (Dirac 1964), one finds the secondary constraints :

$$({}_1H_\perp) \approx 0 \quad , \quad ({}_1H_i) \approx 0 \quad , \quad ({}_1S_A) \approx 0 \quad , \quad ({}_1\overline{S}_{A'}) \approx 0 \quad . \tag{1.3.12}$$

The last two constraints in (1.3.12) are called supersymmetry constraints. Much more details about the Hamiltonian formulation of supergravity can be found in D'Eath 1984 and in Jacobson 1988. A detailed analysis of the algebra of supersymmetry constraints in supersymmetric minisuperspace models can be found in Hughes 1990.

1.4 An Outline of This Work

The most recurring themes of our work are the quantization of constrained systems and one-loop properties of physical theories in the presence of boundaries. We thus insert two more review chapters where these topics are studied. In chapter two Dirac's theory of constraints, his approach to quantization of first- and second-class systems, and the corresponding application to the quantization program for general relativity are presented. The geometrical framework of canonical quantum gravity, i.e. Wheeler's superspace, is described following Fisher 1970.

In chapter three we study some basic tools of perturbative quantum gravity which are frequently used in the rest of our book, i.e. the background-field method, the one-loop approximation, the zeta-function method of regularizing path integrals in curved space-time. A general overview of gravitational-instantons theory is also given. Finally, we describe perturbative renormalization, asymptotic expansions and summability methods in quantum field theory, and we show the nonrenormalizability of quantized general relativity.

At the beginning of each subsequent chapter, we have summarized in detail all results obtained. For the sake of clarity, these results are described again in this section, but emphasizing much more some foundational issues.

In computing the semiclassical approximation of ψ (cf. (1.2.4)) we are writing the four-metric as (Louko 1988a) : $g_{\mu\nu} = g^{c}_{\mu\nu} + \gamma_{\mu\nu}$, where $g^{c}_{\mu\nu}$ is a solution of the classical field equations obeying the HH boundary conditions, and $\gamma_{\mu\nu}$ is a perturbation about $g^{c}_{\mu\nu}$. The action is then expanded as : $I = I^{c} + I^{(2)} + \ldots$, where $I^{c} = I\left(g^{c}_{\mu\nu}\right)$ and $I^{(2)}$ is quadratic in $\gamma_{\mu\nu}$. This is why we find :

$$\psi_{HH} \sim P e^{-\frac{I^{c}}{\hbar}} + \text{higher} - \text{order terms} \quad , \quad P = \int d[\gamma]\, e^{-\left(I^{(2)}(\gamma)\right)} \quad . \tag{1.4.1}$$

Thus the prefactor P is given by the Gaussian integral (1.4.1), which can be regularized using the zeta-function method (cf. appendix B). Efforts have been made in the literature to study the case when (1.4.1) is only taken over the physical degrees of freedom (hereafter often referred to as PDF) of the problem. This can be done using the Hamiltonian formulation of the theory and choosing a gauge condition (in chapters four, five, seven and nine we shall use this method). The alternative method using Faddeev-Popov ghost fields might be more appropriate but presents great technical difficulties, and is studied in chapter six. Writing : $I^{(2)} = \int \gamma^{\mu\nu} A \gamma_{\mu\nu} \sqrt{g} d^4x$ (where A is an elliptic, self-adjoint, positive-definite, second-order differential operator) and : $d[\gamma_{\mu\nu}] = \pi_n (\mu dc_n)$, $\gamma_{\mu\nu} = \sum_n c_n \gamma_{\mu\nu}^{(n)}$, one finds that for bosonic fields (Schleich 1985, Louko 1988a) :

$$P = e^{\left[\frac{1}{2}\zeta(0) ln\left(\pi\mu^2\right) + \frac{1}{2}\zeta'(0)\right]} = D a^{\zeta(0)} \quad , \tag{1.4.2}$$

in perturbing about flat Euclidean space with a three-sphere boundary of radius a. In (1.4.2), D is a constant. In this one-loop calculation, the perturbative modes are required to be regular at the origin, and they or their first derivatives are set to zero on S^3. As $a \rightarrow 0$, the prefactor P had been found to diverge for pure gravity subject to Dirichlet conditions for the perturbed three-metric (Schleich 1985). Thus we have studied the following problem : is the amplitude at one-loop finite in a quantum cosmology based on a supersymmetric theory of gravitation ? Namely, as $a \rightarrow 0$, is P constant in the case of $N = 1$ supergravity or extended supergravities ?

An important comment is now in order. One can perform one-loop calculations paying attention to : (1) S-matrix elements; (2) topological invariants; (3) presence of boundaries. For example, in the case of pure gravity with vanishing cosmological constant : $\Lambda = 0$, it is known that one-loop on-shell S-matrix elements are finite (Duff 1982). This property is also shared by $N = 1$ supergravity when $\Lambda = 0$, and in that theory two-loop on-shell finiteness also holds (Duff 1982). However, when $\Lambda \neq 0$, both pure gravity and $N = 1$ supergravity are no longer one-loop finite in the sense (1) (and (2) as well), because the on-shell one-loop counterterm is given by (Duff 1982) :

$$S_{(1)} = \frac{1}{\epsilon} \left[A\chi - \frac{2BG\Lambda S}{3\pi} \right] \quad . \tag{1.4.3}$$

In (1.4.3), $\epsilon = n - 4$ is the dimensional regularization parameter, χ is the Euler number, S is the classical action on-shell, and one finds (Duff 1982) : $A = \frac{106}{45}$, $B = -\frac{87}{10}$ for pure gravity, and : $A = \frac{41}{24}$, $B = -\frac{77}{12}$ for $N = 1$ supergravity. Thus, $B \neq 0$ is responsible for lack of S-matrix one-loop finiteness, and $A \neq 0$ does not yield topological one-loop finiteness.

In the presence of boundaries, a much larger number of counterterms can be obtained even just at one-loop order in perturbation theory, using the extrinsic-curvature tensor and the Ricci tensor of the boundary (D'Eath 1986). Thus, if any theory of quantum gravity can be studied from the perturbative point of view, boundary effects play a key role in understanding whether it has interesting and useful finiteness properties. The result derived in Schleich 1985 shows that, even when $\Lambda = 0$, pure gravity is no longer PDF one-loop finite if we have a S^3 boundary, but one might hope that in $N = 1$ supergravity, or extended supergravities, one-loop finiteness still holds in the presence of a compact boundary such as a three-sphere. We have found no analytic evidence in favour of this conjecture, and the prefactor P for these theories seems to be still diverging as $a \rightarrow 0$. However, the question of the one-loop finiteness of extended supergravity in the presence of boundaries has not yet been settled, as we will explain later.

In deriving the contribution to P coming from the spin-$\frac{3}{2}$ field which represents the gravitino, it is enlightening to perform at first a similar but simpler calculation for the spin-$\frac{1}{2}$ field. This is why in chapter four we evaluate P for a massless Majorana field. In so doing we pay special attention to the way of dealing with first-order differential operators, and to the asymptotic expansion of the heat kernel using relations derived in Olver 1954 and applied in a simpler case in Stewartson and Waechter 1971. In the Einstein-Dirac theory, the boundary-value problem of interest to quantum cosmology is to find the solution of the Euclidean Dirac equation such that $\phi_A^{(+)}, \chi_A^{(+)}, \widetilde{\phi}_{A'}^{(+)}, \widetilde{\chi}_{A'}^{(+)}$ match prescribed values on the final surface S_F, and $\phi_A, \chi_A, \widetilde{\phi}_{A'}, \widetilde{\chi}_{A'}$ are regular on the interior of S_F. With this notation, the $(+)$ parts correspond to the modes with untwiddled coefficients in the expansion in spinor harmonics on S^3, multiplying unbarred harmonics having positive eigenvalues $\frac{1}{2}(n + \frac{3}{2})$, $\forall n \geq 0$, for the three-dimensional Dirac operator on the bounding S^3. In D'Eath and Halliwell 1987 it was shown that one has to fix the untwiddled variables on S_F so as to have a regular solution in the massless limit. Requiring that all untwiddled coefficients for the spin-$\frac{1}{2}$ field should vanish on S^3, one finds that : $P = Da^{-\frac{11}{360}}$ for a massless Majorana field, which is only described by the pair $\left(\phi_A, \widetilde{\phi}_{A'}\right)$. From now on the term Majorana will only denote this type of spinor which has half as many components as a Dirac spinor (although also a Weyl spinor has this property). Indeed, because in Riemannian space there is no complex conjugation which relates unprimed and primed spinors, *it is not possible* to impose the *reality* conditions which define Majorana spinors in Lorentzian space-time : $\lambda_A = \overline{\lambda}_A \equiv \overline{\lambda_{A'}}$, $\lambda_{A'} = \overline{\lambda}_{A'} \equiv \overline{\lambda_A}$. Thus our terminology has a *different meaning* with respect to the Lorentzian regime, and is chosen so as to make contact with previous papers on one-loop effects, where (though using sometimes a different formalism) the pair $\left(\phi_A, \widetilde{\phi}_{A'}\right)$ has also been studied and called Majorana field. However, the suitable definition of Euclidean Majorana spinors is still receiving careful consideration in the recent literature (Mehta 1991). In Kupsch and Thacker 1990, the authors are able to define massless Euclidean Majorana spinors in four dimensions (and they also study the problem in higher dimensions). Their approach seems to be unique in that it is derived from a systematic decomposition of the algebra of Schwinger functions and of the measure space for fermionic integration, but the formulation of their four-dimensional analysis in two-component spinor language is not yet available.

At the end of chapter four we also introduce Hawking's local boundary conditions on S^3 : $\sqrt{2}\,{}_e n^{AA'} \psi_A = \pm \widetilde{\psi}^{A'}$ (see definition of the Euclidean normal ${}_e n^{AA'}$ after (A.20)), and we show that these conditions lead to a classical boundary-value problem which admits an unique smooth solution inside S^3.

In chapter five, we first extend the method of chapter four to the $N = 1$ supergravity model. The PDF for the prefactor are picked out imposing the supersymmetry constraints (cf. (1.3.12)) and choosing the gauge condition : $e_{AA'}{}^j \psi_j^A = 0$, $e_{AA'}{}^j \widetilde{\psi}_j^{A'} = 0$. Setting the coefficients of these degrees of freedom equal to zero on S^3, we find that the PDF contribution to the prefactor due to the spin-$\frac{3}{2}$ field is $Da^{\frac{289}{360}}$, which does not cancel the one due to the gravitational field. However, one cannot yet reach any final conclusion, because the boundary conditions used are not related by supersymmetry to Dirichlet or other local boundary conditions for the gravitational field.

We then turn our attention to Hawking's local boundary conditions on S^3 involving field strengths and normals : $2^s \ {}_e n^{AA'} \ldots {}_e n^{LL'} \phi_{A \ldots L} = \pm \widetilde{\phi}^{A' \ldots L'}$ for spin $s > 0$, and : $\phi = \pm \overline{\phi}$, ${}_e n^{AA'} \nabla_{AA'} \phi = \mp {}_e n^{BB'} \nabla_{BB'} \overline{\phi}$ for a complex scalar field (Hawking 1983). The consideration of these boundary conditions is suggested by the existence of a one-to-one map between solutions of the massless free-field equations for adjacent spins s and $s + \frac{1}{2}$. This map is obtained using spin-lowering and spin-raising operators (cf. section 5.6), and the condition under which the spin-lowering operator preserves Hawking's boundary conditions is derived. The relevance to gauged supergravity of Hawking's local boundary conditions is as follows. Anti-de Sitter space can be seen as the maximally supersymmetric solution of the $O(N)$-gauged supergravity theories. It has topology $S^1 \times R^3$, and its closed timelike curves can be removed by considering its covering space. The covering space is conformally flat, conformally imbedded into half of the Einstein static universe, and its boundary is the product of the time axis of the Einstein universe and the two-sphere. The solutions of the twistor equation, subject to a suitable boundary condition (Hawking 1983), generate the rigid supersymmetry transformations between massless linearized fields of different spins on an anti-de Sitter background. In Breitenlohner and Freedman 1982 it was shown that the rigid supersymmetry transformations map classical *solutions* of the linearized field equations, subject to local boundary conditions at infinity, to classical *solutions* for an adjacent spin, again obeying the boundary conditions at infinity. The boundary conditions studied in Breitenlohner and Freedman 1982 and in Hawking 1983 are thus in a certain sense supersymmetric. However, rigid supersymmetry transformations do not map *eigenfunctions* of the spin$-s$ wave operators to *eigenfunctions* for adjacent spin $s \pm \frac{1}{2}$ with the *same* eigenvalues. Hence there is no *a priori* reason to expect any cancellation between adjacent spins in a one-loop calculation of the functional determinant about flat space with a S^3 boundary for a supersymmetric theory (D'Eath and Esposito 1991a, Esposito 1991). Indeed, these local boundary conditions are shown to imply that : $\zeta(0) = \frac{7}{45}$ for spin 0, and $\zeta(0) = -\frac{77}{180}$ (magnetic) or $\frac{13}{180}$ (electric) for spin 1. In fact, for a complex scalar field ϕ one is led to study a Dirichlet problem for $Re\ \phi$ and a Neumann problem for $Im\ \phi$ or viceversa, and for the spin-1 field strength one is led to set to zero on S^3 either the components of the magnetic field or the components of the electric field.

For spin $\frac{1}{2}$, $\zeta(0)$ is harder to compute. Indeed, even just the proof that such a number is a well-defined mathematical concept in our case requires a careful study. This result is obtained using the theory of $SU(2)$ spinors in Euclidean four-space and a result due to von Neumann, which finally show that a first-order differential operator for the boundary-value problem exists which is symmetric and has self-adjoint extensions. The corresponding eigenvalue condition is found to be : $J_{n+1}(Ea) = \pm J_{n+2}(Ea)$, $\forall n \geq 0$. One can easily find the solutions to this equation when $n \rightarrow \infty$ and Ea is fixed, or when n is fixed and $| Ea | \rightarrow \infty$, but this information is not sufficient to compute numerically $\zeta(0)$. One of the main problems is that the recurrence relations for generating Bessel functions of higher order become a source of large errors when the argument is comparable with the order. Nevertheless, at this stage we can already remark that, had supersymmetry been *working*, the value of $\zeta(0)$ should be the same for all values of spin. Thus in particular $\zeta(0)$ for spin 0 should be equal to $\zeta(0)$ for spin 1, whereas it turns out this is not true. As a partial justification of this result, we finally prove why, in the comparison spin 0 vs spin $\frac{1}{2}$, the spin-lowering operator does not lead to the same eigenvalues.

In chapter six, we first review the geometric formulation of gauge theories and the problems one meets in trying to quantize them. The Hamiltonian methods for quantization of these theories are essentially (in increasing importance) the reduced phase-space approach, Dirac's method and the Batalin-Fradkin-Vilkovisky theory (Henneaux 1985). As a second step, we apply the extended phase-space method of the Russian authors to the spin-1 field, which is described by a constrained Hamiltonian system with first-class constraints. The charge Q and the gauge-fixed action are derived. The Lorentzian path integral is restricted to the trajectories of the extended phase space which satisfy the boundary conditions which project on the original phase space. This path integral is independent of gauge fixing provided the charge Q remains nilpotent on quantization. As a third step, the Faddeev-Popov formula is derived from the path integral involving the gauge-fixed action, and the relation between extended phase space and reduced phase space is discussed. We then follow the work in Hartle and Schleich 1987 on ghost fields and gauge-fixing terms in the path integral for linearized gravity. The quantum theory is well-defined provided it is expressed in terms of the transverse-traceless modes for the perturbed three-metric. One can formally show that a suitable measure exists such that the gauge-invariant form of the path integral for the ground-state wave function is equal to the one expressed in terms of these PDF only. However, it is not yet clear whether this result can be generalized to the boundary-value problems occurring in quantum cosmology. As an example of this problem, we show how to obtain the Moss and Poletti result, according to which magnetic and electric boundary conditions for spin-1 fields lead to the same result for $\zeta(0)$, thus correcting the PDF values of chapter five.

We finally address the problem of the *direct* $\zeta(0)$ calculation for spin 1 including ghost fields and gauge degrees of freedom. Mixed boundary conditions are then imposed for the PDF and the gauge modes, requiring that self-adjointness of the corresponding differential operators and BRST invariance are respected. The gauge-averaging term $\frac{\Phi^2}{2\alpha}$ is inserted into the action of the Faddeev-Popov formula, where Φ depends linearly on the covariant derivatives of A_0 and A_k, and calculations are finally performed in the $\alpha = 1$ gauge. In fact the final result is known to be independent of α, and the $\alpha = 1$ value leads to a simplification of the Euclidean action quadratic in the gauge modes. However, the gauge modes obey a coupled system of two second-order ordinary differential equations, which is the effect of the $3 + 1$ split of the vector potential. In the $\alpha = 1$ gauge one can easily obtain the corresponding decoupled fourth-order ordinary differential equation, whose solution can be written as $\tau^{\mu} \sum_{k=0}^{\infty} a_{n,k}(n, k, \lambda_n)\tau^k$, $\forall n \geq 2$. With this notation, μ obeys a fourth-order algebraic equation, λ_n are the eigenvalues, and the $a_{n,k}$ coefficients obey recurrence relations. Unfortunately, the recurrence relations for the $a_{n,k}$ are very complicated, so that the exact solution does not have a simple expression in terms of well-known special functions. Thus, a *direct* and gauge-invariant $\zeta(0)$ calculation for gauge fields in the presence of boundaries remains an open problem (although the results of *indirect* calculations in these cases are by now well-known).

In chapter seven, at first we show that the imposition of local boundary conditions on the three-sphere for the Weyl spinor implies that either the electric curvature E_{ij} or the magnetic curvature B_{ij} should vanish on S^3. The physical degrees of freedom of the problem are picked out imposing the transverse-traceless (TT) gauge condition. Taking as background a flat Euclidean space, the perturbed three-metric is expanded in terms of modes $q^n(\tau)$ and of transverse-traceless hyperspherical harmonics. Preserving in time the TT gauge condition one finds that the perturbative part of the lapse, and the shift functions, vanish $\forall\tau$. Working at linear order in the perturbations in the expansion of E_{ij} and B_{ij}, one finds that, if the linearized E_{ij} is vanishing on S^3, this implies that :

$$Q^{(n)}(\tau) = \frac{d^2 q^n}{d\tau^2} - \frac{3}{\tau}\frac{dq^n}{d\tau} + \frac{(n^2-1)}{\tau^2} q^n = 0 \quad \text{on} \quad S^3 \quad , \quad \forall n \geq 3 \quad .$$

However, if $Q^{(n)}$ is required to vanish on S^3, there is no surface term to add to the linearized action integral such that the linearized Einstein equations follow from requiring the action to be stationary. Thus we conclude that fixing the linearized electric curvature on S^3 does not lead to a well-posed classical boundary-value problem.

When the linearized B_{ij} is vanishing on S^3, the first derivatives of the modes must vanish on S^3. Generalizing a technique developed in Moss 1989 for a real scalar field obeying Robin boundary conditions on S^3, this boundary condition is found to imply the PDF result : $\zeta(0) = \frac{112}{45}$. The ultimate consequence of our calculations seems to be that the simplest supersymmetric models are not PDF one-loop finite in the presence of a three-sphere boundary.

In chapter eight, using the theory of canonical products, we prove that the function which occurs in the nonlinear eigenvalue condition for the massless spin-$\frac{1}{2}$ field subject to local boundary conditions on S^3 still obeys a relation of the kind used for $\zeta(0)$ calculations in section 7.3. Using the parameter $x \rightarrow \infty$, after the analytic continuation $x \rightarrow ix$, and defining as usual $\alpha_m \equiv \sqrt{m^2+x^2}$, one can again expand asymptotically a formula of the kind $\log(\Sigma)$ as $\sum_{n=1}^{\infty} \frac{A_n(m,\alpha_m)}{(\alpha_m)^n}$. However, the coefficients A_n are more involved with respect to the bosonic case. In fact they are polynomials with both even and odd powers of $t \equiv \frac{m}{\alpha_m}$. Using the contour formulae of section 7.3 and the uniform asymptotic expansions of the regular Bessel functions J_n and their first derivatives J'_n, we find for a massless Majorana field : $\zeta(0) = \frac{11}{360}$. We also prove that no higher-order terms, i.e. $A_n(m,\alpha_m)$ when $n > 3$, can contribute to our value of $\zeta(0)$. Remarkably, this $\zeta(0)$ value is equal to the one found in chapter four, where we use global boundary conditions.

In chapter nine, we perform a one-loop calculation for the spin-$\frac{3}{2}$ field subject to the following local boundary conditions on S^3 : $\sqrt{2}\ {}_e n_A^{\ A'} \psi_i^A = \pm \widetilde{\psi}_i^{A'}$. The consideration of these boundary conditions is suggested by the work in Luckock and Moss 1989. In fact in that paper it is shown that in simple supergravity the spatial tetrad $e^{AA'}_{\ \ \ i}$ and the projection $\left(\pm \widetilde{\psi}_i^{A'} - \sqrt{2}\ {}_e n_A^{\ A'} \psi_i^A \right)$ formed from the spatial components $\left(\psi_i^A,\ \widetilde{\psi}_i^{A'} \right)$ of the spin-$\frac{3}{2}$ potential, transform into each other under half of the local supersymmetry transformations at the boundary. Moreover, the supergravity action, suitably modified by a boundary term, is invariant under this class of local supersymmetry transformations. One is thus led to specify $e^{AA'}_{\ \ \ i}$ and $\left(\pm \widetilde{\psi}_i^{A'} - \sqrt{2}\ {}_e n_A^{\ A'} \psi_i^A \right)$ on the boundary in computing the quantum amplitude. In our model, the background is again flat Euclidean space, and the PDF are picked out imposing the supersymmetry constraints and choosing the gauge condition : $e_{AA'}^{\ \ \ i} \psi_i^A = 0$, $e_{AA'}^{\ \ \ i} \widetilde{\psi}_i^{A'} = 0$. The boundary conditions are then shown to imply the eigenvalue condition : $[J_{n+2}(E)]^2 - [J_{n+3}(E)]^2 = 0$, $\forall n \geq 0$. We can thus apply the technique of chapter eight. The $\zeta(0)$ value is given by the one for the massless Majorana spin-$\frac{1}{2}$ field plus two other terms, yielding : $\zeta(0) = -\frac{289}{360}$. Thus, for the gravitino field the PDF method leads to a result for $\zeta(0)$ which is equal to the PDF value one obtains setting to zero on S^3 all untwiddled coefficients of ψ_i^A and $\widetilde{\psi}_i^{A'}$. At the end, we show that also extended supergravity theories do not lead to a vanishing result for the total PDF $\zeta(0)$, by using the PDF values we have obtained for bosonic and fermionic fields. However, our PDF results are still preliminary because alternative possibilities such as the effects of antisymmetric tensor fields or pseudoscalars have not yet been considered. Moreover, *direct* $\zeta(0)$ calculations for spin $\frac{3}{2}$ including ghost fields and gauge-fixing terms are still too difficult at present. Thus the *direct* calculation of the *full* $\zeta(0)$ for extended supergravity theories is still an open problem.

The third part of our monograph is devoted to the study of nonperturbative properties of classical gravity. Indeed, the singularity problem in cosmology and theories of gravity with torsion are at the heart of the first chapters of our book, because they motivate quantum cosmology and supergravity. Thus, chapter ten can be seen as the nonperturbative continuation of the first two parts of our work within the framework of classical theories of gravitation. This is why we study the mathematical structures of space-time, presenting at first a description of some aspects of the differentiable, spinor, causal, asymptotic and Hamiltonian structure. For example we present the Poincaré group as the subgroup of the Bondi-Metzner-Sachs group which maps good cuts into good cuts, and we introduce Ashtekar's spinorial variables for canonical gravity. These variables can also be useful in reformulating local boundary conditions for the Weyl spinor.

The gauge theory of the Poincaré group naturally leads to theories with torsion (Hehl et al. 1976, Nester 1983, Awada et al. 1986, Esposito 1989b). After a brief clarification of this point, we finally study the singularity problem in space-times with torsion. A preliminary definition of singularities is proposed which is based on nonspacelike geodesic incompleteness, even though for theories with torsion test particles do not move on geodesics. The study of the geodesic equation for a closed FRW universe with torsion shows that the definition has physical relevance. We then prove how to extend Hawking's singularity theorem without causality assumptions to the space-time of the ECSK theory. For our proof we study the generalized Raychaudhuri equation in the ECSK theory, the conditions for the existence of conjugate points and properties of maximal timelike geodesics.

There are relevant differences between our work and the one in Hehl et al. 1974. Basically, we have a different approach to the theory of geodesics and we emphasize the role played by the full extrinsic-curvature tensor and by the variation formulae. In fact our definition of geodesics as autoparallel curves involves the full connection with torsion, whereas extremal curves used by Hehl et al. only involve Christoffel symbols. Moreover, we keep the field equations of the ECSK theory in their original form, avoiding the introduction of a modified energy-momentum tensor. In so doing (the reader might compare with section 10.7.2.4) we prove that the space-time of the ECSK theory cannot be timelike geodesically complete if :

(a) $\left[R(U,U) - 2\tilde{\omega}^2 - \tilde{\nabla}_a\left(\dot{U}\right)^a\right] \geq 0$ for any nonspacelike vector U;

(b) there exists a compact spacelike three-surface S without edge;

(c) the trace K of the extrinsic-curvature tensor $K(X,Y)$ of S is either everywhere positive or everywhere negative.

In (a), $R(X,Y)$ denotes the Ricci tensor, and $2\tilde{\omega}^2$ is related to the generalized vorticity tensor. This tensor is given by the sum of the vorticity tensor for the Christoffel symbols plus a tensor related to the spin tensor, which is in turn related to the torsion tensor through the field equations of the ECSK theory. A special form of condition (a) can be found in the inequality (10.7.17). Condition (c) also involves torsion, which appears in the full extrinsic-curvature tensor. It is therefore clear why the occurrence of singularities in closed cosmological models based on the ECSK theory can be less generic than in general relativity, even though relevant counterexamples have been given in the literature (cf. Nester and Isenberg 1977; we are very grateful to Professor F. Hehl for bringing this paper on torsion singularities to our attention).

CHAPTER TWO

CANONICAL QUANTUM GRAVITY

Abstract. Dirac's theory of constrained Hamiltonian systems is first described, discussing in detail primary and secondary constraints, first-class and second-class constraints, Dirac brackets, effective Hamiltonian, total Hamiltonian and extended Hamiltonian. On quantization, the operator versions of first-class constraints become supplementary conditions on the wave function, provided these constraints are consistent with one another and with the Schrödinger equation. Second-class constraints, if they cannot be brought into the first class by independent linear combinations, become instead equations between operators in the quantum theory. Moreover, commutation relations are taken to correspond to Dirac-bracket relations, provided it is possible to find an irreducible representation of the Dirac-brackets algebra. Alternative approaches also exist, where second-class constraints play instead a more relevant role in the quantum theory.

Canonical quantum gravity is then studied with emphasis on the original formulation of the theory. The first-class constraint algebra of the theory is studied in some detail following De Witt's work, and geometrical and topological properties of Wheeler's superspace are discussed following the mathematical work of Fisher.

2. Canonical Quantum Gravity

2.1 Hamiltonian Methods in Physics

Since quantum cosmology combines ideas from the canonical and path-integral approach to quantum gravity, it is rather important to describe the ideas underlying these subjects. Thus we here begin by presenting Dirac's theory of constrained Hamiltonian systems, and its application to Einstein's theory of gravitation.

It is quite often the case that theories of interest in modern physics are formulated as constrained systems. This happens whenever the Lagrangian L is singular (we begin by studying the case of a finite number of degrees of freedom for simplicity), so that there is no unique solution of Hamilton's equations of motion expressing the velocities in terms of the canonical coordinates q^i and conjugate momenta p_i (i.e. $det \frac{\partial^2 L}{\partial \dot{q}^i \partial \dot{q}^j} = 0$). One then finds that certain functions $\varphi_m^{(1)} : T^*Q \rightarrow R$ exist such that :

$$\varphi_m^{(1)}(q,p) \approx 0 \quad . \tag{2.1.1}$$

Following Dirac 1964 and Hanson et al. 1976, we say that the primary constraint $\varphi_m^{(1)}$ is weakly zero; in other words, in working out Poisson brackets on the phase space with other canonical variables, $\varphi_m^{(1)}$ can be set to zero only *after* these brackets have been computed, and some of these brackets do not vanish. Note that, on the constraint manifold Σ defined by (2.1.1), no Poisson bracket can be defined, since if some $F(q,p) \approx 0$ on Σ, its gradient does not necessarily vanish weakly on Σ. The problem thus arises to *extend* out of Σ the Legendre transform H_c of L by adding a linear combination of constraints. One is then led to define an effective Hamiltonian $\widetilde{H}$ given by the sum :

$$\widetilde{H} \equiv H_c + u_m(q,p)\varphi_m^{(1)}(q,p) \quad . \tag{2.1.2}$$

Note also that the coefficients u_m of linear combination are not constants, but depend on the canonical variables q,p. In light of (2.1.1), the new equations of motion generated by $\widetilde{H}$ are :

$$\dot{q}^i \equiv \left\{ q^i, \widetilde{H} \right\} \approx \frac{\partial H_c}{\partial p_i} + u_m(q,p)\frac{\partial \varphi_m^{(1)}}{\partial p_i} \quad , \tag{2.1.3}$$

$$\dot{p}_i \equiv \left\{ p_i, \widetilde{H} \right\} \approx -\frac{\partial H_c}{\partial q^i} - u_m(q,p) \frac{\partial \varphi_m^{(1)}}{\partial q^i} \quad , \tag{2.1.4}$$

where curly brackets denote Poisson brackets. We now have to make sure that the primary constraints $\varphi_m^{(1)}$ are preserved in time. This implies that :

$$\dot{\varphi}_m^{(1)} \equiv \left\{ \varphi_m^{(1)}, \widetilde{H} \right\} \approx \left\{ \varphi_m^{(1)}, H_c \right\} + u_j(q,p) \left\{ \varphi_m^{(1)}, \varphi_j^{(1)} \right\} \approx 0 \quad . \tag{2.1.5}$$

Essentially three possibilities occur :

(1) Eq. (2.1.5) already holds by virtue of (2.1.1);

(2) Eq. (2.1.5) can be solved for $u_j = u_j(q,p)$;

(3) Eq. (2.1.5) leads to secondary constraints $\varphi_j^{(2)}(q,p)$ independent of $u_h(q,p)$.

The outlined procedure is then repeated until all secondary constraints, and finally all $u_j = u_j(q,p)$ have been found, so that $\widetilde{H} = \widetilde{H}(q,p)$. The set S_c of all constraints is then given by (we define as secondary all constraints which are not primary) :

$$S_c \equiv \left\{ \varphi_m^{(1)}, m = 1, ..., L; \quad \varphi_j^{(2)}, j = 0, 1, ..., M \right\} \quad . \tag{2.1.6}$$

In geometrical language, one can compute the secondary constraints $\varphi_j^{(2)}(q,p)$ by defining the vector field (cf. (2.1.3-4)) :

$$\Gamma \equiv \left(\frac{\partial H_c}{\partial p_i} + u_m(q,p) \frac{\partial \varphi_m^{(1)}}{\partial p_i} \right) \frac{\partial}{\partial q^i} - \left(\frac{\partial H_c}{\partial q^i} + u_m(q,p) \frac{\partial \varphi_m^{(1)}}{\partial q^i} \right) \frac{\partial}{\partial p_i} \quad ,$$

and then taking the Lie derivative $L_\Gamma \varphi_m^{(1)}$.

A more convenient (and fundamental) division of the set S_c can be, however, obtained. For this purpose, we note that the Poisson bracket of any two elements of S_c may or may not be a linear combination of constraints. This property can be made precise by giving the following definitions (Dirac 1964, Hanson et al. 1976) :

Definition 2.1.1 The function F defined on T^*Q is a first-class quantity if

$$\left\{ F, \varphi \right\} \approx 0 \quad , \quad \forall \varphi \in S_c \quad . \tag{2.1.7}$$

Definition 2.1.2 The function F defined on T^*Q is a second-class quantity if

$$\exists \varphi \in S_c : \left\{ F, \varphi \right\} \not\approx 0 \quad . \tag{2.1.8}$$

Note that a number of second-class constraints might be brought into the first class by means of *independent* linear combinations. We call irreducible those second-class constraints which cannot be brought into the first class. In what follows, we always assume the second-class constraints we deal with are irreducible. First-class constraints are here denoted by $\varphi_m^{(I)}(q,p)$, and second-class constraints by $\varphi_j^{(II)}(q,p)$. An equivalent definition of the set S_c of all constraints is then given by :

$$S_c \equiv \left\{ \varphi_m^{(I)}, m = 0,1,...,L'; \quad \varphi_j^{(II)}, j = 0,1,...,M' \right\} \quad , \tag{2.1.9}$$

where $L' + M' = L + M$. Moreover, with our notation, $\varphi_m^{(1,I)}(q,p)$ and $\varphi_m^{(1,II)}(q,p)$ denote primary first-class and primary second-class constraints, whereas $\varphi_m^{(2,I)}(q,p)$ and $\varphi_m^{(2,II)}(q,p)$ denote secondary first-class and secondary second-class constraints respectively.

As we shall see in our book, relevant examples of first-class Hamiltonian systems are electromagnetism and general relativity (section 2.4, chapters five and six), whereas second-class constraints occur for example in relativistic theories of gravitation with nonvanishing torsion (chapter ten), or after imposing gauge constraints on a first-class system (chapters five and six).

The definition of second-class constraints enables one to understand which quantities can be set to zero also *before* computing any bracket. For this purpose, one first proves (Dirac 1964) that the matrix $C_{lm} = \left\{ \varphi_l^{(II)}, \varphi_m^{(II)} \right\}$ of Poisson brackets of second-class constraints is nonsingular. The second step (Hanson et al. 1976) is to define, for any given $G(q,p)$, a new variable $\widetilde{G}(q,p)$ given by :

$$\widetilde{G} \equiv G - \left\{ G, \varphi_l^{(II)} \right\} C_{lm}^{-1} \varphi_m^{(II)} \quad , \tag{2.1.10}$$

so that $\left\{\widetilde{G},\varphi_r^{(II)}\right\} \approx 0$, $\forall r \in\left\{0,1,...,M'\right\}$. In other words, $\widetilde{G}$ has vanishing Poisson brackets with all second-class constraints, whereas $\left\{\widetilde{G},\varphi_m^{(I)}\right\}$ may be $\not\approx 0$ for some $m \in \left\{0,1,...,L'\right\}$. Thus, defining the Dirac bracket of F_1 and F_2 as :

$$\left\{F_1,F_2\right\}^* \equiv\left\{F_1,F_2\right\}-\left\{F_1,\varphi_l^{(II)}\right\}C_{lm}^{-1}\left\{\varphi_m^{(II)},F_2\right\} \quad , \tag{2.1.11}$$

one finds the fundamental results :

$$\left\{F_1,F_2\right\}^* \approx\left\{\widetilde{F}_1,\widetilde{F}_2\right\} \approx\left\{\widetilde{F}_1,F_2\right\} \approx\left\{F_1,\widetilde{F}_2\right\} \quad , \tag{2.1.12}$$

$$\left\{G,\varphi_n^{(II)}\right\}^* \approx\left\{G,\varphi_n^{(II)}\right\}-\left\{G,\varphi_l^{(II)}\right\}C_{lm}^{-1}C_{mn}=0 \quad . \tag{2.1.13}$$

In other words, second-class constraints are now strongly vanishing, and Dirac brackets are the tool needed to achieve this and deal only with first-class constraints in the classical theory (see also section 2.3).

The Hamiltonians we are interested in are essentially of four kinds :

(1) The canonical Hamiltonian H_c, i.e. the Legendre transform of L. This is the relevant Hamiltonian if L is nonsingular;

(2) The effective Hamiltonian $\widetilde{H}$, i.e.

$$\widetilde{H} \equiv H_c + u_m(q,p)\varphi_m^{(1)}(q,p) = H_c-\left\{H_c,\varphi_l^{(1)}\right\}C_{lm}^{-1}\varphi_m^{(1)} \quad . \tag{2.1.14}$$

This is the relevant Hamiltonian for theories with second-class constraints (section 2.3).

(3) The total Hamiltonian H_T, i.e.

$$H_T \equiv \widetilde{H} + r_m(q,p)\varphi_m^{(1,I)}(q,p) \quad . \tag{2.1.15}$$

(4) The extended Hamiltonian H_E, i.e.

$$\begin{aligned} H_E &\equiv \widetilde{H} + r_m^{(A)}(q,p)\varphi_m^{(1,I)}(q,p) + r_j^{(B)}(q,p)\varphi_j^{(2,I)}(q,p) \\ &= H_T + r_j^{(B)}(q,p)\varphi_j^{(2,I)}(q,p) \quad . \end{aligned} \tag{2.1.16}$$

Thus H_T is given by the effective Hamiltonian $\widetilde{H}$ plus a linear combination of primary first-class constraints, whereas H_E is given by H_T plus a linear combination of secondary first-class constraints. This means that it is possible to include secondary constraints in the Hamiltonian, provided they are first-class.

In fact, as explained in Dirac 1964 (see below), there are certain changes in the $q's$ and $p's$ that do not correspond to a change of state, and which have as generators first-class secondary constraints. One is then led to generalize the equations of motion in order to allow as variation of a dynamical variable also any variation which does not correspond to a change of state. This is obtained using the extended Hamiltonian H_E. The Hamiltonians $\widetilde{H}$, H_T and H_E are sharply distinguished ways of defining a Hamiltonian function on the whole phase space, and they all reduce to H_c on the constraint manifold defined by (2.1.1) (and by (2.1.6) in the case of H_E). In particular, the H_E-formalism is reasonable from the physical point of view (see below), although it does not correspond to the original one deduced from a Lagrangian (Dirac 1964, Marmo et al. 1983).

Note that, in the case of first-class constrained systems, (2.1.15) becomes :

$$H_T \equiv H_c + s_m(q,p)\varphi_m^{(1,I)}(q,p) \quad . \tag{2.1.17}$$

For example, this is what happens for electromagnetism. In chapter six, we shall however use the extended Hamiltonian H_E for electromagnetic fields, following Hanson et al. 1976. An intermediate step is also possible for $U(1)$ gauge theory (and general relativity), following Dirac 1964. Namely, one only includes secondary first-class constraints in the Hamiltonian H_I defined as :

$$H_I \equiv H_c + r_j^{(B)}(q,p)\varphi_j^{(2,I)}(q,p) \quad . \tag{2.1.18}$$

Whenever the extended Hamiltonian H_E is used, the equations of motion take the form :

$$\begin{aligned} \dot{q}^i \equiv \left\{q^i, H_E\right\} \approx \left\{q^i, \widetilde{H}\right\} \\ + r_m^{(A)}(q,p)\left\{q^i, \varphi_m^{(1,I)}\right\} + r_j^{(B)}(q,p)\left\{q^i, \varphi_j^{(2,I)}\right\} \quad , \end{aligned} \tag{2.1.19}$$

$$\dot{p}_i \equiv \Big\{p_i, H_E\Big\} \approx \Big\{p_i, \widetilde{H}\Big\}$$
$$+ r_m^{(A)}(q,p)\Big\{p_i, \varphi_m^{(1,I)}\Big\} + r_j^{(B)}(q,p)\Big\{p_i, \varphi_j^{(2,I)}\Big\} \quad . \tag{2.1.20}$$

The physical state of the system is not affected by the infinitesimal contact transformations generated by the first-class constraints of the theory (Dirac 1964, Hanson et al. 1976). This is a crucial point, and Dirac's original argument is as follows (Dirac 1964).

Given at time t any function $g_t(q,p)$, we study its time evolution. At time $t+\epsilon$, $g_{t+\epsilon}$ is found by definition as :

$$g_{t+\epsilon} = g_t + \epsilon\Big\{g, H_T\Big\} \quad . \tag{2.1.21}$$

Setting $r_m = 0$ in (2.1.15), this yields :

$$g_{t+\epsilon} = g_t + \epsilon\Big\{g, \widetilde{H}\Big\} \quad . \tag{2.1.22}$$

However, we may also take $r_m = c_m \neq 0$. This leads to :

$$\hat{g}_{t+\epsilon} = g_t + \epsilon\Big\{g, \widetilde{H}\Big\} + \epsilon\, c_m \Big\{g, \varphi_m^{(1,I)}\Big\} \quad . \tag{2.1.23}$$

Both choices must correspond to the same physical state at time $t+\delta t$, since the physical state at $t+\epsilon$ is the one arising from the given initial physical state at time t. From (2.1.22-23) we find :

$$\hat{g}_{t+\epsilon} - g_{t+\epsilon} = \epsilon\, c_m \Big\{g, \varphi_m^{(1,I)}\Big\} \quad . \tag{2.1.24}$$

Thus all primary first-class constraints $\varphi_m^{(1,I)}$, regarded as generators of a contact transformation, give rise to a transformation which does not change the physical state.

Dirac's next step is to consider two of these transformations. Denoting by $\chi_a^{(1,I)}$ and $\psi_a^{(1,I)}$ two primary first-class constraints, the first transformation changes g into $g + \epsilon\Big\{g, \chi_a^{(1,I)}\Big\}$, and the second changes $g + \epsilon\Big\{g, \chi_a^{(1,I)}\Big\}$ into

$$g_1 = g + \epsilon\Big\{g, \chi_a^{(1,I)}\Big\} + \epsilon'\Big\{g + \epsilon\Big\{g, \chi_a^{(1,I)}\Big\}, \psi_a^{(1,I)}\Big\} \quad . \tag{2.1.25}$$

He then neglects ϵ^2 and ϵ'^2, but retains $\epsilon\epsilon'$ in the calculation. Even though ϵ^2, ϵ'^2 and $\epsilon\epsilon'$ are of the same order, this approximation is valid, since otherwise, by retaining terms involving ϵ^2 and ϵ'^2, one would obtain an equation holding for all values of ϵ and ϵ'; one has thus to set to zero the coefficients of ϵ^2, of ϵ'^2 and of $\epsilon\epsilon'$. This would lead to three equations, but the first two are trivial so that one is not interested in them. Now, by applying the two transformations in the reverse order, one obtains :

$$g_2 = g + \epsilon\Big\{g, \chi_a^{(1,I)}\Big\} + \epsilon'\Big\{g, \psi_a^{(1,I)}\Big\} + \epsilon\epsilon'\Big\{\Big\{g, \psi_a^{(1,I)}\Big\}, \chi_a^{(1,I)}\Big\} \quad . \tag{2.1.26}$$

In light of (2.1.25), this leads to :

$$g_1 - g_2 = \epsilon\epsilon'\Big\{g, \Big\{\chi_a^{(1,I)}, \psi_a^{(1,I)}\Big\}\Big\} \quad . \tag{2.1.27}$$

Thus, using the group property of all transformations which leave the physical state unchanged, Dirac finds there must be further transformations of this type which do not affect the physical state. He then points out that the only generalization of the argument is that the primary first-class constraints $\chi_a^{(1,I)}$ and $\psi_a^{(1,I)}$ might be replaced by secondary first-class constraints $\rho_a^{(2,I)}$ and $\sigma_a^{(2,I)}$ also leading to an equation formally identical to (2.1.27) and thus generating transformations which do not change the physical state.

However, Dirac made it clear he had not been able to obtain a rigorous mathematical proof that all first-class constraints, whether primary or secondary, do not change the physical state. This has been proved only much later in Costa et al. 1985. As discussed in detail in Costa et al. 1985 and in Zhi 1992, if a function f is first-class, it must be gauge-invariant, and

$$\dot{f}(q,p) \approx \Big\{f, H_T\Big\} \approx \Big\{f, H_E\Big\} \quad . \tag{2.1.28}$$

In other words, the total Hamiltonian H_T and the extended Hamiltonian H_E generate the same time evolution for the gauge-invariant functions $f(q,p)$, and are thus physically equivalent. By contrast, if $f(q,p)$ is gauge-dependent, H_T and H_E generate *different* equations of motion, and the two formalisms cannot be compared. The detailed calculations in Zhi

1992 add evidence in favour of Dirac's and Costa's interpretation of first-class constraints being correct.

2.2 Dirac's Quantization of First-Class Constrained Hamiltonian Systems

Suppose we deal with a theory where all constraints are first-class. The canonical coordinates q^i, with conjugate momenta p_i, are made into operators satisfying canonical commutation relations (hereafter referred to as CCR) corresponding to the Poisson brackets of the classical theory. The mathematically rigorous form of these CCR is the exponentiated Weyl form (Reed and Simon 1972, Isham 1984) :

$$U(a_1)U(a_2) = U(a_1 + a_2) \quad , \tag{2.2.1}$$

$$V(b_1)V(b_2) = V(b_1 + b_2) \quad , \tag{2.2.2}$$

$$U(a)V(b) = e^{i\hbar ab}V(b)U(a) \quad , \tag{2.2.3}$$

where $U(a) = e^{-ia\hat{p}}$ and $V(b) = e^{-ib\hat{q}}$. By virtue of the Stone-von Neumann theorem, the unique (up to unitary equivalence) unitary representation of (2.2.1-3) is :

$$(V(b)\psi)(q) = e^{-ibq}\psi(q) \quad , \tag{2.2.4}$$

$$(U(a)\psi)(q) = \psi(q - \hbar a) \quad . \tag{2.2.5}$$

The generators of $U(a)$ and $V(b)$ are $\hat{p}$ and $\hat{q}$ respectively, and satisfy the familiar relations

$$(\hat{q}\psi)(q) \equiv q\psi(q) \quad , \tag{2.2.6}$$

$$(\hat{p}\psi)(q) \equiv -i\hbar\frac{\partial\psi}{\partial q}(q) \quad , \tag{2.2.7}$$

together with the CCR :

$$\left[\hat{q}, \hat{p}\right] = i\hbar \quad . \tag{2.2.8}$$

This holds on the dense domain of infinitely differentiable functions of compact support. Of course, the more general form of (2.2.8) we are interested in is :

$$\left[\hat{q}^i, \hat{p}_j\right] = i\hbar\delta^i_j \quad . \tag{2.2.9}$$

We then study a Schrödinger equation :

$$i\hbar\frac{\partial\psi}{\partial t} = \hat{H}_T\psi \quad , \tag{2.2.10}$$

where ψ is the wave function, and $\hat{H}_T$ the suitably defined operator corresponding to the first-class Hamiltonian (2.1.17), also denoted by $\hat{H}'$ in the literature (Dirac 1964, Hanson et al. 1976). The next step in the quantization program is to impose *all first-class constraints* as supplementary conditions on the wave function, so that :

$$\hat{\varphi}_l^{(I)}\psi = 0 \quad . \tag{2.2.11}$$

The naturally occurring question is whether the equations (2.2.11) are consistent with one another. Indeed, for $l' \neq l$, we know that :

$$\hat{\varphi}_{l'}^{(I)}\psi = 0 \quad . \tag{2.2.12}$$

We now multiply (2.2.12) by $\hat{\varphi}_l^{(I)}$ and (2.2.11) by $\hat{\varphi}_{l'}^{(I)}$. The subtraction of the resulting equations leads to :

$$\left[\hat{\varphi}_l^{(I)}, \hat{\varphi}_{l'}^{(I)}\right]\psi = 0 \quad . \tag{2.2.13}$$

Note that, in the classical theory, (2.2.13) would be obviously satisfied, since by definition the Poisson bracket of any two first-class constraints is again a linear combination of first-class constraints. In the quantum theory, however, it is not *a priori* obvious that

$$\left[\hat{\varphi}_l^{(I)}, \hat{\varphi}_{l'}^{(I)}\right] = c_{ll'm}(q,p)\hat{\varphi}_m^{(I)}(q,p) \quad . \tag{2.2.14}$$

In other words, for (2.2.13) to be a consequence of (2.2.11-12), the *additional* condition (2.2.14) should hold. Since $c_{ll'm}$ depends on all q's and p's, it does not commute with the

$\hat{\varphi}_m^{(I)}$ in the quantum theory. The problem is thus to make sure that $c_{ll'm}$ appears on the left in (2.2.14), and no extra terms occur.

If (2.2.14) holds, the first-class constraints are consistently imposed as supplementary conditions on ψ. By contrast, if it is not possible to obtain (2.2.14) with the help of suitable factor-ordering prescriptions, the quantum theory we are looking for is ill-defined. In the favourable case, one should go on, and check whether the conditions (2.2.11) are also consistent with the Schrödinger equation (2.2.10). Since $\hat{H}_T$ is a first-class Hamiltonian, this means (following the technique leading to (2.2.13)) one should find :

$$\left[\hat{\varphi}_l^{(I)}, \hat{H}_T\right] = 0 \quad , \tag{2.2.15}$$

which holds provided

$$\left[\hat{\varphi}_l^{(I)}, \hat{H}_T\right] = b_{lm}(q,p)\hat{\varphi}_m^{(I)}(q,p) \quad . \tag{2.2.16}$$

Again, we find a condition which is certainly satisfied in the classical theory, whereas in the quantum theory there may be serious problems, since b_{lm} depends on the $q's$ and $p's$, and thus does not necessarily appear on the left when we compute the commutator $\left[\hat{\varphi}_l^{(I)}, \hat{H}_T\right]$.

2.3 Dirac's Quantization of Second-Class Constrained Hamiltonian Systems

In section 2.1 we remarked that irreducible second-class constraints can be eliminated in the classical theory, after they are set strongly to zero by using the Dirac brackets (2.1.11). Following Dirac 1964, it can be instructive to see what happens in the simplest case, i.e. when two second-class constraints exist of the form :

$$q^l \approx 0 \quad , \quad p_l \approx 0 \quad . \tag{2.3.1}$$

Of course, the corresponding quantum operators cannot be used to impose supplementary conditions on the wave function of the form :

$$\hat{q}^l\psi = 0 \quad , \quad \hat{p}_l\psi = 0 \quad , \quad \Rightarrow \left[\hat{q}^l, \hat{p}_l\right]\psi = 0 \quad ,$$

since these would be inconsistent with the CCR (2.2.9) :

$$\left[\hat{q}^l, \hat{p}_l\right]\psi = i\hbar\psi \quad .$$

Now, by virtue of (2.3.1), one might point out that q^l and p_l are not of interest. One is thus led to define a new bracket $\left\{ \, , \, \right\}^*$ in the classical theory, where the degree of freedom corresponding to the index l has been discarded, so that :

$$\left\{A, B\right\}^* \equiv \sum_{n \neq l} \left(\frac{\partial A}{\partial q^n} \frac{\partial B}{\partial p_n} - \frac{\partial A}{\partial p_n} \frac{\partial B}{\partial q^n} \right) \quad . \tag{2.3.2}$$

One then tries to set up the quantum theory in terms of all degrees of freedom but for the value $n = l$. We are then looking for an operator representation of Dirac brackets, such that second-class constraints are realized strongly, i.e. as equations between operators. In the above example, since Dirac brackets are :

$$\left\{q^l, p_l\right\}^* = 0 \quad , \tag{2.3.3}$$

$$\left\{q^i, p_j\right\}^* = \delta^i_j \quad , \quad \forall i, j \neq l \quad , \tag{2.3.4}$$

an irreducible representation of Dirac brackets is given by :

$$\hat{Q}^l = \hat{p}_l = 0 \quad , \tag{2.3.5}$$

$$\hat{Q}^i \psi = q^i \psi \quad , \quad \forall i \neq l \quad , \tag{2.3.6}$$

$$\hat{p}_i \psi = -i\hbar \frac{\partial \psi}{\partial q^i} \quad , \quad \forall i \neq l \quad , \tag{2.3.7}$$

i.e. the usual Schrödinger representation for q^i and p_i, $\forall i \neq l$, described in section 2.2, and the zero operator for $\hat{Q}^l$ and $\hat{p}_l$. Dirac's quantization program when second-class constraints exist can be thus described as follows.

(1) One picks out irreducible second-class constraints;

(2) Using Dirac brackets (2.1.11), these constraints are set strongly to zero;

(3) Since, for any $g(q,p)$, $\left\{g,H_T\right\}^* \approx \left\{g,H_T\right\}$, the classical equations of motion take the form :

$$\dot{g} \approx \left\{g,H_T\right\}^* \quad ;$$

(4) On quantization, commutation relations are taken to correspond to Dirac-bracket relations, and second-class constraints are realized as equations between operators;

(5) With the exception of some (simple) examples, one has to bear in mind that it may be not possible to find an irreducible representation of the Dirac-brackets algebra. This remains an open problem;

(6) Remaining first-class constraints are imposed as supplementary conditions on the wave function;

(7) For these first-class constraints, one has to check that (2.2.14) and (2.2.16) hold.

In light of points (5) and (7) as above, Dirac's quantization in the presence of second-class constraints is far from being straightforward. It may be thus very helpful to study a nontrivial example of second-class constrained systems. For this purpose, we here describe, following in part Hojman and Shepley 1991 and Kulshreshtha 1992, a second-class system with no secondary constraints. The four primary second-class constraints may be reduced by one. In the corresponding theory there are three primary second-class constraints and one secondary second-class constraint. The theory is then quantized using Dirac's method.

In other words, as done in Kulshreshtha 1992, one begins by studying a theory described by the Lagrangian :

$$L = (q_2 + q_3)\dot{q}_1 + q_4\dot{q}_3 + V(q_2, q_3, q_4) \quad , \tag{2.3.8}$$

where $V(q_2,q_3,q_4) \equiv \frac{1}{2}\left(q_4^2 - 2q_2q_3 - q_3^2\right)$. The study of (2.3.8) is motivated by a generalization of the work by Feynman recently described in Dyson 1990, and we here use, for simplicity of notation, a convention for the indices of q's and p's different from what we have done so far. In light of (2.3.8) and of the definitions $p_i \equiv \frac{\partial L}{\partial \dot{q}_i}$, one finds the four primary constraints :

$$\rho_1 \equiv \left(p_1 - q_2 - q_3\right) \approx 0 \quad , \tag{2.3.9}$$

$$\rho_2 \equiv p_2 \approx 0 \quad , \tag{2.3.10}$$

$$\rho_3 \equiv \left(p_3 - q_4\right) \approx 0 \quad , \tag{2.3.11}$$

$$\rho_4 \equiv p_4 \approx 0 \quad . \tag{2.3.12}$$

As usual, the weak-equality symbol means that the constraints only vanish identically on the constraint manifold Σ, but may have nonvanishing Poisson brackets with some canonical variables. The canonical Hamiltonian H_c, i.e. the Legendre transform of L, is then given by :

$$\begin{aligned} H_c &\equiv \sum_{\alpha=1}^{4} p_\alpha \dot{q}_\alpha - L \\ &= \left(p_1 - q_2 - q_3\right)\dot{q}_1 + p_2\dot{q}_2 + \left(p_3 - q_4\right)\dot{q}_3 \\ &+ p_4\dot{q}_4 - V(q_2, q_3, q_4) \quad . \end{aligned} \tag{2.3.13}$$

However, as we said in section 2.1, we want to extend H_c out of Σ. For this purpose we define the effective Hamiltonian $\widetilde{H}$ (cf. (2.1.2)) on the whole phase space, which coincides with H_c on Σ. In our case, $\widetilde{H}$ becomes :

$$\begin{aligned} \widetilde{H} &\equiv H_c + \Lambda_1\rho_1 + \Lambda_2\rho_2 + \Lambda_3\rho_3 + \Lambda_4\rho_4 \\ &= \lambda_1\rho_1 + \lambda_2\rho_2 + \lambda_3\rho_3 + \lambda_4\rho_4 - V(q_2, q_3, q_4) \quad , \end{aligned} \tag{2.3.14}$$

where $\lambda_i = \lambda_i(q, p) \equiv \left(\Lambda_i + \dot{q}_i\right)$. Interestingly, the preservation in time of the primary constraints (2.3.9-12) leads to no secondary constraints in the theory, since

$$\left\{\rho_1, \widetilde{H}\right\} = -\left(\lambda_2 + \lambda_3\right) \quad , \quad \left\{\rho_2, \widetilde{H}\right\} = \lambda_1 - q_3 \quad , \tag{2.3.15a}$$

$$\left\{\rho_3, \widetilde{H}\right\} = \lambda_1 - \lambda_4 - q_2 - q_3 \quad , \quad \left\{\rho_4, \widetilde{H}\right\} = -\lambda_3 + q_4 \quad , \tag{2.3.15b}$$

which implies :

$$\lambda_1 = q_3 \quad , \quad \lambda_2 = -q_4 \quad , \quad \lambda_3 = q_4 \quad , \quad \lambda_4 = -q_2 \quad . \tag{2.3.16}$$

Moreover, the reader can easily check that the determinant of the 4×4 matrix of Poisson brackets is nonvanishing, so that this matrix is invertible and the constraints are second-class. Using the definition (2.1.11) of Dirac brackets, the nonvanishing Dirac brackets of the theory are found to be :

$$\left\{q_1, q_2\right\}^* = \left\{q_3, q_4\right\}^* = -\left\{q_2, q_4\right\}^* = 1 \quad , \tag{2.3.17}$$

$$\left\{p_3, q_2\right\}^* = -\left\{p_k, q_k\right\}^* = 2 \quad , \quad \forall k = 1,3 \quad . \tag{2.3.18}$$

The calculation expressed by (2.3.17-18) is very important, since nonvanishing Dirac brackets play the key role on quantization (cf. (2.3.4) and (2.3.6-7)).

It may be now instructive to reduce the theory described by (2.3.8) to a three-dimensional one setting $q_4 = constant = k$. One thus deals with a model described by the Lagrangian :

$$\widetilde{L} = \left(q_2 + q_3\right)\dot{q}_1 + k\dot{q}_3 + W(q_2, q_3) \quad , \tag{2.3.19}$$

where $W(q_2, q_3) = V(q_2, q_3, k) = \frac{1}{2}\left(k^2 - 2q_2q_3 - q_3^2\right)$. Using again the definition of canonical momenta p_i, one now finds three primary constraints :

$$\psi_1 \equiv \left(p_1 - q_2 - q_3\right) \approx 0 \quad , \tag{2.3.20}$$

$$\psi_2 \equiv p_2 \approx 0 \quad , \tag{2.3.21}$$

$$\psi_3 \equiv \left(p_3 - k\right) \approx 0 \quad . \tag{2.3.22}$$

This leads to a secondary constraint, i.e. $q_2 \approx 0$, and all constraints are second-class (Corichi 1992). The passage to quantum theory is then made replacing nonvanishing Dirac brackets by the commutators of the corresponding quantum operators, as described in the first part of this section.

However, we should say that progress has been recently made in developing new methods to study similar problems. As pointed out in McMullan 1991, physicists have become interested in theories where the splitting into first- and second-class constraints is not so desirable. The main motivations for taking this point of view seem to be (McMullan 1991) :

(1) In the case of the Green-Schwarz formulation of the super-string, one finds that the requirement of a manifestly supersymmetric description of the system is incompatible with Dirac's treatment of second-class constraints at the classical level.

(2) It may happen that the first-class constraint algebra develops an anomaly upon quantization. In the quantum theory one is then dealing with what is essentially a second-class system, but there is no obvious way to classically remove these constraints, since they were originally first-class. Relevant examples are given by the quantization of the bosonic string, and by the anomaly in Gauss's law when gauge theories are coupled to chiral fermions.

In McMullan 1991, the author finds that one can deal with second-class constraints in the quantum theory, and he shows what are the additional conditions to impose on a state so as to call it a physical state. Attention is there restricted to the case where second-class constraints $\xi_a^{(II)}$ can be decomposed into two first-class subsets $\left(\varphi_a^{(I)}, \chi_b^{(I)}\right)$, where the $\varphi_a^{(I)}$ are linear in momentum and the $\chi_a^{(I)}$ are gauge-fixing terms with no momentum dependence. Interestingly, the example here studied in Eqs. (2.3.1-7) is there studied in a completely different way, finding the two physical states of the second-class system. The details can be found in section 4 of McMullan 1991, but cannot be given here, since this would force us to become very technical, anticipating the content of section 6.2 of our book.

From the point of view of gravitational physics, it appears both desirable and possible to apply McMullan's ideas to the quantization of theories of gravity with torsion, which are relativistic theories of gravitation with second-class constraints (Esposito 1989b and references therein).

2.4 ADM Formalism and Constraints in Canonical Quantum Gravity

In section 2.2 we have studied Dirac's general method to quantize first-class constrained systems, and we are now aiming to apply this technique to general relativity. For this purpose, it may be useful to describe the main ideas of the Arnowitt-Deser-Misner (hereafter referred to as ADM) formalism. This is a canonical formalism for general relativity that enables one to re-write Einstein's field equations in first-order form and explicitly solved with respect to a time variable. For this purpose, one assumes that four-dimensional space-time (M, g) can be foliated by a family of $t = constant$ spacelike surfaces S_t, giving rise to a $3 + 1$ decomposition of the original four-geometry. The basic geometric data of this decomposition are as follows (MacCallum 1975).

(1) The induced three-metric h of the three-dimensional spacelike surfaces S_t. This yields the intrinsic geometry of the three-space. h is also called the first fundamental form of S_t, and is positive-definite with our conventions.

(2) The way each S_t is imbedded in (M, g). This is known once we are able to compute the spatial part of the covariant derivative of the normal n to S_t. Denoting by ∇ the four-connection of (M, g), one is thus led to define the tensor :

$$K_{ij} \equiv -\nabla_j n_i \quad . \tag{2.4.1}$$

Note that K_{ij} is symmetric if and only if ∇ is symmetric (cf. (10.7.18-19) and (10.7.24)). In general relativity, an equivalent definition of K_{ij} is $K_{ij} \equiv -\frac{1}{2}(L_n h)_{ij}$, where L_n denotes the Lie derivative along the normal to S_t. The tensor K is called extrinsic-curvature tensor, or second fundamental form of S_t.

(3) How the coordinates are propagated off the surface S_t. For this purpose one defines the vector $(N, N^1, N^2, N^3)dt$ connecting the point (t, x^i) with the point $(t + dt, x^i)$. Thus, given the surface $x^0 = t$ and the surface $x^0 = t + dt$, $N dt \equiv d\tau$ specifies a displacement normal to the surface $x^0 = t$. Moreover, $N^i dt$ yields the displacement from the point (t, x^i) to the foot of the normal to $x^0 = t$ through $(t + dt, x^i)$. In other words, the N^i arise since

the $x^i = constant$ lines do not coincide in general with the normals to the $t = constant$ surfaces (cf. Fig. 4.2 in MacCallum 1975). According to a well-established terminology, N is the lapse function, and the N^i are the shift functions. They are the tool needed to achieve the desired space-time foliation.

In light of points (1-3) as above, the four-metric g can be locally cast in the form :

$$g = h_{ij}\Big(dx^i + N^i dt\Big)\otimes\Big(dx^j + N^j dt\Big) - N^2 dt \otimes dt \quad , \tag{2.4.2}$$

which coincides with Eq. (1.1.1). This implies that :

$$g_{00} = -\Big(N^2 - N_i N^i\Big) \quad , \tag{2.4.3}$$

$$g_{i0} = g_{0i} = N_i \quad , \tag{2.4.4}$$

$$g_{ij} = h_{ij} \quad , \tag{2.4.5}$$

whereas, using the property $g^{\lambda\nu} g_{\nu\mu} = \delta^{\lambda}{}_{\mu}$, one finds :

$$g^{00} = -\frac{1}{N^2} \quad , \tag{2.4.6}$$

$$g^{i0} = g^{0i} = \frac{N^i}{N^2} \quad , \tag{2.4.7}$$

$$g^{ij} = h^{ij} - \frac{N^i N^j}{N^2} \quad . \tag{2.4.8}$$

Interestingly, the covariant g_{ij} and h_{ij} coincide, whereas the contravariant g^{ij} and h^{ij} differ as shown in (2.4.8). In terms of N, N^i and h, the extrinsic-curvature tensor defined in (2.4.1) takes the form :

$$K_{ij} \equiv \frac{1}{2N}\left(- \frac{\partial h_{ij}}{\partial t} + N_{i|j} + N_{j|i} \right) \quad , \tag{2.4.9}$$

where the stroke | denotes covariant differentiation on the spacelike three-surface S_t, and indices of K_{ij} are raised using h^{il}. Eq. (2.4.9) can be also written as :

$$\frac{\partial h_{ij}}{\partial t} = N_{i|j} + N_{j|i} - 2N K_{ij} \quad . \tag{2.4.10}$$

Eq. (2.4.10) should be supplemented by another first-order equation expressing the time derivative of the momenta π^{ij} conjugate to the three-metric, or equivalently the time evolution of K_{ij} (since π^{ij} is related to K^{ij}). The details can be found for example in MacCallum 1975 and in Ashtekar 1988.

Using the ADM variables described so far, the form of the action integral I for pure gravity that is stationary under variations of the metric vanishing on the boundary is (in $c = 1$ units) :

$$\begin{aligned} I &\equiv \frac{1}{16\pi G}\int_M {}^{(4)}R\,\sqrt{-g}\,d^4x + \frac{1}{8\pi G}\int_{\partial M} K^i_i\,\sqrt{h}\,d^3x \\ &= \frac{1}{16\pi G}\int_M \left[{}^{(3)}R + K_{ij}K^{ij} - \left(K^i_i\right)^2\right] N\sqrt{h}\,d^3x\,dt \quad . \end{aligned} \tag{2.4.11}$$

The boundary term appearing in (2.4.11) is necessary since ${}^{(4)}R$ contains second derivatives of the metric, and integration by parts in the Einstein-Hilbert part

$$I_H \equiv \frac{1}{16\pi G}\int_M {}^{(4)}R\,\sqrt{-g}\,d^4x$$

of the action also leads to a boundary term equal to $-\frac{1}{8\pi G}\int_{\partial M} K^i_i\,\sqrt{h}\,d^3x$. Denoting by G_{ab} the Einstein tensor $G_{ab} \equiv {}^{(4)}R_{ab} - \frac{1}{2}g_{ab}\,{}^{(4)}R$, and defining :

$$\delta\Gamma^d_{ab} \equiv \frac{1}{2}g^{dl}\left[\nabla_a\left(\delta g_{lb}\right) + \nabla_b\left(\delta g_{la}\right) - \nabla_l\left(\delta g_{ab}\right)\right] \quad , \tag{2.4.12}$$

one then finds (York 1986) :

$$(16\pi G)\delta I_H = -\int_M \sqrt{-g}\,G^{ab}\,\delta g_{ab}\,d^4x + \int_{\partial M}\sqrt{-g}\left(g^{ab}\delta^c_d - g^{ac}\delta^b_d\right)\delta\Gamma^d_{ab}\left(d^3x\right)_c \quad , \tag{2.4.13}$$

which clearly shows that I_H is stationary if the Einstein equations hold, and the normal derivatives of the variations of the metric vanish on the boundary ∂M. In other words, I_H is not stationary under *arbitrary* variations of the metric, and stationarity is only achieved after adding to I_H the boundary term appearing in (2.4.11), if δg_{ab} is set to zero on ∂M. Other useful forms of the boundary term can be found in Gibbons and Hawking 1977 and in York 1986. Note also that, strictly, in writing down (2.4.11) one should also take into account a term arising from I_H (De Witt 1967a) :

$$I_t \equiv \frac{1}{8\pi G} \int dt \int_{\partial M} d^3x \, \partial_i \Big[\sqrt{h} \Big(K_l^l \, N^i - h^{ij} \, N_{|j} \Big) \Big] \quad .$$

However, we have not explicitly included I_t since it does not modify the results derived or described hereafter.

We are now ready to apply Dirac's technique to general relativity, so as to get a better understanding of Eqs. (1.1.4-7) and of the following discussion. As we know, consistency of the constraints is proved if one can show that their commutators lead to no new constraints. For this purpose, it may be useful to recall the equal-time commutation relations of the canonical variables :

$$\Big[N(x), \pi(x') \Big] = i\delta(x, x') \quad , \tag{2.4.14}$$

$$\Big[N_i(x), \pi^j(x') \Big] = i\delta_i^{j'} \quad , \tag{2.4.15}$$

$$\Big[h_{ij}, \pi^{k'l'} \Big] = i\delta_{ij}^{\ \ k'l'} \quad . \tag{2.4.16}$$

Note that, following De Witt 1967a, primes have been used, either on indices or on the variables themselves, to distinguish different points of three-space. In other words, we define :

$$\delta_i^{j'} \equiv \delta_i^j \, \delta(x, x') \quad , \tag{2.4.17}$$

$$\delta_{ij}^{\ \ k'l'} \equiv \delta_{ij}^{\ \ kl} \, \delta(x, x') \quad , \tag{2.4.18}$$

$$\delta_{ij}^{\ \ kl} \equiv \frac{1}{2} \Big(\delta_i^k \delta_j^l + \delta_i^l \delta_j^k \Big) \quad . \tag{2.4.19}$$

The reader can easily check that

$$\Big[\pi(x), \pi^i(x') \Big] = \Big[\pi(x), \mathcal{H}^i(x') \Big] = \Big[\pi(x), \mathcal{H}(x') \Big] = \Big[\pi^i(x), \mathcal{H}^j(x') \Big] = \Big[\pi^i(x), \mathcal{H}(x') \Big] = 0 \quad . \tag{2.4.20}$$

It now remains to compute the three commutators $\left[\mathcal{H}_i, \mathcal{H}_{j'}\right]$, $\left[\mathcal{H}_i, \mathcal{H}'\right]$, $\left[\mathcal{H}, \mathcal{H}'\right]$. The first two commutators are obtained using Eq. (1.1.5) and defining $\mathcal{H}_i \equiv h_{ij}\mathcal{H}^j$. Interestingly, $\mathcal{H}_i$ is homogeneous bilinear in the h_{ij} and π^{ij}, with the momenta always to the right. Since the correct version of Eq. (1.1.7) is :

$$\int_{S_t} \mathcal{H}\xi \, d^3x \, \psi = 0 \quad \forall \xi \quad , \tag{2.4.21}$$

$$\int_{S_t} \mathcal{H}_i \xi^i \, d^3x \, \psi = 0 \quad \forall \xi^i \quad , \tag{2.4.22}$$

we begin by computing (De Witt 1967a) :

$$\left[h_{ij}, i \int_{S_t} \mathcal{H}_{k'} \delta\xi^{k'} \, d^3x'\right] = -h_{ij,k} \, \delta\xi^k - h_{kj} \, \delta\xi^k{}_{,i} - h_{ik} \, \delta\xi^k{}_{,j} \quad , \tag{2.4.23}$$

$$\left[\pi^{ij}, i \int_{S_t} \mathcal{H}_{k'} \delta\xi^{k'} \, d^3x'\right] = -\left(\pi^{ij}\delta\xi^k\right)_{,k} + \pi^{kj} \, \delta\xi^i{}_{,k} + \pi^{ik} \, \delta\xi^j{}_{,k} \quad . \tag{2.4.24}$$

This calculation shows that the $\mathcal{H}_i$ are indeed generators of three-dimensional coordinate transformations $\overline{x}^i = x^i + \delta\xi^i$ as we claimed in chapter one. Thus, using the definition of structure constants of the general coordinate-transformation group (De Witt 1967a) :

$$c^{k''}{}_{ij'} \equiv \delta^{k''}{}_{i,l''} \, \delta^{l''}_{j'} - \delta^{k''}{}_{j',l''} \, \delta^{l''}_{i} \quad , \tag{2.4.25}$$

the results (2.4.23-24) may be used to show that :

$$\left[\mathcal{H}_i(x), \mathcal{H}_j(x')\right] = -i \int_{S_t} \mathcal{H}_{k''} \, c^{k''}{}_{ij'} \, d^3x'' \quad , \tag{2.4.26}$$

$$\left[\mathcal{H}_i(x), \mathcal{H}(x')\right] = i\mathcal{H} \, \delta_{,i}(x, x') \quad . \tag{2.4.27}$$

Note that the only term of $\mathcal{H}$ which might lead to difficulties is the one quadratic in the momenta. However, all factors appearing in this term have homogeneous linear transformation laws under the three-dimensional coordinate-transformation group. They thus remain undisturbed in position when commuted with $\mathcal{H}_i$ (De Witt 1967a).

Finally, we have to study the commutator $\Big[\mathcal{H}(x), \mathcal{H}(x')\Big]$. The following remarks are in order :

(i) Terms quadratic in momenta contain no derivatives of h_{ij} or π^{ij} with respect to three-space coordinates. Hence they commute;

(ii) The terms $\sqrt{h(x)}\Big({}^{(3)}R(x)\Big)$ and $\sqrt{h(x')}\Big({}^{(3)}R(x')\Big)$ contain no momenta, so that they also commute;

(iii) The only commutators we are left with are the cross commutators, and they can be evaluated using the variational formula (De Witt 1967a) :

$$\delta\Big(\sqrt{h}\,{}^{(3)}R\Big) = \sqrt{h}\,h^{ij}h^{kl}\Big(\delta h_{ik,jl} - \delta h_{ij,kl}\Big) - \sqrt{h}\left[{}^{(3)}R^{ij} - \frac{1}{2}h^{ij}\Big({}^{(3)}R\Big)\right]\delta h_{ij} \quad , \tag{2.4.28}$$

which leads to :

$$\left[\int_{S_t} \mathcal{H}\,\xi_1\,d^3x, \int_{S_t} \mathcal{H}\,\xi_2\,d^3x\right] = i\int_{S_t} \mathcal{H}^l\Big(\xi_1\,\xi_{2,l} - \xi_{1,l}\,\xi_2\Big)\,d^3x \quad . \tag{2.4.29}$$

The commutators (2.4.26-27) and (2.4.29) clearly show that the constraint equations of canonical quantum gravity are first-class. As we said in section 1.1, the Wheeler-De Witt equation (2.4.21) is an equation on the superspace $S(\mathcal{M}) \equiv Riem(\mathcal{M})/Diff(\mathcal{M})$. In Wheeler's superspace-based hybrid scheme the spatial diffeomorphisms are factored, but the operator constraint (2.4.21) is retained (Isham 1984). Two very useful classical formulae frequently used in Lorentzian canonical gravity are :

$$\mathcal{H} \equiv (16\pi G)G_{ijkl}p^{ij}p^{kl} - \frac{\sqrt{h}}{16\pi G}\Big({}^{(3)}R\Big) \quad , \tag{2.4.30}$$

$$\mathcal{H} \equiv (16\pi G)^{-1}\Big[G^{ijml}K_{ij}K_{ml} - \sqrt{h}\Big({}^{(3)}R\Big)\Big] \quad , \tag{2.4.31}$$

where :

$$G_{ijkl} \equiv \frac{1}{2\sqrt{h}}\Big(h_{ik}h_{jl} + h_{il}h_{jk} - h_{ij}h_{kl}\Big) \quad , \tag{2.4.32}$$

$$G^{ijkl} \equiv \frac{\sqrt{h}}{2}\Big(h^{ik}h^{jl} + h^{il}h^{jk} - 2h^{ij}h^{kl}\Big) \quad , \tag{2.4.33}$$

and p^{ij} is here defined as $-\frac{\sqrt{h}}{16\pi G}\Big(K^{ij} - h^{ij}K\Big)$. Note that the factor -2 multiplying $h^{ij}h^{kl}$ in (2.4.33) is needed so as to obtain the identity :

$$G_{ijmn}G^{mnkl} = \frac{1}{2}\Big(\delta_i^k\delta_j^l + \delta_i^l\delta_j^k\Big) \quad . \tag{2.4.34}$$

Eq. (2.4.30) clearly shows that $\mathcal{H}$ contains a part quadratic in the momenta and a part proportional to ${}^{(3)}R$ (cf. (1.1.4)). On quantization, it is then hard to give a well-defined meaning to the second functional derivative $\frac{\delta^2}{\delta h_{ij}\delta h_{kl}}$ (cf. (1.2.2)), whereas the occurrence of ${}^{(3)}R$ makes it even more difficult to solve exactly the Wheeler-De Witt equation.

Nevertheless, since our book is mainly devoted to canonical quantum gravity within the old-variables approach (cf. section 10.6), our next aim is to describe the relevant geometrical and topological features of Wheeler's superspace.

2.5 Mathematical Theory of Wheeler's Superspace

Let $\mathcal{M}$ be a compact, connected, orientable, Hausdorff, C^∞ three-manifold without boundary. Following Fisher 1970, we say $\mathcal{M}$ is a *superspatial*. If g is a Riemannian (i.e. positive-definite) C^∞ metric on a superspatial $\mathcal{M}$, the pair $(\mathcal{M}, g)$ is also called a superspatial. For a given superspatial $\mathcal{M}$, one denotes by $Riem(\mathcal{M})$ the space of Riemannian C^∞ metrics on $\mathcal{M}$, and by $Diff(\mathcal{M})$ the group of C^∞ orientation-preserving diffeomorphisms of $\mathcal{M}$. $Diff(\mathcal{M})$ acts as a transformation group on $Riem(\mathcal{M})$; its action maps (f, g) to f^*g, where $f \in Diff(\mathcal{M})$ and $g \in Riem(\mathcal{M})$. As we said in section 1.1, the space of all orbits of $Diff(\mathcal{M})$:

$$S(\mathcal{M}) \equiv Riem(\mathcal{M})/Diff(\mathcal{M}) \quad , \tag{2.5.1}$$

is called *superspace*. In other words, $\forall g \in Riem(\mathcal{M})$, we consider all metrics obtained from g by the action of elements $f \in Diff(\mathcal{M})$. If two metrics g and $\overline{g}$ are on the same orbit, a diffeomorphism f of $\mathcal{M}$ onto itself exists such that :

$$f^* g = \overline{g} \quad , \tag{2.5.2}$$

which implies that g and $\overline{g}$ are isometric. Two metrics are isometric if and only if they lie on the same orbit. $S(\mathcal{M})$ is thus the set of geometries of $\mathcal{M}$, which are equivalence classes of isometric Riemannian metrics.

An important topological property of superspace, which is necessary to prove theorems on its structure, is the following result (Fisher 1970) :

Theorem 2.5.1 Metrization Theorem for Superspace: The superspace $S(\mathcal{M})$ is a connected, second-countable, metrizeable space (i.e. a countable basis of open sets exists for its topology, and there also exists a metric on $S(\mathcal{M})$ inducing on $S(\mathcal{M})$ the given topology).

Note that, since there are symmetric geometries on $\mathcal{M}$, there are neighbourhoods of $S(\mathcal{M})$ not homeomorphic to neighbourhoods of nonsymmetric geometries. This implies the superspace defined in (2.5.1) cannot be a manifold. However, all geometries with the same kind of symmetry have homeomorphic neighbourhoods and they are thus a manifold. Two theorems hold which enable one to understand how these manifolds can be put together to give rise to superspace. For us to be able to state these theorems, some further definitions are in order (Fisher 1970).

(i) Let G be a compact subgroup of $Diff(\mathcal{M})$, and let us denote by (G) all compact subgroups of $Diff(\mathcal{M})$ that are conjugate to G by an element in $Diff(\mathcal{M})$, i.e.

$$(G) \equiv \left\{ fGf^{-1} \mid f \in Diff(\mathcal{M}) \right\} \quad . \tag{2.5.3}$$

If some element of (H) is included in some element of (G), we say that $(H) \leq (G)$. The relation $\leq$ is a partial ordering. The partially-ordered set of conjugacy classes of compact subgroups of $Diff(\mathcal{M})$ is used to index a partition of $S(\mathcal{M})$.

(ii) A partition of a second-countable Hausdorff space X, is a set of nonempty subspaces $\left\{X_\alpha\right\}$ such that :

$$X = \bigcup_\alpha X_\alpha \quad , \quad \alpha \in A \quad , \tag{2.5.4}$$

$$X_\alpha \cap X_\beta \neq \phi \quad \Rightarrow \alpha = \beta \quad . \tag{2.5.5}$$

The subspace X_α is Hausdorff, second-countable, and its components

$$\left\{\left\{X_\alpha^i\right\} \mid i \in C_\alpha\right\}$$

yield a partition of X_α indexed by C_α. If $\left\{X_\alpha\right\}$ is a partition for X, $\left\{X_\alpha^i\right\}$ is the complete partition of X, indexed by

$$\left\{(\alpha, i) \mid \alpha \in A,\, i \in C_\alpha\right\} = \prod_\alpha C_\alpha \quad .$$

A partition is said to be a manifold partition if each X_α is a manifold.

(iii) A *stratification* (respectively, an *inverted stratification*) of a connected, second-countable, Hausdorff topological space X, is a countable, partially-ordered, manifold partition of X whose complete partition has the frontier property (respectively, inverted frontier property) :

$$X_\alpha^i \cap \overline{X_\beta^j} \neq \phi \quad , \quad \alpha \neq \beta \left(\Leftrightarrow (\alpha, i) \neq (\beta, j)\right)$$

$$\Rightarrow X_\alpha^i \subset \overline{X_\beta^j} \quad \textit{and} \quad \alpha < \beta \left(\Leftrightarrow (\alpha, i) < (\beta, j)\right)$$

(respectively : $\beta < \alpha$ ($\Leftrightarrow$ $(\beta, j) < (\alpha, i)$)).

The manifolds X_α are said the *strata* of the stratification, and the manifolds $\left\{X_\alpha^i\right\}$ are the connected strata.

We are now in a position to state the following theorems (Fisher 1970) :

Theorem 2.5.2 Decomposition Theorem for Superspace: The decomposition of $S(\mathcal{M})$ by the subspaces $\left\{S_G(\mathcal{M})\right\}$ is a countable, partially-ordered, C^∞-Fréchet manifold partition [A locally convex space which is metrizeable and complete is called a Fréchet space, and manifolds can be modeled on any linear space in which one has a theory of differential calculus. Fréchet manifolds are thus differentiable manifolds whose charts have values in a Fréchet space].

Theorem 2.5.3 Stratification Theorem for Superspace: The manifold partition $\left\{S_G(\mathcal{M})\right\}$ of $S(\mathcal{M})$ is an inverted stratification indexed by the symmetry type.

This means that geometries with a given symmetry are completely contained within the boundary of less symmetric geometries. For example, the so-called minisuperspace models studied in the literature (cf. section 1.2) only consider certain strata of superspace.

Since physicists are interested in writing down and studying differential equations, it would be of much help if superspace could be extended to a suitable manifold. This extended superspace is obtained from $Riem(\mathcal{M})$ by the action of a subgroup of $Diff(\mathcal{M})$. With the help of a suitable choice of such subgroup, the resulting orbit space is a manifold. One can then prove what follows (Fisher 1970).

Theorem 2.5.4 Extension Theorem for Superspace: For every n-dimensional superspatial $\mathcal{M}$, the superspace $S(\mathcal{M})$ can be extended to a suitable manifold $S^{ext}(\mathcal{M})$, such that :

$$dim\left(S^{ext}(\mathcal{M})/S(\mathcal{M})\right) = n(n+1) \quad . \tag{2.5.6}$$

CHAPTER THREE

PERTURBATIVE QUANTUM GRAVITY

Abstract. Perturbative quantum gravity can be formulated in terms of amplitudes of going from a three-metric and a matter-field configuration on a spacelike surface Σ to a three-metric and a field configuration on a spacelike surface Σ'. The Wick-rotated quantum amplitudes are here studied under the assumption that the analytic continuation to the real Riemannian section of the complexified space-time is possible, but this is not a generic property. Within the background-field method, one then expands both the four-metric g and matter fields ϕ about a configuration (g_0, ϕ_0) which is a solution of the classical equations of motion. If the one-loop approximation holds, the part of the action quadratic in the fluctuations about (g_0, ϕ_0) gives the dominant contribution to the quantum amplitudes. This leads to Gaussian integrals and to formally divergent amplitudes, since the one-loop result involves the determinant of second-order elliptic operators.

The corresponding divergences are regularized using the zeta-function method. For this purpose, following Hawking, one first defines a generalized zeta-function $\zeta(s)$ obtained from the eigenvalues of the elliptic operator $\mathcal{B}$ appearing in the calculation. Such $\zeta(s)$ can be analytically extended to a meromorphic function which only has poles at some finite values of s. The values of ζ and its first derivative at the origin enable one to express the one-loop quantum amplitudes, whose scaling properties only depend on $\zeta(0)$ under suitable assumptions on the measure in the path integral. Although it frequently happens that the eigenvalues of $\mathcal{B}$ cannot be computed exactly, the regularized $\zeta(0)$ value can be obtained by studying the heat equation for the elliptic operator $\mathcal{B}$. The corresponding integrated heat kernel $G(\tau)$ has an asymptotic expansion as $\tau \rightarrow 0^+$ for those boundary conditions which ensure self-adjointness of $\mathcal{B}$. The $\zeta(0)$ value is then given by the constant term in the asymptotic form of $G(\tau)$, and it also determines the one-loop divergences of physical theories.

Some relevant examples of gravitational background fields are then studied. These gravitational instantons are complete, four-dimensional Riemannian manifolds whose metric solves the Einstein equations with cosmological constant : $R(X,Y) - \Lambda g(X,Y) = 0$. The possible boundary conditions are : asymptotically Euclidean, asymptotically locally Euclidean, asymptotically flat, asymptotically locally flat, compact without boundary.

Perturbative renormalization, asymptotic expansions and summability methods in quantum field theory are finally described, and the standard argument showing the non-renormalizability of quantized general relativity is presented.

3.1 The One-Loop Approximation

In section 1.1 we have described the assumptions and problems of the path-integral approach to quantum gravity. We here examine perturbative properties of this theory within the background-field method, which will play a key role in the following chapters.

In the one-loop approximation (also called stationary phase or WKB method) one first expands *both* the metric g and matter fields ϕ about a metric g_0 and a field ϕ_0 which are solutions of the classical field equations :

$$g = g_0 + \overline{g} \quad , \tag{3.1.1}$$

$$\phi = \phi_0 + \overline{\phi} \quad . \tag{3.1.2}$$

One then assumes that the fluctuations $\overline{g}$ and $\overline{\phi}$ are so small that the dominant contribution to the path integral (1.1.17) comes from the quadratic order in the Taylor-series expansion of the action about the background fields g_0 and ϕ_0 (Hawking 1979b) :

$$I_E[g,\phi] = I_E[g_0,\phi_0] + I_2[\overline{g},\overline{\phi}] + \text{ higher} - \text{order terms} \quad , \tag{3.1.3}$$

so that the logarithm of the quantum-gravity amplitude $\widetilde{A}$ can be expressed as :

$$\log\left(\widetilde{A}\right) \sim -I_E[g_0,\phi_0] + \log \int D[\overline{g},\overline{\phi}] e^{-I_2[\overline{g},\overline{\phi}]} \quad . \tag{3.1.4}$$

Various properties of background fields will be studied in section 3.3. For our present purposes we are instead interested in the second term appearing on the right-hand side of (3.1.4). An useful factorization is obtained if ϕ_0 can be set to zero. One then finds that $I_2[\overline{g},\overline{\phi}] = I_2[\overline{g}] + I_2[\overline{\phi}]$, which implies (Hawking 1979b) :

$$\log\left(\widetilde{A}\right) \sim -I_E[g_0] + \log \int D[\phi] e^{-I_2[\phi]} + \log \int D[\overline{g}] e^{-I_2[\overline{g}]} \quad . \tag{3.1.5}$$

The one-loop term for matter fields with various spins (and boundary conditions) is the main object studied in the next chapters. We here recall some basic results, following again Hawking 1979b.

A familiar form of $I_2[\phi]$ is :

$$I_2[\phi] = \frac{1}{2} \int \phi B \phi \sqrt{g_0} \, d^4x \quad , \tag{3.1.6}$$

where the elliptic differential operator B depends on the background metric g_0. Note that B is a second-order operator for bosonic fields, whereas it is first-order for fermionic fields. In light of (3.1.6) it is clear we are interested in the eigenvalues $\left\{\lambda_n\right\}$ of B, with corresponding eigenfunctions $\left\{\phi_n\right\}$. If boundaries are absent, it is sometimes possible to know explicitly the eigenvalues with their degeneracies. This is what happens for example in de Sitter space (Allen 1983a-b). If boundaries are present, however, very little is known about the detailed form of the eigenvalues, once boundary conditions have been imposed.

We here assume for simplicity to deal with bosonic fields subject to (homogeneous) Dirichlet conditions on the boundary surface : $\phi = 0$ on ∂M, and $\phi_n = 0$ on ∂M, $\forall n$. It is in fact well-known that the Laplace operator subject to Dirichlet conditions has a positive-definite spectrum (Chavel 1984, page 9). The field ϕ can then be expanded in terms of the eigenfunctions ϕ_n of B as :

$$\phi = \sum_{n=n_0}^{\infty} y_n \phi_n \quad , \tag{3.1.7}$$

where the eigenfunctions ϕ_n are normalized so that :

$$\int \phi_n \phi_m \sqrt{g_0} \, d^4x = \delta_{nm} \quad . \tag{3.1.8}$$

Another formula we need is the one expressing the measure on the space of all fields ϕ as :

$$D[\phi] = \prod_{n=n_0}^{\infty} \mu \, dy_n \quad , \tag{3.1.9}$$

where the normalization parameter μ has dimensions of mass or $(\text{length})^{-1}$. Note that, if gauge fields appear in the calculation, the choice of gauge-fixing and the form of the measure in the path integral are not a trivial problem. The reader should always bear in mind this remark, and he will find a detailed (although incomplete) study of these issues in the following chapters.

Using well-known results about Gaussian integrals, the one-loop matter amplitudes $\widetilde{A}_{\phi}^{(1)}$ can be now obtained as :

$$\begin{aligned}\widetilde{A}_{\phi}^{(1)} &\equiv \int D[\phi] e^{-I_2[\phi]} \\ &= \prod_{n=n_0}^{\infty} \int \mu \, dy_n \, e^{-\frac{\lambda_n}{2} y_n^2} \\ &= \prod_{n=n_0}^{\infty} \left(2\pi\mu^2 \lambda_n^{-1}\right)^{\frac{1}{2}} \\ &= \frac{1}{\sqrt{\det\left(\frac{1}{2}\pi^{-1}\mu^{-2}\mathcal{B}\right)}} \quad . \end{aligned} \tag{3.1.10}$$

In the particular (and relevant) case of a complex scalar field in a complex or real Riemannian space-time, the complex conjugate ϕ^* of ϕ can be replaced by its analytic continuation $\widetilde{\phi}$, where $\widetilde{\phi}$ is now completely independent of ϕ. The formula (3.1.6) for the one-loop term is then replaced by :

$$I_2[\phi, \widetilde{\phi}] = \frac{1}{2} \int \widetilde{\phi} \mathcal{B} \phi \sqrt{g_0} \, d^4x \quad . \tag{3.1.11}$$

The adjoint operator $\mathcal{B}^\dagger$ has now eigenfunctions $\widetilde{\phi}_n$, and the field $\widetilde{\phi}$ can be expanded in terms of these $\widetilde{\phi}_n$ as :

$$\widetilde{\phi} = \sum_{n=n_0}^{\infty} \widetilde{y}_n \, \widetilde{\phi}_n \quad , \tag{3.1.12}$$

whereas the measure $D[\phi, \widetilde{\phi}]$ takes the form :

$$D[\phi, \widetilde{\phi}] = \prod_{n=n_0}^{\infty} \mu^2 \, dy_n \, d\widetilde{y}_n \quad . \tag{3.1.13}$$

Since one has to integrate over y_n and $\widetilde{y}_n$ independently, one now finds (cf. (3.1.10)) :

$$\widetilde{A}_{\phi}^{(1)} = \frac{1}{\det\left(\frac{1}{2}\pi^{-1}\mu^{-2}\mathcal{B}\right)} \quad . \tag{3.1.14}$$

This result will be very useful in chapter six, where complex scalar fields in real Riemannian space-time are studied at the end.

When fermionic fields appear in the path integral (1.1.17), one deals with a first-order elliptic operator, the Dirac operator, acting on independent spinor fields ψ and $\widetilde{\psi}$. These are anticommuting Grassmann variables obeying Berezin integration rules :

$$\int dw = 0 \quad , \quad \int w \, dw = 1 \quad . \tag{3.1.15}$$

The formulae (3.1.15) are all what we need, since powers of w greater than or equal to 2 vanish in light of the anticommuting property. The reader can then check that the one-loop amplitude for fermionic fields is :

$$\widetilde{A}_{\psi}^{(1)} = \det\left(\frac{1}{2}\mu^{-2}\mathcal{B}\right) \quad . \tag{3.1.16}$$

The main difference with respect to bosonic fields is the direct proportionality to the determinant. The following comments can be useful in understanding the meaning of (3.1.16).

Let us denote by γ^{μ} the gamma matrices, and by λ_i the eigenvalues of the Dirac operator $\gamma^{\mu}D_{\mu}$, and suppose that no zero-modes exist. More precisely, the eigenvalues of $\gamma^{\mu}D_{\mu}$ occur in equal and opposite pairs : $\pm\lambda_1, \pm\lambda_2$, ..., whereas the eigenvalues of the Laplace operator on spinors occur as $(\lambda_1)^2$ twice, $(\lambda_2)^2$ twice, and so on. For Dirac fermions (D) one thus finds :

$$\det{}_D\left(\gamma^{\mu}D_{\mu}\right) = \left(\prod_{i=1}^{\infty} \mid \lambda_i \mid\right)\left(\prod_{i=1}^{\infty} \mid \lambda_i \mid\right) = \prod_{i=1}^{\infty} \mid \lambda_i \mid^2 \quad , \tag{3.1.17}$$

whereas in the case of Majorana spinors (M), for which the number of degrees of freedom is halved (cf. section 1.4), one finds :

$$\det_M\Big(\gamma^\mu D_\mu\Big) = \prod_{i=1}^{\infty} \mid \lambda_i \mid = \sqrt{\det_D\Big(\gamma^\mu D_\mu\Big)} \quad . \tag{3.1.18}$$

In chapters four, five, eight and nine we shall apply several times Eq. (3.1.18).

3.2 Zeta-Function Regularization of Path Integrals

The formal expression (3.1.10) for the one-loop quantum amplitude clearly diverges since the eigenvalues λ_n increase without bound, and a regularization is thus necessary. For this purpose, the following technique has been described and applied by many authors (Hawking 1977, Hawking 1979b).

Bearing in mind that Riemann's zeta-function $\zeta_R(s)$ is defined as :

$$\zeta_R(s) \equiv \sum_{n=1}^{\infty} n^{-s} \quad , \tag{3.2.1}$$

one first defines a generalized zeta-function $\zeta(s)$ obtained from the (positive) eigenvalues of the second-order, self-adjoint operator $\mathcal{B}$. As already explained in appendix B (some repetition is unavoidable and indeed also useful), such $\zeta(s)$ can be defined as :

$$\zeta(s) \equiv \sum_{n=n_0}^{\infty} \sum_{m=m_0}^{\infty} d_m(n) \lambda_{n,m}^{-s} \quad . \tag{3.2.2}$$

This means that all the eigenvalues are completely characterized by two integer labels n and m, while their degeneracy d_m only depends on n. This is the case studied in the following chapters of our book. Note that formal differentiation of (3.2.2) at the origin yields :

$$\det\Big(\mathcal{B}\Big) = e^{-\zeta'(0)} \quad . \tag{3.2.3}$$

This result can be given a sensible meaning since, in four dimensions, $\zeta(s)$ converges for $Re(s) > 2$, and one can perform its analytic extension to a meromorphic function of s which only has poles at $s = \frac{1}{2}, 1, \frac{3}{2}, 2$. Since $\det\left(\mu \mathcal{B}\right) = \mu^{\zeta(0)} \det\left(\mathcal{B}\right)$, one finds the useful formula :

$$\log\left(\widetilde{A}_{\phi}\right) = \frac{1}{2}\zeta'(0) + \frac{1}{2}\log\left(2\pi\mu^2\right)\zeta(0) \quad . \tag{3.2.4}$$

As we said following (3.1.6), it may happen quite often that the eigenvalues appearing in (3.2.2) are unknown, since the eigenvalue condition, i.e. the equation leading to the eigenvalues by virtue of the boundary conditions, is a complicated equation which cannot be solved exactly for the eigenvalues. However, since the scaling properties of the one-loop amplitude are still given by $\zeta(0)$ (and $\zeta'(0)$) as shown in (3.2.4), efforts have been made to compute the regularized $\zeta(0)$ also in this case. The various steps of this program are as follows (Hawking 1977).

(1) One first studies the heat equation for the operator $\mathcal{B}$:

$$\frac{\partial}{\partial \tau} F(x, y, \tau) + \mathcal{B} F(x, y, \tau) = 0 \quad , \tag{3.2.5}$$

where the Green's function F satisfies the initial condition $F(x, y, 0) = \delta(x, y)$.

(2) Assuming completeness of the set $\left\{\phi_n\right\}$ of eigenfunctions of $\mathcal{B}$, the field ϕ can be expanded as

$$\phi = \sum_{n=n_i}^{\infty} a_n \phi_n \quad .$$

(3) The Green's function $F(x, y, \tau)$ is then given by :

$$F(x, y, \tau) = \sum_{n=n_0}^{\infty} \sum_{m=m_0}^{\infty} e^{-\lambda_{n,m}\tau} \phi_{n,m}(x) \otimes \phi_{n,m}(y) \quad . \tag{3.2.6}$$

(4) The corresponding (integrated) heat kernel is then :

$$G(\tau) = \int_M Tr\ F(x, x, \tau)\sqrt{g}\ d^4x = \sum_{n=n_0}^{\infty} \sum_{m=m_0}^{\infty} e^{-\lambda_{n,m}\tau} \quad . \tag{3.2.7}$$

(5) In light of (3.2.2) and (3.2.7), the generalized zeta-function can be also obtained as an integral transform of the integrated heat kernel :

$$\zeta(s) = \frac{1}{\Gamma(s)} \int_0^\infty \tau^{s-1} G(\tau)\, d\tau \quad . \tag{3.2.8}$$

(6) The hard part of the analysis is now to prove that $G(\tau)$ has an asymptotic expansion as $\tau \rightarrow 0^+$ (Greiner 1971). This property has been proved for all boundary conditions such that the Laplace operator is self-adjoint. The corresponding asymptotic expansion of $G(\tau)$ can be written as :

$$G(\tau) \sim b_1 \tau^{-2} + b_2 \tau^{-\frac{3}{2}} + b_3 \tau^{-1} + b_4 \tau^{-\frac{1}{2}} + b_5 + O\left(\sqrt{\tau}\right) \quad , \tag{3.2.9}$$

which implies :

$$\zeta(0) = b_5 \quad . \tag{3.2.10}$$

The result (3.2.10) is proved splitting the integral in (3.2.8) into an integral from 0 to 1 and an integral from 1 to ∞. The asymptotic expansion of $\int_0^1 \tau^{s-1} G(\tau)\, d\tau$ is then obtained using (3.2.9).

In other words, for a given second-order self-adjoint elliptic operator, we study the corresponding heat equation, and the integrated heat kernel $G(\tau)$. The regularized $\zeta(0)$ value is then given by the constant term appearing in the asymptotic expansion of $G(\tau)$ as $\tau \rightarrow 0^+$. As anticipated in section 1.4, the regularized $\zeta(0)$ value also yields the one-loop divergences of the theory for bosonic and fermionic fields. These divergences will be computed, in the presence of boundaries, in the second part of our book.

3.3 Gravitational Instantons

This section is devoted to the study of the background gravitational fields appearing in Eqs. (3.1.1-5). As we said in section 1.1, these gravitational instantons are complete four-geometries solving the Einstein equations $R(X,Y) - \Lambda g(X,Y) = 0$ when the four-metric g

has signature +4 (i.e. it is positive-definite, and thus called Riemannian). Following Pope 1981, essentially three cases can be studied.

3.3.1.1 Asymptotically Locally Euclidean Instantons

Even though it might seem natural to define first the asymptotically Euclidean instantons, it turns out there is not much choice in this case, since the only asymptotically Euclidean instanton is flat space. It is in fact well-known that the action of an asymptotically Euclidean metric with vanishing scalar curvature is ≥ 0, and it vanishes if and only if the metric is flat. Suppose now such a metric is a solution of the Einstein equations $R(X,Y) = 0$. Its action should be thus stationary also under constant conformal rescalings $g \rightarrow k^2 g$ of the metric. However, the whole action rescales then as $I_E \rightarrow k^2 I_E$, so that it can only be stationary and finite if $I_E = 0$. By virtue of the theorem previously mentioned, the metric g must then be flat (Gibbons and Pope 1979, Le Brun 1988).

In the asymptotically locally Euclidean case, however, the boundary at infinity has topology S^3/Γ rather than S^3, where Γ is a discrete subgroup of the local tetrad rotation group $SO(4)$. Many examples can then be found. The simplest was discovered by Eguchi and Hanson, and corresponds to $\Gamma = Z_2$ and $\partial M = RP^3$. This instanton is conveniently described using three left-invariant one-forms $\left\{\omega_i\right\}$ on the three-sphere, satisfying the $SU(2)$ algebra : $d\omega_i = -\frac{1}{2}\epsilon_i^{\ jk}\, \omega_j \wedge \omega_k$, and parametrized by Euler angles as follows :

$$\omega_1 = (\cos\psi)d\theta + (\sin\psi)(\sin\theta)d\phi \quad , \tag{3.3.1}$$

$$\omega_2 = -(\sin\psi)d\theta + (\cos\psi)(\sin\theta)d\phi \quad , \tag{3.3.2}$$

$$\omega_3 = d\psi + (\cos\theta)d\phi \quad , \tag{3.3.3}$$

where $\theta \in [0,\pi]$, $\phi \in [0,2\pi]$. The metric of the Eguchi-Hanson instanton may be thus written in the Bianchi-IX form (Pope 1981) :

$$g_1 = \left(1 - \frac{a^4}{r^4}\right) dr \otimes dr + \frac{r^2}{4}\left[(\omega_1)^2 + (\omega_2)^2 + \left(1 - \frac{a^4}{r^4}\right)(\omega_3)^2\right] \quad , \tag{3.3.4}$$

where $r \in [a, \infty[$. The singularity of g_1 at $r = a$ is only a coordinate singularity. We may get rid of it defining $4\frac{\rho^2}{a^2} \equiv 1 - \frac{a^4}{r^4}$, so that, as $r \rightarrow a$, the metric g_1 is approximated by the metric :

$$g_2 = d\rho \otimes d\rho + \rho^2 \Big[d\psi + (\cos\theta)d\phi\Big]^2 + \frac{a^2}{4}\Big[d\theta \otimes d\theta + (\sin\theta)^2 d\phi \otimes d\phi\Big] \quad . \tag{3.3.5}$$

Regularity of g_2 at $\rho = 0$ is then guaranteed provided one identifies ψ with period 2π. This implies in turn the local surfaces $r = constant$ have topology RP^3 rather than S^3, as we claimed. Note that at $r = a \Rightarrow \rho = 0$ the metric becomes that of a two-sphere of radius $\frac{a}{2}$. Following Gibbons and Hawking 1979, we say that $r = a$ is a bolt, where the action of the Killing vector $\frac{\partial}{\partial \psi}$ has a two-dimensional fixed-point set (Pope 1981).

A whole family of multi-instanton solutions is obtained taking the group $\Gamma = Z_k$. They all have a self-dual Riemann-curvature tensor (Hawking 1979b), and their metric takes the form :

$$g = V^{-1}\Big(d\tau + \underline{\gamma} \cdot \underline{dx}\Big)^2 + V\, \underline{dx} \cdot \underline{dx} \quad . \tag{3.3.6}$$

Following Pope 1981, $V = V(\underline{x})$ and $\underline{\gamma} = \underline{\gamma}(\underline{x})$ on an auxiliary flat three-space with metric $\underline{dx} \cdot \underline{dx}$. This metric g solves the Einstein vacuum equations provided $grad\ V = rot \underline{\gamma}$, which implies $\nabla^2 V = 0$. If we take :

$$V = \sum_{i=1}^{n} \frac{1}{\mid \underline{x} - \underline{x}_i \mid} \quad , \tag{3.3.7}$$

we obtain the desired asymptotically locally Euclidean multi-instantons. In particular, if $n = 1$ in (3.3.7), g describes flat space, whereas $n = 2$ leads to the Eguchi-Hanson instanton. If $n > 2$, there are $3n - 6$ arbitrary parameters, related to the freedom to choose the positions $\underline{x}_i$ of the singularities in V. These singularities correspond actually to coordinate singularities in (3.3.5), and can be removed using suitable coordinate transformations (Pope 1981).

3.3.1.2 Asymptotically Flat Instantons

This name is chosen since the underlying idea is to deal with metrics in the path integral which tend to the flat metric in three directions but are periodic in the Euclidean-time dimension. The basic example is provided by the Riemannian version $g_R^{(1)}$ (also called Euclidean) of the Schwarzschild solution :

$$g_R^{(1)} = \left(1 - 2\frac{M}{r}\right) d\tau \otimes d\tau + \left(1 - 2\frac{M}{r}\right)^{-1} dr \otimes dr + r^2 \Omega \quad , \tag{3.3.8}$$

where $\Omega = d\theta \otimes d\theta + (\sin\theta)^2 d\phi \otimes d\phi$ is the metric on a unit two-sphere. It is indeed well-known that, in the Lorentzian case (see section 10.2), the metric g_L is more conveniently written using Kruskal-Szekeres coordinates :

$$g_L = 32M^3 r^{-1} e^{-\frac{r}{2M}} \Big(- dz \otimes dz + dy \otimes dy \Big) + r^2 \Omega \quad , \tag{3.3.9}$$

where z and y obey the relations :

$$-z^2 + y^2 = \left(\frac{r}{2M} - 1\right) e^{\frac{r}{2M}} \quad , \tag{3.3.10}$$

$$\frac{(y+z)}{(y-z)} = e^{\frac{t}{2M}} \quad . \tag{3.3.11}$$

In the Lorentzian case, the coordinate singularity at $r = 2M$ can be thus avoided, whereas the curvature singularity at $r = 0$ remains and is described by the surface $z^2 - y^2 = 1$. However, if we set $\zeta = iz$, the analytic continuation to the section of the complexified space-time where ζ is real yields the positive-definite (i.e. Riemannian) metric :

$$g_R^{(2)} = 32M^3 r^{-1} e^{-\frac{r}{2M}} \Big(d\zeta \otimes d\zeta + dy \otimes dy \Big) + r^2 \Omega \quad , \tag{3.3.12}$$

where :

$$\zeta^2 + y^2 = \left(\frac{r}{2M} - 1\right) e^{\frac{r}{2M}} \quad . \tag{3.3.13}$$

It is now clear that also the curvature singularity at $r = 0$ has disappeared, since the left-hand side of (3.3.13) is ≥ 0, whereas the right-hand side of (3.3.13) would be equal to

-1 at $r = 0$. Note also that, setting $z = -i\zeta$ and $t = -i\tau$ in (3.3.11), and writing $\zeta^2 + y^2$ as $(y + i\zeta)(y - i\zeta)$ in (3.3.13), one finds :

$$y + i\zeta = e^{\frac{i\tau}{4M}} \sqrt{\frac{r}{2M} - 1} \; e^{\frac{r}{4M}} \quad , \tag{3.3.14}$$

$$y = \cos\left(\frac{\tau}{4M}\right) \sqrt{\frac{r}{2M} - 1} \; e^{\frac{r}{4M}} \quad , \tag{3.3.15}$$

which imply that the Euclidean time τ is periodic with period $8\pi M$. This periodicity on the Euclidean section leads to the interpretation of the Riemannian Schwarzschild solution as describing a black hole in thermal equilibrium with gravitons at a temperature $(8\pi M)^{-1}$ (Pope 1981). Moreover, the fact that any matter-field Green's function on this Schwarzschild background is also periodic in imaginary time leads to some of the thermal-emission properties of black holes. This is one of the greatest conceptual revolutions in modern gravitational physics.

There is also a local version of the asymptotically flat boundary condition in which ∂M has the topology of a nontrivial S^1-bundle over S^2, i.e. S^3/Γ, where Γ is a discrete subgroup of $SO(4)$. However, unlike the asymptotically Euclidean boundary condition, the S^3 is distorted and expands with increasing radius in only two directions rather than three (Pope 1981). The simplest example of an asymptotically locally flat instanton is the self-dual Taub-NUT solution, which can be regarded as a special case of the two-parameter Taub-NUT metrics :

$$g = \frac{(r+M)}{(r-M)} dr \otimes dr + 4M^2 \frac{(r-M)}{(r+M)} (\omega_3)^2 + \left(r^2 - M^2\right) \left[(\omega_1)^2 + (\omega_2)^2\right] \quad , \tag{3.3.16}$$

where the $\left\{\omega_i\right\}$ have been defined in (3.3.1-3). The main properties of the metric (3.3.16) are :

(I) $r \in [M, \infty[$, and $r = M$ is a removable coordinate singularity provided ψ appearing in (3.3.1-3) is identified modulo 4π;

(II) the $r = constant$ surfaces have S^3 topology;

(III) $r = M$ is a point at which the isometry generated by the Killing vector $\frac{\partial}{\partial\psi}$ has a zero-dimensional fixed-point set.

In other words, $r = M$ is a nut, using the terminology of section 1.1 (Gibbons and Hawking 1979).

There is also a family of asymptotically locally flat multi-Taub-NUT instantons. Their metric takes the form (3.3.6), but one should bear in mind that the formula (3.3.7) is replaced by :

$$V = 1 + \sum_{i=1}^{n} \frac{2M}{\mid \underline{x} - \underline{x}_i \mid} \quad . \tag{3.3.17}$$

Again, the singularities at $\underline{x} = \underline{x}_i$ can be removed, and the instantons are all self-dual.

3.3.1.3 Compact Instantons

Compact gravitational instantons occur in the course of studying the topological structure of the gravitational vacuum. This can be done by first of all normalizing all metrics in the functional integral to have a given four-volume V, and then evaluating the instanton contributions to the partition function as a function of their topological complexity. One then sends the volume V to infinity at the end of the calculation. If one wants to constrain the metrics in the path integral to have a volume V, this can be obtained by adding a term $\frac{\Lambda}{8\pi}V$ to the action. The stationary points of the modified action are solutions of the Einstein equations with cosmological constant Λ : $R(X,Y) - \Lambda g(X,Y) = 0$.

The few compact instantons that are known can be described as follows (Pope 1981).

(1) The four-sphere S^4, i.e. the Riemannian version of de Sitter space obtained by analytic continuation to positive-definite metrics. Setting to 3 for convenience the cosmological constant, the metric on S^4 takes the conformally-flat form (Pope 1981) :

$$g_I = d\beta \otimes d\beta + \frac{1}{4}(\sin\beta)^2 \Big[(\omega_1)^2 + (\omega_2)^2 + (\omega_3)^2\Big] \quad , \tag{3.3.18}$$

where $\beta \in [0, \pi]$. The apparent singularities at $\beta = 0, \pi$ can be made into regular nuts, provided the Euler angle ψ appearing in (3.3.1-3) is identified modulo 4π. The $\beta =$

constant surfaces are topologically S^3, and the isometry group of the metric (3.3.18) is $SO(5)$.

(2) If in C^3 we identify (z_1, z_2, z_3) and $(\lambda z_1, \lambda z_2, \lambda z_3)$, $\forall \lambda \in C - \{0\}$, we obtain, by definition, CP^2. For this two-dimensional complex space one can find a real four-dimensional metric, which solves the Einstein equations with cosmological constant Λ. If we set Λ to 6 for convenience, the metric of CP^2 takes the form (Pope 1981) :

$$g_{II} = d\beta \otimes d\beta + \frac{1}{4}(\sin\beta)^2 \Big[(\omega_1)^2 + (\omega_2)^2 + (\cos\beta)^2(\omega_3)^2\Big] \quad , \tag{3.3.19}$$

where $\beta \in \left[0, \frac{\pi}{2}\right]$. A bolt exists at $\beta = \frac{\pi}{2}$, where $\frac{\partial}{\partial\psi}$ has a two-dimensional fixed-point set. The isometry group of g_{II} is locally $SU(3)$, which has a $U(2)$ subgroup acting on the three-spheres $\beta = constant$.

(3) The Einstein metric on the product manifold $S^2 \times S^2$ is obtained as the direct sum of the metrics on 2 two-spheres :

$$g = \frac{1}{\Lambda}\sum_{i=1}^{2}\Big(d\theta_i \otimes d\theta_i + (\sin\theta_i)^2 d\phi_i \otimes d\phi_i\Big) \quad . \tag{3.3.20}$$

The metric (3.3.20) is invariant under the $SO(3) \times SO(3)$ isometry group of $S^2 \times S^2$, but is not of Bianchi-IX type as (3.3.18-19). This can be achieved by a coordinate transformation leading to (Pope 1981) :

$$g_{III} = d\beta \otimes d\beta + (\cos\beta)^2(\omega_1)^2 + (\sin\beta)^2(\omega_2)^2 + (\omega_3)^2 \quad , \tag{3.3.21}$$

where $\Lambda = 2$ and $\beta \in \left[0, \frac{\pi}{2}\right]$. Regularity at $\beta = 0, \frac{\pi}{2}$ is obtained provided that ψ is identified modulo 2π (cf. (3.3.18)). Remarkably, this is a regular Bianchi-IX Einstein metric in which the coefficients of ω_1, ω_2 and ω_3 are all different.

(4) The nontrivial S^2-bundle over S^2 has a metric which, setting $\Lambda = 3$, may be cast in the form (Pope 1981) :

$$g_{IV} = \left(1+\nu^2\right)\left[\frac{\left(1-\nu^2x^2\right)}{\left(3-\nu^2-\nu^2\left(1+\nu^2\right)x^2\right)\left(1-x^2\right)} dx \otimes dx + \frac{\left(1-\nu^2x^2\right)}{\left(3+6\nu^2-\nu^4\right)}\left((\omega_1)^2+(\omega_2)^2\right) + \frac{\left(3-\nu^2-\nu^2\left(1+\nu^2\right)x^2\right)}{\left(3-\nu^2\right)\left(1-\nu^2x^2\right)}\left(1-x^2\right)(\omega_3)^2\right] , \tag{3.3.22}$$

where $x \in [0,1]$, and ν is the positive root of :

$$w^4 + 4w^3 - 6w^2 + 12w - 3 = 0 \quad . \tag{3.3.23}$$

The isometry group corresponding to (3.3.22) may be shown to be $U(2)$.

(5) Another compact instanton of fundamental importance is the $K3$ surface, whose explicit metric has not yet been found. $K3$ is defined as the compact complex surface whose first Betti number and first Chern class are vanishing. The mathematically-oriented reader might now like to recall the following basic concepts.

C1. The p-th Betti number B_p can be seen as the number of independent closed p-surfaces that are not boundaries of some $(p+1)$-surface (Hawking 1979b).

C2. A complex structure on a real manifold M is a tensor field $J^\mu{}_\nu$ such that $J^\mu{}_\nu J^\nu{}_\sigma = -\delta^\mu{}_\sigma$, and satisfying an integrability condition that enables one to introduce local complex coordinates z^j on M so that transition functions between different coordinate patches are holomorphic.

C3. Given a complex structure $J^\mu{}_\nu$, a Hermitian metric is a Riemannian metric g such that $J^\mu{}_\rho J^\nu{}_\sigma \, g_{\mu\nu} = g_{\rho\sigma}$.

C4. Let $g_{j\overline{k}}$ be a Hermitian metric, and consider the real $(1,1)$ form $J \equiv i g_{j\overline{k}} \, dz^j \wedge d\overline{z}^{\overline{k}}$. A Kähler metric is by definition a Hermitian metric with $dJ = 0$. If $g_{j\overline{k}}$ is Kähler, the corresponding closed form J is called the Kähler form. A Hermitian metric is Kähler if and only if $J^{\mu}{}_{\nu}$ is covariantly constant with respect to the connection defined by g. This means that, for Kähler metrics, the Riemannian structure is compatible with the complex structure.

C5. The Ricci tensor of a Kähler metric (Chern 1979) is a (real, symmetric) bilinear form of type $(1,1)$, and the associated two-form is the Ricci form ρ.

C6. The first real Chern class c_1 is represented by the two-form $\frac{\rho}{2\pi}$.

C7. A Kähler metric on a complex manifold is said to be Kähler-Einstein if the Ricci form ρ is proportional to the Kähler form $\omega : \rho = \lambda\omega$.

C8. According to Calabi's conjecture, given a compact Kähler manifold M, its Kähler form ω, its real first Chern class $c_1(M)$, then any closed (real) two-form of type $(1,1)$ belonging to $2\pi c_1(M)$ is the Ricci form of one and only one Kähler metric in the Kähler class of ω.

So far twistor theory has given important contributions to the $K3$-metric problem, since it provides a method so as to obtain explicit approximations to the $K3$ metric (Hitchin 1984), and it leads to a proof of the existence of Kähler-Einstein metrics on $K3$ (Topiwala 1987a-b). We now focus on the latter development, trying to explain its importance, its limits and what remains to be done.

Indeed, Yau's proof (Yau 1978) of Calabi's conjecture (Calabi 1954) already leads to an existence theorem for Kähler-Einstein metrics on $K3$, but it does not lead to explicit calculations. Topiwala obtained the same result using the following idea (Topiwala 1987a-b) : if a Kähler-Einstein metric on $K3$ exists, then it gives rise to a canonical one-parameter deformation of the complex structure. Before going on, it is worth saying that a one-parameter deformation is a set of deformations indexed by t, where t can either be real or complex. The deformation is said to be canonical if it depends only on the Kähler-Einstein metric, and not on any choice of coordinate used to express structures (e.g. a coordinate description of metrics). The advantage of Topiwala's method is that it shows under which conditions the twistor space for $K3$ is biholomorphic to the one for the Eguchi-Hanson metric (3.3.4). One thus gets an explicit result, proving Page's conjecture (Topiwala 1985).

However, it should be emphasized that this method is a substitute for the existence theorem obtained using partial differential equations, and does not lead to the knowledge of the metric. Thus it seems we are left with two main possibilities :

(a) To go on with twistor theory. Remarkably, in order to know *a priori* the twistor space (cf. section 5.6 and references therein), one should know all complex structures on $K3$. A number of them has already been studied (Hitchin 1984), but the task seems to be very hard;

(b) To use deformation theory (Barth et al. 1984, Kodaira 1986).

Two topological invariants exist which may be used to characterize the various gravitational instantons studied so far. These invariants are the Euler number χ and the Hirzebruch signature τ. The Euler number can be defined as an alternating sum of Betti numbers :

$$\chi \equiv B_0 - B_1 + B_2 - B_3 + B_4 \quad . \tag{3.3.24}$$

The Hirzebruch signature can be defined as :

$$\tau \equiv B_2^+ - B_2^- \quad , \tag{3.3.25}$$

where B_2^+ is the number of self-dual harmonic two-forms (Ward and Wells 1990), and B_2^- is the number of anti-self-dual harmonic two-forms [in terms of the Hodge-star operator ${}^*F_{ab} \equiv \frac{1}{2}\epsilon_{abcd}F^{cd}$, self-duality of a two-form F is expressed as ${}^*F = F$, and anti-self-duality as ${}^*F = -F$]. In the case of compact four-dimensional manifolds without boundary, χ and τ can be expressed as integrals of the curvature (Hawking 1979b) :

$$\chi = \frac{1}{128\pi^2}\int_M R_{abcd}\, R_{efgh}\, \epsilon^{abef}\, \epsilon^{cdgh}\, \sqrt{g}\, d^4x \quad , \tag{3.3.26}$$

$$\tau = \frac{1}{96\pi^2}\int_M R_{abcd}\, R^{ab}{}_{ef}\, \epsilon^{cdef}\, \sqrt{g}\, d^4x \quad . \tag{3.3.27}$$

For the instantons previously listed one finds (Pope 1981) :

Eguchi-Hanson : $\chi = 2$, $\tau = 1$.

Asymptotically locally Euclidean multi-instantons : $\chi = n$, $\tau = n - 1$.

Schwarzschild : $\chi = 2$, $\tau = 0$.

Taub-NUT : $\chi = 1$, $\tau = 0$.

Asymptotically locally flat multi-Taub-NUT instantons : $\chi = n$, $\tau = n - 1$.

S^4 : $\chi = 2$, $\tau = 0$.

CP^2 : $\chi = 3$, $\tau = 1$.

$S^2 \times S^2$: $\chi = 4$, $\tau = 0$.

S^2-bundle over S^2 : $\chi = 4$, $\tau = 0$.

$K3$: $\chi = 24$, $\tau = 16$.

3.4 Perturbative Renormalization of Quantum Field Theories

The one-loop properties described at the beginning of this chapter are just a part of the renormalizability problem in quantum field theory and quantum gravity. This is why we conclude our introduction to perturbative quantum gravity with a discussion of renormalization, asymptotic expansions and summability methods.

Following Collins 1984, we distinguish three types of renormalizable theory :

(3.4.1) Finite : no counterterms needed at all.

(3.4.2) Super-renormalizable : only a finite set of graphs need overall counterterms.

(3.4.3) Strictly renormalizable : infinitely many graphs need overall counterterms.

Note that, if a theory fails to be one-loop finite, one may find such a proliferation of counterterms that the theory also fails to be renormalizable according to definitions (3.4.2-3). This remark plays an important role in understanding the ultimate consequence of the perturbative calculations appearing in our book. If a theory is not renormalizable, this means that physical quantities are infinite, or, if they are finite, an infinite set of parameters is needed to specify the finite parts of the counterterms. This is why it does not seem possible to make sense of a nonrenormalizable theory.

As explained in Collins 1984, in the case of super-renormalizable theories, the series for a bare mass or coupling has a finite number of terms and converges. The rigorous proof amounts then to showing that, in summing the perturbative series to a finite order, the error is correctly estimated by the first term omitted. In the case of strictly renormalizable theories, the perturbative series is (at the very best) asymptotic rather than convergent, and the error obtained if one uses a truncated form of the series is always divergent.

It may be now instructive to study a bit more in detail the asymptotic and summability techniques available in the literature, following Wightman 1979. It is not yet clear whether many renormalized perturbative series studied in the literature are convergent or at least asymptotic. The latter concept (cf. (1.2.4) and (3.1.4)) was introduced by Poincaré in his 1886 paper on the irregular integrals of linear equations.

3.4.1.1 Asymptotic Expansions

Definition 3.4.1 Given a function g, defined on an open interval $]a,b[$, one says that a function g_N defined on the same interval is asymptotic to g of order N at a if :

$$\lim_{x \rightarrow a} \frac{\Big[g(x) - g_N(x)\Big]}{(x-a)^N} = 0 \quad . \tag{3.4.1}$$

This holds in particular when $g_N(x) = \sum_{n=0}^{N} a_n (x-a)^n$.

Definition 3.4.2 Given a sequence $\{a_n\}_{n \in \mathcal{N}}$, if $\sum_{n=0}^{N} a_n (x-a)^n$ is asymptotic to g of order N at a $\forall N = 0,1,2,...$, one then says that the series $\sum_{n=0}^{\infty} a_n (x-a)^n$ is asymptotic to g at a. This is written as :

$$g(x) \sim \sum_{n=0}^{\infty} a_n (x-a)^n \quad . \tag{3.4.2}$$

Note that, if (3.4.1) holds, this implies that

$$\lim_{x \to a} \frac{\Big[g(x) - g_N(x)\Big]}{(x-a)^k} = 0 \quad , \quad \forall k = 0, 1, ..., N-1 \quad .$$

Thus, if $\sum_{n=0}^{N} a_n (x-a)^n$ is asymptotic to g of order N at a, one finds the useful relations (Wightman 1979) :

$$\lim_{x \to a} g(x) = a_0 = g(a^+) \quad , \tag{3.4.3}$$

$$\lim_{x \to a} \frac{\Big[g(x) - g(a^+)\Big]}{(x-a)} = a_1 = g'(a^+) \quad , \tag{3.4.4}$$

$$\lim_{x \to a} \frac{\Big[g(x) - g(a^+) - g'(a^+)(x-a)\Big]}{(x-a)^2} = a_2 = \frac{g''(a^+)}{2!} \quad , \tag{3.4.5}$$

$$\lim_{x \to a} \left[g(x) - \sum_{n=0}^{N-1} \frac{g^{(n)}(a^+)}{n!}(x-a)^n \right] (x-a)^{-N} = a_N = \frac{g^{(N)}(a^+)}{N!} \quad . \tag{3.4.6}$$

In other words, when $\sum_{n=0}^{N} a_n (x-a)^n$ is asymptotic to g of order N at a, g has N derivatives from the right at a, and the coefficients a_n are *uniquely* determined as Taylor coefficients :

$$a_n = \frac{g^{(n)}(a^+)}{n!} \quad , \quad n = 0, 1, ..., N \quad . \tag{3.4.7}$$

For example, if the renormalized perturbative series $\sum_{n=1}^{\infty} a_n (\alpha/\pi)^n$ for the gyromagnetic anomaly $\frac{(\gamma-2)}{2}$ is asymptotic, this means that $\frac{(\gamma-2)}{2}$ is defined for α (i.e. the fine-structure constant) $\in]0, \alpha_0[$, and has derivatives of all orders from the right at the origin. The a_n coefficients are then uniquely determined as shown in (3.4.7).

3.4.1.2 Summability Methods

In the case of summability methods, the idea is to operate on a divergent (or also convergent) infinite series, so as to obtain a convergent series of functions. Following Wightman 1979, we here describe :

(1) Cesàro Summability. This converts $\sum_{n=0}^{\infty} a_n$ to :

$$\lim_{n\to\infty}\left[\frac{1}{(n+1)}\sum_{l=0}^{n}\sum_{k=0}^{l} a_k\right] \quad .$$

For example, in the case of the series $\sum_{n=0}^{\infty}(-1)^n$, the Cesàro method yields $\frac{1}{2}$.

(2) Abel Summability. This converts $\sum_{n=0}^{\infty} a_n$ to :

$$\lim_{r\to 1^-}\sum_{n=0}^{\infty} a_n r^n \quad .$$

If applied to $\sum_{n=0}^{\infty}(-1)^n$, this yields $\lim_{r\to 1^-}\frac{1}{(1+r)}=\frac{1}{2}$.

(3) Borel Summability. This is a more powerful method, and is also applied in quantum electrodynamics (Feldman et al. 1988). When applied to a formal power series $\sum_{n=0}^{\infty} a_n z^n$, it yields :

$$\frac{1}{x}\int_0^{\infty} e^{-z/x}\left(\sum_{n=0}^{\infty}\frac{a_n z^n}{n!}\right)\, dz \quad .$$

Note that analytic continuation may be necessary to define the sum under the integral sign. Moreover, it is still very difficult to sum the renormalized perturbative series using a summability method, so that one tries instead to prove properties of the nonperturbative solutions which ensure that a *particular* summability method yields the unique, correct answer. This may be obtained as follows (Wightman 1979).

Definition 3.4.3 A formal power series $\sum_{n=0}^{\infty} a_n z^n$ is a strong asymptotic series for a function g if and only if :

(a) g is analytic in a region

$$R_{\epsilon,B}\equiv\left\{z :\mid arg\ z\mid<\frac{\pi}{2}+\epsilon\ ,\ \mid z\mid\in\,]0,B[\right\} \quad , \qquad (3.4.8)$$

for some ϵ and $B>0$.

(b) For some A and B, $\forall N = 0, 1, ..., \forall z \in R_{\epsilon,B}$:

$$\left| g(z) - \sum_{n=0}^{N} a_n z^n \right| \leq AB^{n+1}(N+1)! | z |^{N+1} \quad . \tag{3.4.9}$$

The important result is that Borel's summability always works for strongly asymptotic series, since the following theorem holds.

Theorem 3.4.1 Let g be analytic in a neighbourhood $R_{\epsilon,B}$ defined as in (3.4.8), and satisfy the estimate (3.4.9), then g has derivatives of all orders $n = 0, 1, 2, ...$ from the right at the origin. Its formal Taylor series has an associated series :

$$\sum_{n=0}^{\infty} \frac{g^{(n)}(0^+)}{(n!)^2} z^n \quad .$$

This series converges for $| z | < B$ and defines an analytic function G which may be continued analytically to the sector $| arg \ z | < \epsilon$. g can be then recovered from G using the integral formula :

$$g(z) = \frac{1}{z} \int_0^{\infty} G(x) e^{-x/z} \, dx \quad . \tag{3.4.10}$$

If a quantum field theory can be renormalized, but the renormalized perturbative series does not converge with the help of a suitable summability method, it remains unclear whether this theory exists in a mathematical sense. This is why many authors claim that quantized gauge theories do not exist (e.g. end of chapter five of Ward and Wells 1990). Even though the problem remains open, progress has been recently made in applying new techniques so as to prove the renormalizability of (Euclidean) quantum electrodynamics in four dimensions (hereafter referred to as QED_4).

As pointed out in Feldman et al. 1988, no clear proof existed in a vast literature that, for QED_4, only gauge-invariant counterterms are required. Interestingly, these authors have successfully generalized to QED_4 a previous technique by Gallavotti and Nicolò (GN). The GN tree expansion enables one to choose counterterms and to renormalize scale by scale without introducing overlapping divergences (i.e. when two divergent subgraphs intersect, and neither is a subgraph of the other) or combinatorial complexities. However, if the GN method is applied to gauge theories, it is not *a priori* clear that renormalization is proved using only gauge-invariant counterterms. Still, Feldman et al. have been able to prove that, when gauge-dependent counterterms are summed up over all scales, they add up to zero by virtue of the Ward identities. They can thus define a renormalized effective potential where only gauge-invariant counterterms are employed, whereas cancellations are performed in a gauge-dependent way (Feldman et al. 1988).

The reader should bear in mind that the choice of review topics in this chapter is motivated by the second part of our book (starting in chapter four), dealing with new research results, where asymptotic analysis and perturbative concepts are frequently applied. It may be thus useful to conclude this sèction by showing why the quantized version of general relativity cannot be renormalized, following De Witt 1967b and Duff 1982.

Since the scalar curvature contains second derivatives of the metric, the corresponding momentum-space vertex functions behave like p^2, and the propagator like p^{-2}. In d dimensions each loop integral contributes p^d, so that with L loops, V vertices and P internal lines, the superficial degree D of divergence of a Feynman diagram is given by :

$$D = dL + 2V - 2P \quad . \tag{3.4.11}$$

Moreover, a topological relation holds :

$$L = 1 - V + P \quad , \tag{3.4.12}$$

which leads to :

$$D = (d-2)L + 2 \quad . \tag{3.4.13}$$

In other words, D increases with increasing loop order for $d > 2$, so that it clearly leads to a nonrenormalizable theory according to the definitions given in this section.

PART II:

ONE-LOOP QUANTUM COSMOLOGY

CHAPTER FOUR

GLOBAL BOUNDARY CONDITIONS AND $\zeta(0)$ VALUE FOR THE MASSLESS SPIN-$\frac{1}{2}$ FIELD

Abstract. The prefactor of the semiclassical approximation of the wave function is evaluated in quantum cosmological models where a massless Majorana spin-$\frac{1}{2}$ field is regarded as a perturbation around a flat Euclidean background bounded by a three-sphere of radius a. At first we outline the one-loop calculation in Schleich 1985 for pure gravity, and we discuss in some detail the global boundary conditions used in D'Eath and Halliwell 1987 and their relation to the spectral theory of elliptic operators. In performing the one-loop calculation, all untwiddled coefficients of the spin-$\frac{1}{2}$ field, multiplying unbarred harmonics having positive eigenvalues for the three-dimensional Dirac operator, are set to zero on S^3. This means that half of the fermionic field, corresponding to harmonics of the intrinsic three-dimensional Dirac operator with positive eigenvalues, is required to vanish on the three-sphere boundary. The corresponding $\zeta(0)$ value is obtained studying the Laplace transform of the heat equation for the squared Dirac operator, and finally deriving the asymptotic expansion of the inverse Laplace transform, i.e. the heat kernel. This squared operator arises from the study of the coupled system of first-order eigenvalue equations for the perturbative modes, subject to the eigenvalue condition $J_{n+1}(Ea) = 0$, $\forall n \geq 0$, with degeneracy $2(n+2)(n+1)$. We compute in detail the infinite sums occurring in the asymptotic expansion of the heat kernel. The global boundary conditions previously described are shown to imply that the prefactor is still diverging in the limit of small three-geometry, because the generalized ζ-function is such that $\zeta(0) = \frac{11}{360}$ for massless Majorana fields.

Finally, Hawking's local boundary conditions involving the spin-$\frac{1}{2}$ field and the spinor version of the timelike future-pointing normal vector to S^3 are introduced. The conditions under which one can get a classical boundary-value problem with a unique regular solution inside S^3 are thus described.

4.1 Physical-Degrees-of-Freedom One-Loop Results for Pure Gravity

As already remarked in section 1.4, the semiclassical calculation of the wave function of the universe has played so far a vital role, because its exact evaluation is impossible. The first accurate calculation of the prefactor has been done in Schleich 1985 for pure gravity, and we now outline her calculation because her technique will be applied again in the remaining part of this chapter and in chapter five (although for fermionic fields we shall initially study global boundary conditions, whereas for pure gravity Schleich studied a *local* boundary-value problem).

We want to study pure gravity at one-loop about flat Euclidean space with a three-sphere boundary of radius a. We thus write the four-metric as : $g_{ab} = s_{ab} + \gamma_{ab}$, where s_{ab} is the metric of flat Euclidean space, and γ_{ab} is a perturbation regular at the origin and which vanishes on the boundary. The prefactor of the semiclassical wave function is thus given by :

$$P(a) = \int D[\gamma] e^{-I_2[\gamma]} \quad , \tag{4.1.1}$$

which is a path integral over all metric perturbations satisfying the above-mentioned boundary conditions. The measure in (4.1.1) is defined integrating over the physical degrees of freedom which are found using the Hamiltonian formulation of the theory with the following choice of gauge :

$$D^i \gamma_{ij} = 0 \quad , \tag{4.1.2}$$

$$\gamma_k^k = 0 \quad . \tag{4.1.3}$$

The relations (4.1.2-3) pick out the transverse-traceless tensor hyperspherical harmonics $G_{ij}^{(n)}(\phi^k)$, multiplied by functions of the radial coordinate t. Thus we have :

$$\gamma_{ij} = \gamma_{ij}^{TT} = \sum_{n=3}^{\infty} q^n(t) G_{ij}^{(n)}(\phi^k) \quad , \tag{4.1.4}$$

$$P(a) = \int D\left[\gamma^{TT}\right] e^{-I_2\left[\gamma^{TT}\right]} \quad , \tag{4.1.5}$$

where $q^n(t=a) = q^n(t=0) = 0$. Denoting by μ a normalization parameter having dimensions of inverse length, by l_p the Planck length and by $-\nabla_f \nabla^f$ the Laplacian operator, one finds that $P(a)$ is formally given by :

$$P(a) = \frac{1}{\sqrt{\det\left(\frac{-\nabla_f \nabla^f}{4\pi l_p^2 \mu^2}\right)}} \quad , \tag{4.1.6}$$

and the ζ-function technique together with a scaling argument (Schleich 1985) show that P is proportional to $a^{\zeta(0)}$. Thus, the task remains to study the heat equation for the Laplacian acting on transverse-traceless tensors. Indeed, the Laplace transform of this equation is (Schleich 1985) :

$$\left(\sigma^2 - \nabla_f \nabla^f\right) G_{ab,cd}\left(x, x', \sigma^2\right) = \delta_{ab,cd}(x, x') \quad , \tag{4.1.7}$$

where σ^2 is the parameter of the transform. The Green's function $G_{ab,cd}(x, x', \sigma^2)$ of transverse-traceless tensors can be expanded according to :

$$\begin{aligned} G_{ab,cd}\left(x, x', \sigma^2\right) &= \sum_{n=3}^{\infty} d(n) f^{(n)}(t, t') G_{ab}^{(n)}(\phi^i) G_{cd}^{(n)}(\phi'^i) \\ &= G_{ab,cd}^{F}\left(x, x', \sigma^2\right) + G_{ab,cd}^{int}\left(x, x', \sigma^2\right) \quad . \end{aligned} \tag{4.1.8}$$

In (4.1.8), we have used the property that $f^{(n)}(t, t')$ is a solution of the equation :

$$\left(\frac{\partial^2}{\partial t^2} - \frac{1}{t}\frac{\partial}{\partial t} - \sigma^2 - \frac{(n^2-1)}{t^2}\right) f^{(n)}(t, t') = -\frac{\delta(t-t')}{t^3} \quad . \tag{4.1.9}$$

Therefore, requiring that $f^{(n)}(a, t') = f^{(n)}(t, a) = 0$, we find :

$$f^{(n)}(t, t') = t_< t_> \frac{I_n(\sigma t_<)}{I_n(\sigma a)} \Big[I_n(\sigma a) K_n(\sigma t_>) - I_n(\sigma t_>) K_n(\sigma a) \Big] \quad , \tag{4.1.10}$$

(where : $t_< = min(t, t')$, $t_> = max(t, t')$), which implies that :

$$G_{ab,cd}^{F}\left(x, x', \sigma^2\right) = \sum_{n=3}^{\infty} d(n) t_< t_> I_n(\sigma t_<) K_n(\sigma t_>) G_{ab}^{(n)}(\phi^i) G_{cd}^{(n)}(\phi'^i) \quad , \tag{4.1.11}$$

$$G_{ab,cd}^{int}(x,x',\sigma^2) = -\sum_{n=3}^{\infty} d(n) t_< t_> \frac{K_n(\sigma a)}{I_n(\sigma a)} I_n(\sigma t_<) I_n(\sigma t_>) G_{ab}^{(n)}(\phi^i) G_{cd}^{(n)}(\phi'^i) . \quad (4.1.12)$$

In (4.1.8) and (4.1.11), the label F reminds us that $G_{ab,cd}^F$ is the *free* part of $G_{ab,cd}$ in that it is regular at the origin and it vanishes as $t_> \rightarrow \infty$, whereas $d(n) = 2(n^2-4)$ is the degeneracy of the eigenvalues. In order to compute the heat kernel we must now take the inverse Laplace transform of (4.1.11-12), set $t=t'$, $\phi^i = \phi'^i$, and integrate over the volume interior to the three-sphere of radius a. Thus we can define :

$$G^F(t,t',\tau) \equiv \sum_{n=3}^{\infty} \Big[2\left(n^2-4\right)\Big] G_{(n)}^F(t,t',\tau) \quad , \quad (4.1.13)$$

where :

$$G_{(n)}^F(t,t',\tau) = \frac{tt'}{2\tau} I_n\left(\frac{tt'}{2\tau}\right) \exp\left[-\frac{1}{4\tau}\left(t^2+t'^2\right)\right] \quad , \quad (4.1.14)$$

which implies :

$$G_{(n)}^F(\tau) = \frac{1}{2}\int_0^{\frac{a^2}{2\tau}} I_n(y) e^{-y} dy \quad . \quad (4.1.15)$$

Therefore, using the integral representation of Bessel functions, the identity : $I_0'(y) = I_1(y)$, and the rules for expressing the Dirac delta and its derivatives (Gelfand and Shilov 1964), one finds :

$$G^F(\tau) = \frac{a^4}{16\tau^2} - \frac{a^2}{\tau} + 3\left[e^{\frac{-a^2}{2\tau}} I_0\left(\frac{a^2}{2\tau}\right) - 1\right] + \frac{5a^2}{2\tau} e^{\frac{-a^2}{2\tau}} \left[I_0\left(\frac{a^2}{2\tau}\right) + I_1\left(\frac{a^2}{2\tau}\right)\right] . \quad (4.1.16)$$

The *interacting* part $G^{int}(\tau)$ is found by taking the inverse Laplace transform of :

$$G^{int}(\sigma^2) = -a^2 \sum_{n=3}^{\infty} \left(n^2-4\right) f(n;\sigma a) \quad , \quad (4.1.17)$$

where $f(n;\sigma a)$ is an even function of n, given by (Stewartson and Waechter 1971, Schleich 1985) :

$$f(n;\sigma a) \equiv \left(1+\frac{n^2}{\sigma^2 a^2}\right) I_n(\sigma a) K_n(\sigma a) - I_n'(\sigma a) K_n'(\sigma a) - \frac{I_n'(\sigma a)}{\sigma a I_n(\sigma a)} . \quad (4.1.18)$$

Indeed, the series appearing in (4.1.17) leads to divergences in the calculation. As explained in section 4.2 of Kennedy 1979, this problem arises because we have attempted to take the Laplace transform of a function which is singular at $\tau = 0$. Thus we replace the $\sum_{n=3}^{\infty}$ by $\sum_{n=3}^{N}$, where N acts as an intermediate cut-off which is only removed ($N \rightarrow \infty$) *after* inversion of (4.1.17). Since the operations of inverse Laplace transform L_I and $\sum_{n=3}^{N}$ commute for any finite N, the removal of cut-off *after* inversion means we are actually obtaining the interacting heat kernel as $-a^2 \sum_{n=3}^{\infty} L_I \Big[(n^2-4)f(n;\sigma a)\Big]$. Note that, when each term in the asymptotic expansion of $L_I \Big[(n^2-4)f(n;\sigma a)\Big]$ is summed from $n = 3$ to N, the resulting function of τ does not always converge uniformly to the infinite sum $\sum_{n=3}^{\infty}$ in a neighbourhood $\tau \in (0, \delta)$ (D'Eath and Esposito 1991b). Nevertheless, a study of the error terms shows that it is valid to take the limit $N \rightarrow \infty$, and then examine the small-τ behaviour of the resulting contributions to $G^{int}(\tau)$, so that, setting $a = 1$ for simplicity and defining $r \equiv \frac{n}{\sqrt{n^2+\sigma^2}}$:

$$\lim_{N \rightarrow \infty} \sum_{n=3}^{N} L_I \Big[(n^2-4)f(n;\sigma)\Big] \sim \sum_{n=3}^{\infty} L_I \left[\frac{1}{n^2}\frac{r^4}{8}(1-12r^2+15r^4)+...\right] + \sum_{n=3}^{\infty} L_I \left[\frac{r^2}{2} - \frac{r^5}{2n}\right] \quad , \tag{4.1.19}$$

where the second infinite sum on the right-hand side of (4.1.19) is the one for which pointwise (rather than uniform) convergence is obtained when the cut-off is removed.

A powerful way of performing this calculation is to introduce the Watson transform :

$$G^{int}(\sigma^2) = -\frac{a^2}{4i} \int_{C'-Q} (\nu^2-4)\, f(\nu;\sigma a) \cot(\pi\nu) d\nu \quad . \tag{4.1.20}$$

In (4.1.20), C' is a contour enclosing all poles of the integrand along the real axis, whereas Q encloses anti-clockwise the poles at $\nu = 0, \pm 1, \pm 2$. In fact the sum over all n only starts from $n = 3$, and the functions (n^2-4) and $f(n;\sigma a)$ are even functions of n, so that also -1 and -2 must be taken into account. In the case of fermionic fields, there will be both a contribution of the kind (4.1.20) and a contribution involving an odd function of n. The derivation of (4.1.17) is not performed here because a similar one, even more involved and original, has been described in section 4.4 of our monograph. However, it should be recalled that, in computing (4.1.20), one expands $\cot(\pi\nu)$ and one finds that only a constant term equal to 1 contributes as $\sigma^2 \rightarrow \infty$. This holds for an S^{p+1} bounding an R^{p+2}, but only when p is even (Kennedy 1978-1979). In Schleich 1985 it is shown that, as $\tau \rightarrow 0^+$:

$$G^{int}(\tau) \sim -\frac{a^3\sqrt{\pi}}{8\tau^{\frac{3}{2}}} + \frac{a^2}{4\tau} + \frac{245}{256}\frac{a\sqrt{\pi}}{\sqrt{\tau}} - \frac{143}{45} + O(\sqrt{\tau}) \quad . \tag{4.1.21}$$

Thus (4.1.16) and (4.1.21) yield :

$$G(\tau) \sim \frac{a^4}{16\tau^2} - \frac{a^3\sqrt{\pi}}{8\tau^{\frac{3}{2}}} - \frac{3a^2}{4\tau} + \left(\frac{5}{\sqrt{\pi}} + \frac{245}{256}\sqrt{\pi}\right)\frac{a}{\sqrt{\tau}} - \frac{278}{45} + O(\sqrt{\tau}) \quad , \tag{4.1.22}$$

which in turn implies that (see appendix B, equations (B.5-6)) :

$$\zeta(0) = -\frac{278}{45} \quad , \quad P = Da^{\zeta(0)} = Da^{-\frac{278}{45}} \quad . \tag{4.1.23}$$

However, we warn the reader that the results in Moss and Poletti 1990a and Poletti 1990, where ghost fields have been explicitly introduced using the Faddeev-Popov technique, are in disagreement with (4.1.23) derived from Schleich (for further details see section 6.4).

4.2 Mathematical Foundations of Global Boundary Conditions

In this section we first illustrate the mathematical foundations of global boundary conditions (Atiyah et al. 1975). Finally, we discuss global boundary conditions for the quantum cosmology of the Einstein-Dirac theory (D'Eath and Halliwell 1987), and for one-loop calculations with massless Majorana spin-$\frac{1}{2}$ fields.

We here denote by $H^p(X;R)$ the p-th cohomology group of X with real coefficients. We should recall that a fundamental algebraic structure in cohomology theory is the cup product $\cup$. It associates to an element of $H^p(X;R) \times H^q(X;R)$ an element of $H^{p+q}(X;R)$

through the relation : $[\omega] \cup [\rho] = [\omega \wedge \rho]$, where $[\omega] \in H^p(X;R)$ and $[\rho] \in H^q(X;R)$. Now, if X is a closed, oriented, four-dimensional Riemannian manifold, in addition to the Gauss-Bonnet formula there is another formula which relates cohomological invariants with curvature. In fact, it is known that the signature (i.e. number of positive eigenvalues minus number of negative eigenvalues) of the quadratic form on $H^2(X;R)$ given by the cup product is expressed by :

$$sign(X) = \frac{1}{3} \int_X p_1 \quad . \tag{4.2.1}$$

In (4.2.1), p_1 is the differential four-form which represents the first Pontrjagin class (Milnor and Stasheff 1974, Nash and Sen 1983), and is equal to : $p_1 = (2\pi)^{-2} Tr\left(R^2\right)$, where R is the curvature matrix (Atiyah et al. 1975). However, (4.2.1) does not hold in general for manifolds with boundary, so that one has :

$$sign(X) - \frac{1}{3} \int_X p_1 = f(Y) \neq 0 \quad , \tag{4.2.2}$$

where $Y = \partial X$. Therefore, if X' is another manifold such that : $Y = \partial X'$, one has :

$$sign(X) - \frac{1}{3} \int_X p_1 = sign(X') - \frac{1}{3} \int_{X'} p_1' \quad . \tag{4.2.3}$$

Thus, we are looking for a continuous function f of the metric on Y such that $f(-Y) = -f(Y)$. Atiyah et al. were able to prove that $f(Y)$ is a spectral invariant, computed in the following way. We look at the Laplace operator $\triangle$ acting on forms as well as on scalar functions. This operator $\triangle$ is the square of the self-adjoint first-order operator $B = \pm(d * - * d)$, where d is the exterior-derivative operator, and $*$ is the Hodge-star operator which maps p-forms into $(l-p)$-forms in l dimensions. Thus, if λ is an eigenvalue of B, the eigenvalues of $\triangle$ are of the form λ^2. But the eigenvalues of B can be both negative and positive. We take this result into account by defining the η-function :

$$\eta(s) = \sum_{\lambda \neq 0} d(\lambda)(sign\lambda) \mid \lambda \mid^{-s} \quad , \tag{4.2.4}$$

where $d(\lambda)$ is the multiplicity of the eigenvalue λ. It is important to remark that, since B involves the $*$ operator, in reversing the orientation of Y we change B into $-B$, and hence $\eta(s)$ into $-\eta(s)$. The main result of Atiyah et al. 1975 states therefore that :

$$f(Y) = \frac{1}{2}\eta(0) \quad . \tag{4.2.5}$$

Now, for a manifold X with boundary Y, if one tries to set up an elliptic boundary-value problem for the signature operator, one finds there is no local boundary condition for this operator. But for global boundary conditions (i.e. expressed by the vanishing of a given integral computed on Y) one has a good elliptic theory and a finite index (recall that for any elliptic differential operator Γ on X, one defines index $\Gamma \equiv$ dim $N(\Gamma)$ - dim $N(\Gamma^{\dagger})$, where $N(\Gamma)$ is the null space of Γ, and $\Gamma^{\dagger}$ is the adjoint of Γ). This index can be identified with $sign(X)$ appearing in (4.2.2). Therefore we have to consider the theorem expressed by (4.2.5) in the context of index theorems for global boundary conditions. Atiyah et al. were also able to derive the relation between the index of the Dirac operator on X with a global boundary condition and $\eta(0)$, where η is the η-function of the Dirac operator on Y.

Let us now look at global boundary conditions in quantum cosmology. The Lorentzian action of the Einstein-Dirac theory is (D'Eath and Halliwell 1987) :

$$I \equiv I_V + I_B \quad . \tag{4.2.6}$$

In (4.2.6), the volume part is given by :

$$\begin{aligned} I_V \equiv & \frac{1}{2\kappa^2}\int eR \, d^4x \\ & - \frac{i}{2}\int e\left[\bar{\phi}^{A'} e^{\mu}_{AA'} D_\mu \phi^A + \bar{\chi}^{A'} e^{\mu}_{AA'} D_\mu \chi^A\right] d^4x + H.C. \\ & - \frac{m}{\sqrt{2}}\int e\left[\chi_A\phi^A + \bar{\phi}^{A'}\bar{\chi}_{A'}\right] d^4x \quad , \end{aligned} \tag{4.2.7}$$

and the boundary part can be assumed to be :

$$\begin{aligned} I_B \equiv & \frac{1}{\kappa^2}\left[\int_{S_F} - \int_{S_I}\right] K\sqrt{h}\, d^3x \\ & + \frac{i}{2}\left[\int_{S_F} + \int_{S_I}\right]\left(\bar{\phi}^{A'} n_{AA'}\phi^A + \bar{\chi}^{A'} n_{AA'}\chi^A\right)\sqrt{h}\, d^3x \quad . \end{aligned} \tag{4.2.8}$$

In (4.2.7-8) we use two-component spinor notation (appendix A) so as to denote the spin-$\frac{1}{2}$ fields ϕ^A, χ^A which, together with their complex conjugates $\overline{\phi}^{A'}$ and $\overline{\chi}^{A'}$ describe the Dirac field, and the spinor-valued one-form $e^{AA'}{}_{\mu}$ which represents the tetrad. The spinor covariant derivative is D_μ, and $n^{AA'}$ is the spinor version of the unit future-pointing normal to the three-surface. Moreover, R is the scalar curvature, h the determinant of the three-metric, K the trace of the extrinsic-curvature tensor, and m is the mass of fermions. We can now focus our attention on the Dirac equations obtained by requiring the stationarity of I :

$$e^{\mu}_{AA'} D_\mu \phi^A = \frac{im}{\sqrt{2}} \bar{\chi}_{A'} \quad , \tag{4.2.9}$$

$$e^{\mu}_{AA'} D_\mu \chi^A = \frac{im}{\sqrt{2}} \bar{\phi}_{A'} \quad , \tag{4.2.10}$$

$$e^{\mu}_{AA'} D_\mu \bar{\phi}^{A'} = -\frac{im}{\sqrt{2}} \chi_A \quad , \tag{4.2.11}$$

$$e^{\mu}_{AA'} D_\mu \bar{\chi}^{A'} = -\frac{im}{\sqrt{2}} \phi_A \quad . \tag{4.2.12}$$

Following D'Eath and Halliwell, we can now make the decomposition $\phi_A = \phi_A^{(+)} + \phi_A^{(-)}$, with similar relations for χ_A, $\bar{\phi}_{A'}$ and $\bar{\chi}_{A'}$, where the $(+)$ and $(-)$ parts correspond to the modes with unbarred and barred coefficients respectively in the expansion in spinor harmonics on S^3 :

$$\phi_A = \frac{e^{-\frac{3\alpha}{2}}}{2\pi} \sum_{npq} \alpha_n^{pq} \Big[m_{np}(t) \rho_A^{nq}(x) + \bar{r}_{np}(t) \bar{\sigma}_A^{nq}(x) \Big] \quad . \tag{4.2.13}$$

The modes with unbarred (barred) coefficients multiply harmonics having positive (negative) eigenvalues for the three-dimensional Dirac operator ${}_e n_{AB'} e^{BB'j} \left({}^{(3)}D_j \right)$ on the bounding S^3 (cf. section 1.4). In the summation, n runs from 0 to ∞, p and q from 1 to $(n+1)(n+2)$. The α_n^{pq} are a collection of matrices introduced for convenience, where, for each n, α_n^{pq} is block-diagonal in the indices pq, with blocks $\begin{pmatrix} 1 & 1 \\ 1 & -1 \end{pmatrix}$. In Euclidean four-space, we have untwiddled and twiddled coefficients not related by complex conjugation, rather than unbarred and barred coefficients. However, unbarred and barred harmonics still appear in the expansion on S^3 of the spin-$\frac{1}{2}$ field. The boundary-value problem (D'Eath and Halliwell 1987) of interest to quantum cosmology is to find the solution of the Euclidean Dirac equation such that $\phi_A^{(+)}, \chi_A^{(+)}, \widetilde{\phi}_{A'}^{(+)}, \widetilde{\chi}_{A'}^{(+)}$ match prescribed values on the final surface S_F, and ϕ_A, χ_A, $\widetilde{\phi}_{A'}$ and $\widetilde{\chi}_{A'}$ are regular on the interior of S_F. The authors chose to fix the untwiddled variables on S_F so as to have a regular solution in the massless limit. In fact, in that limit the general solution for the twiddled variables is singular at $\tau = 0$, and one is forced to take them to be identically zero.

In the study of the eigenvalues for the quantum theory of the massless Majorana spin-$\frac{1}{2}$ field, we still fix the untwiddled variables (m_{np}, r_{np}) on S^3, bearing in mind the only correct choice of boundary data for the classical theory. In the case of a general manifold with boundary, knowledge of the spectrum of the intrinsic three-dimensional Dirac operator is necessary if one wishes to compute the η-invariant (cf. (4.2.4)) which gives a boundary contribution to the index of the Dirac operator for the manifold with boundary. In the generic case, the index does not vanish, and the classical boundary-value problem is not well-posed (D'Eath and Esposito 1991b), since the solution is not unique.

In one-loop calculations about flat Euclidean backgrounds, we require that m_{np} and r_{np} should vanish on S^3. More precisely, the complete set of boundary conditions obeyed by the fields summed over in the Hartle-Hawking path integral are spectral conditions on S^3 :

$$m_{np}(a) = r_{np}(a) = 0 \quad , \quad \forall n,p \quad , \tag{4.2.14}$$

and regularity at the origin :

$$m_{np}(0) = r_{np}(0) = \widetilde{m}_{np}(0) = \widetilde{r}_{np}(0) = 0 \quad , \quad \forall n,p \quad . \tag{4.2.15}$$

However, in sections 4.6 and 5.7-8 we shall see that also local boundary conditions can be chosen in the quantum cosmology of fermionic fields.

4.3 How to Deal with First-Order Differential Operators

We are now aiming to compute the prefactor of the semiclassical approximation of the wave function of the universe in the case of massless Majorana spin-$\frac{1}{2}$ fields regarded as perturbation around a flat Euclidean background. In their paper on fermionic fields in quantum cosmology, D'Eath and Halliwell did show that the expansion in harmonics of the Lorentzian action of a Dirac field yields :

$$I = \sum_{np} \Big[I_n(m_{np}, \overline{m}_{np}, s_{np}, \overline{s}_{np}) + I_n(t_{np}, \overline{t}_{np}, r_{np}, \overline{r}_{np}) \Big] \quad , \tag{4.3.1}$$

where :

$$I_n[x, \overline{x}, y, \overline{y}] = \int dt \; N \left[\frac{i}{2N}(\overline{x}\dot{x} + x\dot{\overline{x}} + \overline{y}\dot{y} + y\dot{\overline{y}}) + e^{-\alpha}\left(n + \frac{3}{2}\right)(\overline{x}x + \overline{y}y) - m(yx + \overline{x}\overline{y}) \right]$$

$$+ \frac{i}{2}(\overline{x}x + \overline{y}y)_{S_F} + \frac{i}{2}(\overline{x}x + \overline{y}y)_{S_I} \quad . \tag{4.3.2}$$

In (4.3.1-2), the m_{np}, r_{np}, s_{np}, t_{np} are functions of t only and obey anticommutation relations. We are now interested in the Euclidean version $\widetilde{I}$ of this action integral. The transformation rules are :

$$\widetilde{I} = -iI \quad , \quad N \rightarrow -iN \quad , \tag{4.3.3}$$

so that, for flat backgrounds, (4.3.2) becomes :

$$\widetilde{I}_n[x, \widetilde{x}, y, \widetilde{y}] = \frac{1}{2}(\widetilde{x}x + \widetilde{y}y)_{S_F} + \frac{1}{2}(\widetilde{x}x + \widetilde{y}y)_{S_I}$$

$$+ \frac{1}{2}\int_0^a d\tau \left(\widetilde{x}\dot{x} + x\dot{\widetilde{x}} + \widetilde{y}\dot{y} + y\dot{\widetilde{y}} \right)$$

$$- \int_0^a d\tau N \left[\frac{1}{\tau}\left(n + \frac{3}{2}\right)(\widetilde{x}x + \widetilde{y}y) - m(yx + \widetilde{x}\widetilde{y}) \right] \quad , \tag{4.3.4}$$

subject to regularity at the origin and spectral boundary conditions (cf. section 4.2) :

$$x(0) = y(0) = \widetilde{x}(0) = \widetilde{y}(0) = 0 \quad , \quad x(a) = x' \quad , \quad y(a) = y' \quad . \tag{4.3.5}$$

4. Global Boundary Conditions and $\zeta(0)$ Value for the Massless Spin-$\frac{1}{2}$ Field

We are here interested in the massless case (which is the most relevant also in chapters five, eight and nine), and we introduce the notation $\nu = \frac{(n+\frac{3}{2})}{\tau}$ for future calculations. The prefactor P for a massless Majorana field, which only involves two sets of modes $(x, \widetilde{x})$ and $(y, \widetilde{y})$, can be computed by means of the path-integral formula :

$$
\begin{aligned}
P &= \int d[x]d[\widetilde{x}]d[y]d[\widetilde{y}] \exp\left(- \sum_{n=0}^{\infty} \widetilde{I}_n[x, \widetilde{x}, y, \widetilde{y}] \right) \\
&= \left\{ \int d[x]d[\widetilde{x}] \exp\left(- \sum_{n=0}^{\infty} \widetilde{I}_n[x, \widetilde{x}] \right) \right\}^2 \quad . \qquad (4.3.6)
\end{aligned}
$$

The problem now arises to compute a functional integral involving first-order differential operators. The known regularization techniques can be especially helpful if we are able to transform the problem into one involving a second-order differential operator. This can be achieved after a more careful consideration of the action integral (4.3.4) when $m = 0$ and $x' = y' = 0$. In fact the corresponding Lagrangian for each set of modes has the form :

$$
\frac{1}{2} \left(\widetilde{x}\dot{x} + x\dot{\widetilde{x}} - \nu \widetilde{x} x - \nu \widetilde{x} x \right) \quad .
$$

This leads to the one-dimensional version (the spatial integration having been carried out in the action (4.3.4)) of the variational principle for the eigenvalues :

$$
\left(\frac{d}{d\tau} - \nu \right) x = E \widetilde{x} \quad , \qquad (4.3.7)
$$

$$
\left(-\frac{d}{d\tau} - \nu \right) \widetilde{x} = E x \quad . \qquad (4.3.8)
$$

The operator on the left-hand side of (4.3.8) can be easily shown to be the adjoint $A^{\dagger}$ of the operator A on the left-hand side of (4.3.7) : $(Au, v) = (u, A^{\dagger}v)$, provided we require the vanishing of u and v on the boundary and at the origin (the latter condition ensures their regularity). In so doing, we are defining : $(u, v) \equiv \int_0^1 u(t)v(t)dt$. Thus, *squaring up* A means we have to require these boundary conditions and then to take the second-order operator $D = A^{\dagger}A$ (other details about this procedure can be found in Atiyah 1988a, pp 336-337). In light of (4.3.7-8), we find that :

$$\left(-\frac{d}{d\tau}-\nu\right)\left(\frac{d}{d\tau}-\nu\right)x=\left(-\frac{d^2x}{d\tau^2}+x\frac{d\nu}{d\tau}+\nu^2x\right)=E^2x \quad . \tag{4.3.9a}$$

Equation (4.3.9a) is the one we are looking for. It can also be written in the form :

$$\left[\frac{d^2}{d\tau^2}-\frac{((n+1)^2-\frac{1}{4})}{\tau^2}+E^2\right]x=0 \quad , \quad \forall n\geq 0 \quad , \tag{4.3.9b}$$

using the definition of ν. The regular solutions of (4.3.9b) as $\tau \rightarrow 0$ are of the form : $x = A\sqrt{\tau}J_{n+1}(E\tau)$ (whereas it may be shown that : $\widetilde{x} = -A\sqrt{\tau}J_{n+2}(E\tau)$). In fact the Bessel function J_{n+1} is regular at $\tau = 0$, and allows x to vanish on the boundary, having well-known zeros.

We are now in a position to compute the path integral (4.3.6) using zeta-function regularization. In fact, if we denote by Ω a normalization constant with dimensions of mass, we can formally write (Allen 1983a) :

$$\int d[x]d[\widetilde{x}]\exp\left[-\sum_{n=0}^{\infty}\widetilde{I}_n[x,\widetilde{x}]\right]=\prod_{m=1}^{\infty}\frac{|\lambda_m|}{\Omega}=\sqrt{\prod_{m=1}^{\infty}\Omega^{-2}|\lambda_m|^2} \quad , \tag{4.3.10}$$

where the $|\lambda_m|^2$ are the E^2 appearing in (4.3.9b). The notation used in (4.3.10) is loose since m stands for 2 degeneracy labels, say n and k, which characterize the eigenvalues. Moreover, in writing the infinite product, the dimensionless ratios $\left(\frac{|\lambda_m|}{\Omega}\right)$ should be thought as raised to an integer power given by the degeneracy of the eigenvalues (see discussion following (4.5.3-4)). The Berezin integration rules imply that one should only include those values of k, say $k = 1, 2, \ldots$, which correspond to positive values of $E_{n,k}$ (the n,k indices are omitted for simplicity of notation in (4.3.7-9)).

From now on, we put $\tau = t$, and $\widetilde{\tau}$ is just a parameter. The heat equation for the operator appearing on the left-hand side of (4.3.9a) is :

$$\left[\frac{\partial}{\partial\widetilde{\tau}}-\frac{\partial^2}{\partial t^2}+\frac{\left((n+1)^2-\frac{1}{4}\right)}{t^2}\right]G_n(t,t',\widetilde{\tau})=\delta(t-t')\delta(\widetilde{\tau}) \quad . \tag{4.3.11}$$

Taking the Laplace transforms of both members of (4.3.11), with σ^2 as parameter of the transform, we get, $\forall n \geq 0$:

$$\left[\frac{\partial^2}{\partial t^2} - \sigma^2 - \frac{\left((n+1)^2 - \frac{1}{4}\right)}{t^2}\right] G_n(t,t',\sigma^2) = -\delta(t-t') \quad . \tag{4.3.12}$$

When $t \neq t'$, the solutions of (4.3.12) regular at the origin and that vanish on the boundary are of the form : $G_n = A\sqrt{t} I_{n+1}(\sigma t)$, by virtue of the relation : $J_n(iz) = e^{\frac{n\pi i}{2}} I_n(z)$ when $-\pi < arg\ z \leq \frac{\pi}{2}$. However, we are interested in the general solution of (4.3.12) so as to compute the heat kernel, and in so doing we must also take into account the contribution of the Bessel functions K_n, and write the suitable normalization of the eigenfunctions. Thus we find for a single set of modes $(m, \widetilde{m})$ or $(r, \widetilde{r})$ (compare with section 4.1) :

$$G(t,t',\sigma^2) = \sum_{n=0}^{\infty} (n+1)(n+2) G_n(t,t',\sigma^2) = \sum_{n=1}^{\infty} n(n+1) G_{n-1}(t,t',\sigma^2)$$

$$= \sum_{n=1}^{\infty} n(n+1) \sqrt{t_<}\sqrt{t_>} \frac{I_n(\sigma t_<)}{I_n(\sigma a)}$$

$$\Big[I_n(\sigma a) K_n(\sigma t_>) - I_n(\sigma t_>) K_n(\sigma a)\Big] \quad . \tag{4.3.13}$$

One may point out that (4.1.11-12) and (4.3.13) differ because the powers of $t_<$ and $t_>$ multiplied by $I_n(\sigma t_<)$ are not the same. However, the integration over t involved in the calculation of the heat kernel gets rid of this difference.

4.4 Detailed Calculation of the Infinite Sums

The comparison with (4.1.13-15) shows that in our case the free part of the heat kernel becomes :

$$G^F(\widetilde{\tau}) = \int_0^{\frac{a^2}{2\tau}} \sum_{n=1}^{\infty} \Big[n(n+1)\Big] \frac{I_n(y)}{2} e^{-y}\, dy \quad . \tag{4.4.1}$$

Now, using again the integral representation of Bessel functions, one finds the identity : $\sum_{n=1}^{\infty} n^2 I_n(y) = \frac{y}{2} e^y$, which implies :

$$\sum_{n=1}^{\infty} \int_0^{\frac{a^2}{2\widetilde{\tau}}} n^2 I_n(y) e^{-y} \, dy = \frac{a^4}{16\widetilde{\tau}^2} \quad . \tag{4.4.2}$$

Moreover, using recurrence relations obeyed by Bessel functions, we find :

$$\begin{aligned}
\frac{1}{2} \int_0^{\frac{a^2}{2\widetilde{\tau}}} \sum_{n=1}^{\infty} n I_n(y) e^{-y} \, dy &= \frac{1}{4} \int_0^{\frac{a^2}{2\widetilde{\tau}}} \sum_{n=1}^{\infty} \Big[I_{n-1}(y) - I_{n+1}(y) \Big] y e^{-y} \, dy \\
&= \frac{1}{4} \int_0^{\frac{a^2}{2\widetilde{\tau}}} y e^{-y} I_0(y) \, dy + \frac{1}{4} \int_0^{\frac{a^2}{2\widetilde{\tau}}} y e^{-y} \frac{dI_0}{dy} \, dy \\
&= \frac{a^2}{8\widetilde{\tau}} e^{-\frac{a^2}{2\widetilde{\tau}}} I_0 \left(\frac{a^2}{2\widetilde{\tau}} \right) \\
&+ \frac{1}{4} \int_0^{\frac{a^2}{2\widetilde{\tau}}} e^{-y} (2y - 1) I_0(y) dy \quad .
\end{aligned} \tag{4.4.3}$$

The problem now arises to compute (4.4.3), and finally turn our attention to the interacting part of the heat kernel. Indeed one has :

$$\begin{aligned}
\int y e^{-y} I_0(y) \, dy &= \int \left[y e^{-y} (I_0 + I_1) - y^2 e^{-y} (I_0 + I_1) + y^2 e^{-y} \left(I_1 + I_0 - \frac{I_1}{y} \right) \right] dy \\
&= y^2 e^{-y} (I_0 + I_1) - \int y e^{-y} (I_0 + I_1) \, dy \quad ,
\end{aligned} \tag{4.4.4}$$

which implies :

$$3 \int y e^{-y} I_0(y) \, dy = y^2 e^{-y} (I_0 + I_1) - y e^{-y} I_0 + \int I_0 e^{-y} \, dy \quad . \tag{4.4.5}$$

In addition :

$$\begin{aligned}
\int I_0 e^{-y} \, dy &= \int \left[y e^{-y} \left(I_1 + I_0 - \frac{I_1}{y} \right) + (I_0 + I_1) \left(e^{-y} - y e^{-y} \right) \right] dy \\
&= y e^{-y} (I_0 + I_1) \quad ,
\end{aligned} \tag{4.4.6}$$

where in deriving (4.4.4) and (4.4.6) we have used the identities : $I_0' = I_1$; $I_1' = I_0 - \frac{I_1}{y}$. The integral on the right-hand side of (4.4.3) is thus found to be :

$$\int e^{-y}(2y-1)I_0(y)\,dy = e^{-y}\left\{\frac{2}{3}\left[y^2 I_0 + (y^2+y)I_1\right] - y(I_0+I_1)\right\} \quad . \tag{4.4.7}$$

The relations (4.4.1-3) and (4.4.7) imply in turn that, as $\widetilde{\tau} \rightarrow 0^+$:

$$G^F(\widetilde{\tau}) \sim \frac{a^4}{32}\widetilde{\tau}^{-2} + \frac{a^3}{12\sqrt{\pi}}\widetilde{\tau}^{-\frac{3}{2}} - \frac{a}{16\sqrt{\pi}}\widetilde{\tau}^{-\frac{1}{2}} + O(\sqrt{\widetilde{\tau}}) \quad . \tag{4.4.8}$$

In deriving (4.4.8), we have used the following asymptotic expansions (cf. appendix C, formulae (C.8-9)) valid as $z \rightarrow \infty$:

$$e^{-z}z^2 I_0(z) \sim \frac{z^{\frac{3}{2}}}{\sqrt{2\pi}} + \frac{1}{8}\frac{\sqrt{z}}{\sqrt{2\pi}} + O\left(\frac{1}{\sqrt{z}}\right) \quad , \tag{4.4.9}$$

$$e^{-z}z^2 I_1(z) \sim \frac{z^{\frac{3}{2}}}{\sqrt{2\pi}} - \frac{3}{8}\frac{\sqrt{z}}{\sqrt{2\pi}} + O\left(\frac{1}{\sqrt{z}}\right) \quad . \tag{4.4.10}$$

The Laplace transform of the kernel of the interacting part is given by :

$$G^{int}(\sigma^2) = -\frac{a^2}{2}\sum_{n=1}^{\infty} n^2 f(n;\sigma a) - \frac{a^2}{2}\sum_{n=1}^{\infty} n f(n;\sigma a) \quad , \tag{4.4.11}$$

where $f(n;\sigma a)$ has already been defined in (4.1.18), and *from now on* the notation using infinite sums for Laplace transforms of heat kernels is actually subject to a cut-off procedure, as explained in section 4.1. From the study of scalar fields we already know that the contribution to $\zeta(0)$ due to $-\frac{a^2}{2}\sum_{n=1}^{\infty} n^2 f(n;\sigma a)$ is given by $-\frac{1}{180}$ (Stewartson and Waechter 1971). The task is now to compute the effect of $\sum_{n=1}^{\infty} n f(n;\sigma a)$. In our case the Watson transform used in Schleich 1985 is a source of complications, because $n f(n;\sigma a)$ is not an even function of n. However, as explained following (4.1.17-18), we can use the asymptotic expansion of $f(n;\sigma a)$ valid uniformly with respect to n at large σa (Olver 1954), and compute at first the asymptotic form of the inverse Laplace transform of $nf(n;\sigma a)$. The second step of the calculation will consist in computing infinite sums using the Euler-Maclaurin formula for the difference between an infinite sum and the corresponding integral. Finally we can take the limit as $\widetilde{\tau} \rightarrow 0^+$, and pick out the constant contribution to the asymptotic form of $G^{int}(\widetilde{\tau})$. Let us now put $\tau \equiv \frac{n}{\sqrt{n^2+\sigma^2 a^2}}$. According to Stewartson and Waechter 1971, one has that :

$$nf(n;\sigma a) \sim \frac{n\sqrt{n^2+\sigma^2 a^2}}{\sigma^2 a^2}\left[\frac{\tau}{2}\frac{(1-\tau^2)}{n} - \frac{\tau^4}{2}\frac{(1-\tau^2)}{n^2} + \frac{\tau^3}{8}\frac{(1-\tau^2)(1-12\tau^2+15\tau^4)}{n^3} + \frac{\tau^4}{16}\frac{(1-\tau^2)(2-53\tau^2+168\tau^4-125\tau^6)}{n^4} + \ldots\right] . \tag{4.4.12}$$

This asymptotic expansion will play an important role also in the next chapter; thus it is here described in detail. The modified Bessel functions $I_\nu(z)$, $K_\nu(z)$ and their first derivatives have the following asymptotic expansions which are uniformly valid in the order ν for $| \, arg \nu \, | < \frac{\pi}{2}$, as the argument $z \rightarrow \infty$ (Olver 1954, Abramowitz and Stegun 1964, Stewartson and Waechter 1971) :

$$I_\nu(z) \sim \frac{e^z}{\sqrt{2\pi\nu}\left(1+\frac{z^2}{\nu^2}\right)^{\frac{1}{4}}} \sum_{s=0}^{\infty} \frac{U_s}{\nu^s} \quad , \tag{4.4.13}$$

$$K_\nu(z) \sim \sqrt{\frac{\pi}{2\nu}} \frac{e^{-z}}{\left(1+\frac{z^2}{\nu^2}\right)^{\frac{1}{4}}} \sum_{s=0}^{\infty} (-1)^s \frac{U_s}{\nu^s} \quad , \tag{4.4.14}$$

$$I'_\nu(z) \sim \frac{\left(1+\frac{z^2}{\nu^2}\right)^{\frac{1}{4}}}{\left(\frac{z}{\nu}\right)} \frac{e^z}{\sqrt{2\pi\nu}} \sum_{s=0}^{\infty} \frac{V_s}{\nu^s} \quad , \tag{4.4.15}$$

$$K'_\nu(z) \sim -\sqrt{\frac{\pi}{2\nu}} \frac{\left(1+\frac{z^2}{\nu^2}\right)^{\frac{1}{4}}}{\left(\frac{z}{\nu}\right)} e^{-z} \sum_{s=0}^{\infty} (-1)^s \frac{V_s}{\nu^s} \quad . \tag{4.4.16}$$

In (4.4.13-16), the polynomials U_s and V_s can be computed by means of recurrence relations. Defining $u \equiv \frac{\nu}{\sqrt{\nu^2+z^2}}$, the first few of them are given by (Olver 1954) :

$$U_0 = 1 \quad , \quad U_1 = \frac{u}{8} - \frac{5}{24}u^3 \quad , \tag{4.4.17}$$

$$U_2 = \frac{9}{128}u^2 - \frac{77}{192}u^4 + \frac{385}{1152}u^6 \quad , \tag{4.4.18}$$

$$U_3 = \frac{75}{1024}u^3 - \frac{4563}{5120}u^5 + \frac{17017}{9216}u^7 - \frac{85085}{82944}u^9 \quad , \tag{4.4.19}$$

$$V_0 = 1 \quad , \quad V_1 = -\frac{3}{8}u + \frac{7}{24}u^3 \quad , \tag{4.4.20}$$

$$V_2 = -\frac{15}{128}u^2 + \frac{99}{192}u^4 - \frac{455}{1152}u^6 \quad , \tag{4.4.21}$$

$$V_3 = -\frac{105}{1024}u^3 + \frac{5577}{5120}u^5 - \frac{6545}{3072}u^7 + \frac{95095}{82944}u^9 \quad . \tag{4.4.22}$$

It is the insertion of (4.4.13-22) into (4.1.18) which finally yields (4.4.12). Let us now denote by L_I the inverse Laplace transform. One has therefore :

$$\begin{aligned} L_I\Big[nf(n;\sigma a)\Big] \sim\ & \frac{1}{2a^2}ne^{-\frac{n^2\tilde{\tau}}{a^2}} - \frac{2\tilde{\tau}^{\frac{3}{2}}}{3\sqrt{\pi}a^5}n^3e^{-\frac{n^2\tilde{\tau}}{a^2}} \\ & + \frac{\tilde{\tau}}{8a^4}ne^{-\frac{n^2\tilde{\tau}}{a^2}} - \frac{3\tilde{\tau}^2}{4a^6}n^3e^{-\frac{n^2\tilde{\tau}}{a^2}} \\ & + \frac{5}{16}\frac{\tilde{\tau}^3}{a^8}n^5e^{-\frac{n^2\tilde{\tau}}{a^2}} + \frac{1}{6\sqrt{\pi}}\frac{\tilde{\tau}^{\frac{3}{2}}}{a^5}ne^{-\frac{n^2\tilde{\tau}}{a^2}} \\ & - \frac{53}{30}\frac{\tilde{\tau}^{\frac{5}{2}}}{\sqrt{\pi}a^7}n^3e^{-\frac{n^2\tilde{\tau}}{a^2}} + \frac{168}{105}\frac{\tilde{\tau}^{\frac{7}{2}}}{\sqrt{\pi}a^9}n^5e^{-\frac{n^2\tilde{\tau}}{a^2}} \\ & - \frac{50}{189}\frac{\tilde{\tau}^{\frac{9}{2}}}{\sqrt{\pi}a^{11}}n^7e^{-\frac{n^2\tilde{\tau}}{a^2}} + \ldots \quad , \end{aligned} \tag{4.4.23}$$

and the asymptotic expansion of the corresponding part of the interacting heat kernel is obtained from $-\frac{a^2}{2}\sum_{n=1}^{\infty} L_I[nf(n;\sigma a)]$. Indeed, when each term on the right-hand side of

(4.4.23) is summed from $n=1$ to N, the resulting function of $\widetilde{\tau}$ does not always converge uniformly to the infinite sum $\sum_{n=1}^{\infty}$ in a neighbourhood $\widetilde{\tau} \in (0,\delta)$. However, as already remarked following (4.1.18), the study of the error terms shows that it is valid to take the limit $N \rightarrow \infty$ as in (4.4.29-31) (D'Eath and Esposito 1991b), so that we can write (setting here $a=1$ for simplicity) :

$$\lim_{N\to\infty} \sum_{n=1}^{N} L_I\Big[nf(n;\sigma)\Big] \sim \sum_{n=1}^{\infty} L_I\left[-\frac{\tau^5}{2n^2}+\ldots\right] + \sum_{n=1}^{\infty} L_I\left(\frac{\tau^2}{2n}\right) \quad . \tag{4.4.24}$$

In order to compute sums of the type : $\sum_{n=1}^{\infty} n^{(2m+1)} e^{-\frac{n^2\widetilde{\tau}}{a^2}}$, where $m = 0,1,2,\ldots$, we use the Euler-Maclaurin formula (Jeffreys and Jeffreys 1946, Wong 1989) :

$$\frac{1}{2}F(0)+F(1)+F(2)+\ldots-\int_0^{\infty} F(y)dy = -\frac{B_2}{2}F'(0)$$

$$-\frac{B_4}{4!}F'''(0)-\frac{B_6}{6!}F'''''(0)\ldots , \tag{4.4.25}$$

for the function $F(y) = ye^{-\frac{y^2\widetilde{\tau}}{a^2}}$. In (4.4.25), the B_i denote the Bernoulli numbers. Relevant details on the right-hand side of (4.4.25) can be found in appendix C. Hence we obtain the formula :

$$\sum_{n=1}^{\infty} ne^{-\frac{n^2\widetilde{\tau}}{a^2}} - \int_0^{\infty} xe^{-\frac{x^2\widetilde{\tau}}{a^2}}dx$$

$$= \sum_{n=1}^{\infty} ne^{-\frac{n^2\widetilde{\tau}}{a^2}} - \frac{a^2}{2\widetilde{\tau}}$$

$$= -\frac{B_2}{2}F'(0)$$

$$-\frac{B_4}{4!}F'''(0)-\frac{B_6}{6!}F'''''(0)\ldots \quad . \tag{4.4.26}$$

Let us now point out that :

$$F'(n) = e^{-\frac{n^2\widetilde{\tau}}{a^2}} - \frac{2n^2\widetilde{\tau}}{a^2}e^{-\frac{n^2\widetilde{\tau}}{a^2}} \quad , \tag{4.4.27}$$

$$F''(n) = \left(-\frac{6n\widetilde{\tau}}{a^2} + \frac{4n^3\widetilde{\tau}^2}{a^4}\right) e^{-\frac{n^2\widetilde{\tau}}{a^2}} \quad . \tag{4.4.28}$$

Thus $F'(0) = 1$, whereas all derivatives of F higher than F'' contain, when computed at $n = 0$, contributions which are at least linear in $\widetilde{\tau}$, so that we may write :

$$\sum_{n=1}^{\infty} n e^{-\frac{n^2\widetilde{\tau}}{a^2}} = \frac{a^2}{2\widetilde{\tau}} - \frac{1}{12} - \frac{\widetilde{\tau}}{120a^2} - \frac{\widetilde{\tau}^2}{504a^4} + \dots \quad . \tag{4.4.29}$$

The other sums are obtained differentiating (4.4.29) with respect to $\widetilde{\tau}$. This yields :

$$\sum_{n=1}^{\infty} n^3 e^{-\frac{n^2\widetilde{\tau}}{a^2}} = \frac{a^4}{2\widetilde{\tau}^2} + \frac{1}{120} + \frac{\widetilde{\tau}}{252a^2} + \dots \quad , \tag{4.4.30}$$

$$\sum_{n=1}^{\infty} n^5 e^{-\frac{n^2\widetilde{\tau}}{a^2}} = \frac{a^6}{\widetilde{\tau}^3} - \frac{1}{252} + \dots \quad , \tag{4.4.31}$$

and so on.

4.5 The Heat Kernel and the Prefactor

The constant appearing in the asymptotic expansion of $\sum_{n=1}^{\infty} L_I[nf(n;\sigma a)]$ is found using (4.4.29-31) plus other formulae for infinite sums with higher odd powers of n. This leads to :

$$-\frac{B_2}{4a^2} + \frac{\widetilde{\tau}}{8a^4}\left(\frac{a^2}{2\widetilde{\tau}}\right) + \frac{5\widetilde{\tau}^3}{16a^8}\frac{a^6}{\widetilde{\tau}^3} - \frac{3\widetilde{\tau}^2}{4a^6}\frac{a^4}{2\widetilde{\tau}^2} = -\frac{1}{24a^2} \quad . \tag{4.5.1}$$

In light of (4.4.8), (4.4.11) and (4.5.1) we find that for a single set of modes $\zeta(0)$ is given by :

$$\zeta_{el}(0) = -\frac{a^2}{2}\left(-\frac{1}{24a^2}\right) - \frac{1}{180} = \frac{11}{720} \quad . \tag{4.5.2}$$

Therefore the behaviour of the prefactor P in (4.3.6) for the complete massless Majorana field in the limit of small three-geometry is given by :

$$P = Da^{-2\zeta_{el}(0)} = Da^{-\zeta_{tot}(0)} = Da^{-\frac{11}{360}} \quad . \tag{4.5.3}$$

In other words, using the notation in (4.3.9-10), the full degeneracy of the eigenvalues λ_m is here $2(n+1)(n+2)$, $\forall n \geq 0$. The $| \lambda_m |^2 = E^2$ have degeneracy $4(n+1)(n+2)$, but, in order to obtain $\zeta(0)$ for a Majorana field, one has to divide the final result by 2. This is equivalent to the result obtained from the zeta-function :

$$\zeta_M(s) \equiv \sum_{n=0}^{\infty} \sum_{k=1}^{\infty} \Big[2(n+1)(n+2)\Big] \left(E_{n,k}^2\right)^{-s} \quad , \tag{4.5.4}$$

provided one does not divide by 2 the corresponding $\zeta(0)$. The advantage of using (4.5.4) is that the formal product $\prod_{m=1}^{\infty} | \lambda_m |$ is then expressed in terms of $\zeta'(0)$ and $\zeta(0)$ in a way entirely analogous to bosonic calculations (cf. (1.4.2)), apart from a sign change in the exponent for the prefactor, since fermionic fields form an anticommuting Grassmann algebra. This is why a $(-)$ sign appears in (4.5.3). In general, the value $\zeta_{tot}(0) = \frac{11}{360}$ in (4.5.3) should be multiplied by the number N of massless Majorana fields present in the theory.

It is also worth deriving the asymptotic expansion of the whole kernel $\hat{G}(\widetilde{\tau})$ of the heat equation in the limit of small $\widetilde{\tau}$, for the complete Majorana field. Indeed one has : $\hat{G}(\widetilde{\tau}) = \hat{G}^F(\widetilde{\tau}) + \hat{G}^{int}(\widetilde{\tau}) = 2G^F(\widetilde{\tau}) + 2G^{int}(\widetilde{\tau})$, where the relations (4.4.11-31) show that :

$$\begin{aligned}
\hat{G}^{int}(\widetilde{\tau}) \sim & \left(-\frac{a^3\sqrt{\pi}}{8\widetilde{\tau}^{\frac{3}{2}}} + \frac{a^2}{4\widetilde{\tau}} - \frac{11\sqrt{\pi}a}{256\sqrt{\widetilde{\tau}}}\right) - \frac{1}{90} \\
& - a^2\left[\left(\frac{1}{2a^2} + \frac{\widetilde{\tau}}{8a^4}\right)\left(\frac{a^2}{2\widetilde{\tau}} - \frac{1}{12} - \frac{\widetilde{\tau}}{120a^2} - \frac{\widetilde{\tau}^2}{504a^4}\right)\right. \\
& + \frac{5\widetilde{\tau}^3}{16a^8}\left(\frac{a^6}{\widetilde{\tau}^3} - \frac{1}{252}\right) - \frac{3\widetilde{\tau}^2}{4a^6}\left(\frac{a^4}{2\widetilde{\tau}^2} + \frac{1}{120} + \frac{\widetilde{\tau}}{252a^2}\right) \\
& + \frac{\widetilde{\tau}^{\frac{3}{2}}}{6\sqrt{\pi}a^5}\left(\frac{a^2}{2\widetilde{\tau}} - \frac{1}{12} - \frac{\widetilde{\tau}}{120a^2} - \frac{\widetilde{\tau}^2}{504a^4}\right)
\end{aligned}$$

$$
- \left(\frac{2\tilde{\tau}^{\frac{3}{2}}}{3\sqrt{\pi}a^5} + \frac{53}{30} \frac{\tilde{\tau}^{\frac{5}{2}}}{\sqrt{\pi}a^7} \right) \left(\frac{a^4}{2\tilde{\tau}^2} + \frac{1}{120} + \frac{\tilde{\tau}}{252a^2} \right)
$$

$$
+ \frac{168}{105} \frac{\tilde{\tau}^{\frac{7}{2}}}{\sqrt{\pi}a^9} \left(\frac{a^6}{\tilde{\tau}^3} - \frac{1}{252} \right)
$$

$$
\left. - \frac{50}{189} \frac{\tilde{\tau}^{\frac{9}{2}}}{\sqrt{\pi}a^{11}} \frac{3a^8}{\tilde{\tau}^4} \right] + \mathrm{O}(\sqrt{\tilde{\tau}}) \quad . \tag{4.5.5}
$$

From (4.4.8) and (4.5.5) we find that, as $\tilde{\tau} \rightarrow 0^+$, the whole heat kernel $\hat{G}(\tilde{\tau})$ has the asymptotic expansion :

$$
\hat{G}(\tilde{\tau}) \sim \frac{a^4}{16}\tilde{\tau}^{-2} + a^3 \left(\frac{1}{6\sqrt{\pi}} - \frac{\sqrt{\pi}}{8} \right) \tilde{\tau}^{-\frac{3}{2}} + a \left(\frac{5}{24\sqrt{\pi}} - \frac{11\sqrt{\pi}}{256} \right) \tilde{\tau}^{-\frac{1}{2}} + \frac{11}{360} + \mathrm{O}(\sqrt{\tilde{\tau}}) \, . \tag{4.5.6}
$$

4.6 From Global to Local Boundary Conditions

In section 4.2, following D'Eath and Halliwell 1987, we chose to fix the untwiddled variables on the final surface S_F and the twiddled variables on the initial surface S_I so as to ensure that a regular solution exists in the massless limit. We now show that also the *local* classical boundary-value problem is well-posed for a massless Majorana spin-$\frac{1}{2}$ field $\left(\psi^A, \tilde{\psi}^{A'}\right)$ obeying the Weyl equation. We begin by writing the decompositions :

$$
\psi^A = \psi^A_{(+)} + \psi^A_{(-)} \quad , \tag{4.6.1}
$$

$$
\tilde{\psi}^{A'} = \tilde{\psi}^{A'}_{(+)} + \tilde{\psi}^{A'}_{(-)} \quad . \tag{4.6.2}
$$

With our notation, the $(+)$ parts correspond to the modes with untwiddled coefficients in the expansion in spinor harmonics on S^3, multiplying unbarred harmonics having positive eigenvalues $\frac{1}{2}(n+\frac{3}{2})$ for the three-dimensional Dirac operator on the bounding S^3. The $(-)$ parts correspond instead to twiddled modes multiplying barred harmonics having negative eigenvalues $-\frac{1}{2}(n+\frac{3}{2})$ for this operator. Moreover, denoting by P_a and P_b complementary projection operators (i.e. such that $P_a+P_b=1$), the formulae (4.6.1-2) can be re-expressed as :

$$\psi^A = (P_a+P_b)\psi^A_{(+)} + (P_a+P_b)\psi^A_{(-)} \quad , \tag{4.6.3}$$

$$\widetilde{\psi}^{A'} = (P_a+P_b)\widetilde{\psi}^{A'}_{(+)} + (P_a+P_b)\widetilde{\psi}^{A'}_{(-)} \quad . \tag{4.6.4}$$

We now choose :

$$P_a\psi^A_{(-)} = -P_b\psi^A_{(-)} \quad , \tag{4.6.5}$$

$$P_a\widetilde{\psi}^{A'}_{(-)} = -P_b\widetilde{\psi}^{A'}_{(-)} \quad , \tag{4.6.6}$$

in order to get rid of the singular part of (4.6.3-4). Moreover, setting $\epsilon=\pm 1$, we specify on the boundary $\sqrt{2}\ {}_e n_A{}^{A'}\psi^A - \epsilon\ \widetilde{\psi}^{A'} = \lambda^{A'}$. Thus, in light of the relations between unbarred and barred harmonics (D'Eath and Halliwell 1987), the boundary conditions can be expressed as follows :

$$\sqrt{2}\ {}_e n_A{}^{A'}(P_a+P_b)\psi^A_{(+)} = \lambda_1^{A'} \quad \text{on} \quad S^3 \quad , \tag{4.6.7}$$

$$-\epsilon(P_a+P_b)\widetilde{\psi}^{A'}_{(+)} = \lambda_2^{A'} \quad \text{on} \quad S^3 \quad , \tag{4.6.8}$$

where $\lambda_1^{A'}$ and $\lambda_2^{A'}$ are the parts of $\lambda^{A'}$ related to the $\overline{\rho}$ and σ harmonics respectively. In other words, since (4.6.5-6) hold, we find a smooth solution of the Weyl equation; the parts $\left[P_a\psi^A_{(+)}\right]$ and $\left[P_b\psi^A_{(+)}\right]$, $\left[P_a\widetilde{\psi}^{A'}_{(+)}\right]$ and $\left[P_b\widetilde{\psi}^{A'}_{(+)}\right]$, are related on the boundary in the way specified by (4.6.7-8). Note also that, if $\lambda^{A'}$ does not vanish, the classical Euclidean action should be supplemented by a boundary term proportional to

$$\int_{\partial M} \lambda^{A'}\ {}_e n_{AA'}\psi^A \sqrt{h}\ d^3x \quad .$$

In particular, if $\lambda_1^{A'} = \lambda_2^{A'} = 0$, we find :

$$\sum_{npq} \alpha_n^{pq} m_{np}(a)\ {}_e n_A{}^{A'} \rho^A_{nq} = 0 \quad , \tag{4.6.9}$$

$$\sum_{npq} \alpha_n^{pq} r_{np}(a) \sigma_{nq}^{A'} = 0 \quad . \tag{4.6.10}$$

Because the harmonics appearing in (4.6.9-10) are linearly independent, these relations imply : $m_{np}(a) = r_{np}(a) = 0$, $\forall n, p$, where a is the S^3 radius.

In section 5.8 and chapter eight we will study one-loop calculations with the local boundary conditions : $\sqrt{2}\, {}_e n_A^{\ A'} \psi^A - \epsilon\, \widetilde{\psi}^{A'} = 0$ on S^3. In that case, since one studies the eigenvalue equations for the quantum theory, $\psi_{(-)}^A$ and $\widetilde{\psi}_{(-)}^{A'}$ are no longer singular. Hence we do not have to impose (4.6.5-6). However, as far as the classical theory is concerned, our argument can be easily extended to a massive Dirac field described by the pairs $\left(\phi^A, \widetilde{\phi}^{A'}\right)$ and $\left(\chi^A, \widetilde{\chi}^{A'}\right)$. Since the field is massive, $\phi_{(-)}^A$, $\widetilde{\phi}_{(-)}^{A'}$, $\chi_{(-)}^A$ and $\widetilde{\chi}_{(-)}^{A'}$ are not singular, and the *local* boundary-value problem is well-posed.

CHAPTER FIVE

CHOICE OF BOUNDARY CONDITIONS IN ONE-LOOP QUANTUM COSMOLOGY

Abstract. The problem of boundary conditions in a supersymmetric theory of quantum cosmology is studied. For fermionic fields one has a choice of local or nonlocal boundary conditions. $N = 1$ supergravity is the simplest supersymmetric model, which we study in a background cosmological model which is flat Euclidean space bounded by a three-sphere. We pick out the physical degrees of freedom by imposing the supersymmetry constraints and choosing a gauge condition. A set of naturally occurring nonlocal boundary conditions for fermions is such that the coefficients of the physical degrees of freedom are regular at the origin, and half of them are set equal to zero on S^3. However, supersymmetry does not relate them to Dirichlet or other local boundary conditions for the gravitational field. Studying $N = 1$ supergravity at 1 loop about flat space, this approach is shown to yield a nonvanishing result for the PDF $\zeta(0)$. Namely, the PDF contribution to the prefactor due to the spin-$\frac{3}{2}$ field is proportional to $a^{\frac{289}{360}}$ (a being the three-sphere radius), which does not cancel $a^{-\frac{278}{45}}$ due to the gravitational field subject to Dirichlet boundary conditions for the perturbed three-metric.

We therefore study possible local boundary conditions for both bosons and fermions. One set, related by supersymmetry, was originally introduced by Breitenlohner, Freedman and Hawking for gauged supergravity theories in anti-de Sitter space. It involves the normal to the boundary and field strengths for spins $s = 1, \frac{3}{2}, 2$, while for $s = 0$ it involves a complex scalar field and its normal derivative, and for $s = \frac{1}{2}$ the undifferentiated field. Using twistor theory in flat space, the existence is proved of a form of the spin-lowering operator which preserves these local boundary conditions required on S^3 for solutions to the massless free-field equations for adjacent spins s and $s + \frac{1}{2}$.

The eigenvalue condition implied by these boundary conditions for the massless spin-$\frac{1}{2}$ field is found to be : $J_{n+1}(Ea) = \pm J_{n+2}(Ea)$, $\forall n \geq 0$, with degeneracy $(n+2)(n+1)$. Using the formalism of $SU(2)$ spinors in Euclidean four-space and a theorem due to von Neumann, we also prove that a first-order differential operator for this boundary-value problem exists which is symmetric and has self-adjoint extensions.

Moreover, in the case of the spin-1 field strength these local boundary conditions are shown to lead to the following PDF values of $\zeta(0)$: $-\frac{77}{180}$ (magnetic) and $\frac{13}{180}$ (electric), whereas for a complex scalar field one finds : $\zeta(0) = \frac{7}{45}$. Thus the values of $\zeta(0)$ are not equal for all spins. As a partial justification of this result, we prove why, in the comparison spin 0 vs spin $\frac{1}{2}$, the spin-lowering operator does not lead to the same eigenvalues. Finally, we discuss in detail the issue of the preservation in time of the gauge constraint used in evaluating $\zeta(0)$.

5.1 General Form of the Action of the Spin-$\frac{3}{2}$ Field

In the previous chapter, we studied one-loop properties of massless Majorana spin-$\frac{1}{2}$ fields using global boundary conditions. Thus we are here interested in generalizing this technique to the case of the gravitino field occurring in supergravity, as a first step in our analysis of various possible choices of boundary conditions in one-loop quantum cosmology. The first part of our analysis describes some properties of supergravity in the language of Lorentzian geometry, to make more easily contact with relevant parts of section 1.3 and of our basic references (D'Eath 1984, Hughes 1990). Our language will then become the one of Riemannian geometry in section 5.5, devoted to a PDF one-loop calculation.

The volume part I of the action of the spin-$\frac{3}{2}$ field is given by :

$$I \equiv I_L + H.C.(I_L) \quad , \tag{5.1.1}$$

where :

$$I_L \equiv \frac{1}{2}\int_M d^4x \; \epsilon^{\mu\nu\rho\sigma}\left(\overline{\psi}_\mu^{A'} e_{AA'\nu} D_\rho \psi_\sigma^A\right) = I_A + I_B \quad . \tag{5.1.2}$$

In (5.1.2) we have used the notation :

$$I_A \equiv \frac{1}{2}\int_M d^4x \; \epsilon^{ilj}\overline{\psi}_i^{A'} e_{AA'l}\dot{\psi}_j^A - \frac{1}{2}\int_M d^4x \; N\epsilon^{ilj}\overline{\psi}_i^{A'} n_{AA'}\left({}^{(4)}D_l\psi_j^A\right) \quad , \tag{5.1.3}$$

$$I_B \equiv \frac{1}{2}\int_M d^4x \; \epsilon^{ilj}\overline{\psi}_0^{A'} e_{AA'i}\left({}^{(4)}D_l\psi_j^A\right) - \frac{1}{2}\int_M d^4x \; \epsilon^{ilj}\overline{\psi}_i^{A'} e_{AA'l}\left({}^{(4)}D_j\psi_0^A\right) \quad . \tag{5.1.4}$$

In (5.1.4), ψ_0^A represents the unprimed part of a spin-$\frac{1}{2}$ field. It may be thus expanded on a family of three-spheres centred on the origin as :

$$\psi_0^A = \frac{e^{-\frac{3\alpha}{2}}}{2\pi}\sum_{npq}\alpha_n^{pq}\left(u_{np}\rho^{nqA} + \overline{v}_{np}\overline{\sigma}^{nqA}\right) \quad . \tag{5.1.5}$$

In (5.1.5), α is the logarithm of the three-sphere radius, and the properties of the α_n^{pq} and of the harmonics are all discussed in detail in section 4.2 and in appendix A. The expansion of the spin-$\frac{3}{2}$ field is more elaborated. In fact ψ_j^D has both a symmetric and an antisymmetric part, so that :

$$\psi_j^D = \left(\psi^{(DB)B'} + \epsilon^{DB}\overline{\psi}^{B'}\right) e_{BB'j} \quad . \tag{5.1.6}$$

If we insert into (5.1.6) a relation of the type (5.1.5) for ψ^B and the formulae for the expansion of $\psi^{(DB)B'}$ on the three-sphere, we get :

$$\begin{aligned}\psi_j^D &= \frac{e^{-\frac{3\alpha}{2}}}{2\pi}\sum_{npq}\alpha_n^{pq}\left(m_{np}^{(\alpha)}\alpha^{nqDBB'} + m_{np}^{(\beta)}\beta^{nqDBB'} + m_{np}^{(\gamma)}\gamma^{nqDBB'}\right)e_{BB'j}\\ &+ \frac{e^{-\frac{3\alpha}{2}}}{2\pi}\sum_{npq}\alpha_n^{pq}\left(\overline{r}_{np}^{(\lambda)}\overline{\lambda}^{nqDBB'} + \overline{r}_{np}^{(\mu)}\overline{\mu}^{nqDBB'} + \overline{r}_{np}^{(\nu)}\overline{\nu}^{nqDBB'}\right)e_{BB'j}\\ &+ \frac{e^{-\frac{3\alpha}{2}}}{2\pi}\sum_{npq}\alpha_n^{pq}\left(\overline{s}_{np}\overline{\rho}^{nqB'} + t_{np}\sigma^{nqB'}\right)\epsilon^{DB}e_{BB'j} \quad ,\end{aligned} \tag{5.1.7}$$

plus the obvious correspondent relation for $\overline{\psi}_j^{D'}$. For the sake of clarity, it is better to write explicitly the behaviour of $\alpha^{nqACC'}$, $\beta^{nqACC'}$, $\gamma^{nqACC'}$, $\lambda^{nqA'B'B}$, $\mu^{nqA'B'B}$ and $\nu^{nqA'B'B}$ as functions of the harmonics and of the Lorentzian normal (Hughes 1990 and our appendix A) :

$$\alpha^{nqACC'} = \rho^{nq(AC)C'} - \frac{4i}{3}\rho^{n+1,q(A}{}_{n}{}^{C)C'} \quad , \tag{5.1.8}$$

$$\beta^{nqACC'} = -\rho^{nq(ACD)}n_D{}^{C'} \quad , \tag{5.1.9}$$

$$\gamma^{nqACC'} = \rho^{nq(A}{}_{n}{}^{C)C'} \quad , \tag{5.1.10}$$

$$\lambda^{nqA'B'B} = \sigma^{nq(A'B')B} - \frac{4i}{3}\sigma^{n+1,q(A'}{}_{n}{}^{BB')} \quad , \tag{5.1.11}$$

$$\mu^{nqA'B'B} = -\sigma^{nq(A'B'C')}n^B{}_{C'} \quad , \tag{5.1.12}$$

$$\nu^{nqA'B'B} = \sigma^{nq(A'}{}_{n}{}^{BB')} \quad . \tag{5.1.13}$$

From (5.1.3-4) and (5.1.7) it is clear that we also need an identity for $\epsilon^{ijk}e_{AA'i}e_{BB'j}$ and $\epsilon^{ijk}e_{AA'i}e_{BB'j}e_{CC'k}$. Indeed we have that :

$$\epsilon^{ijk}e_{AA'i}e_{BB'j} = \epsilon^{ijk}\Big[e_{[AA'i}e_{B]B'j} + e_{B[A'i}e_{AB']j}\Big] \quad , \tag{5.1.14}$$

$$2e_{[AA'i}e_{B]B'j} = -\frac{1}{2}h_{ij}\epsilon_{AB}\epsilon_{A'B'} - i\epsilon_{ijs}\sqrt{h}\,\epsilon_{AB}n_{EA'}e^{E}_{B'}{}^{s} \quad , \tag{5.1.15}$$

$$2e_{B[A'i}e_{AB']j} = -\frac{1}{2}h_{ij}\epsilon_{A'B'}\epsilon_{BA} + i\epsilon_{ijs}\sqrt{h}\,\epsilon_{A'B'}n_{BE'}e_{A}^{E's} \quad . \tag{5.1.16}$$

Thus (5.1.14) becomes :

$$\epsilon^{ijk}e_{AA'i}e_{BB'j} = -i\sqrt{h}\left(\epsilon_{AB}n_{EA'}e^{E}_{B'}{}^{k} - \epsilon_{A'B'}n_{BE'}e_{A}^{E'k}\right) \quad . \tag{5.1.17}$$

Moreover, we know the identity :

$$e_{AA'\mu}e^{BB'\mu} = -n_{AA'}n^{BB'} + e_{AA'i}e^{BB'i} = -\epsilon_{A}^{\ B}\epsilon_{A'}^{\ B'} \quad , \tag{5.1.18}$$

which in turn implies that :

$$\epsilon^{ijk}e_{AA'i}e_{BB'j}e_{CC'k} = -i\sqrt{h}\,\epsilon_{AB}n_{EA'}\epsilon_{D'B'}\left(-\epsilon_{C}^{\ E}\epsilon_{C'}^{\ D'} + n_{CC'}n^{ED'}\right)$$
$$+ i\sqrt{h}\,\epsilon_{A'B'}n_{BE'}\epsilon_{DA}\left(-\epsilon_{C}^{\ D}\epsilon_{C'}^{\ E'} + n_{CC'}n^{DE'}\right) \quad . \tag{5.1.19}$$

We are now in a position to expand I_A and I_B appearing in (5.1.3-4).

5.2 Hamiltonian Form of the Action and Supersymmetry Constraints

In section 1.3 we have learned that ψ_0^A and $\overline{\psi}_0^{A'}$ play the role of Lagrange multipliers in the Hamiltonian of supergravity :

$$H = \int d^3x\Big[N({}_1H_\perp) + N^i({}_1H_i) + \psi_0^A({}_1S_A)$$
$$+({}_1\overline{S}_{A'})\overline{\psi}_0^{A'} + M_{AB}J^{AB} + \overline{M}_{A'B'}\overline{J}^{A'B'}\Big] \quad , \tag{5.2.1}$$

i.e. ψ_0^A and $\tilde{\psi}_0^{A'}$ multiply right-handed and left-handed supersymmetry constraints respectively. This implies that we must set $I_B + H.C.(I_B) = 0$ in (5.1.4). We now expand I_A and I_B to quadratic order. This implies that no torsion terms will appear in our action integral (cf. (1.3.1)). Let us put : $I_B = I_3 - I_4$, where :

$$I_3 \equiv \frac{1}{2} \int_M d^4x \; \epsilon^{ilj} \overline{\psi}_0^{A'} e_{AA'i} \left({}^{(4)}D_l \psi_j^A \right) \quad , \tag{5.2.2}$$

$$I_4 \equiv \frac{1}{2} \int_M d^4x \; \epsilon^{ilj} \overline{\psi}_i^{A'} e_{AA'l} \left({}^{(4)}D_j \psi_0^A \right) \quad . \tag{5.2.3}$$

In view of the properties of the $\rho's$ and $\sigma's$, we can focus our attention just on that part of I_3 and I_4 which is related to the $\rho's$, and here denoted by $I_3^{(\rho\tilde{\rho})}$ and $I_4^{(\rho\tilde{\rho})}$. Entirely analogous results will then hold for the $\sigma's$. We must now insert into (5.2.2-3) the expansions (5.1.5), (5.1.7) and the identity (5.1.17). Moreover, denoting by f a real-valued function of r and $\dot{r}$, we know for example that for a spin-$\frac{1}{2}$ field in a closed FRW space-time one has :

$${}^{(4)}D_j \phi^A = \frac{1}{r} \left[{}^{(3)}D_j \phi^A \right] + f(r, \dot{r}) n^{AB'} e_{BB'j} \phi^B \quad , \tag{5.2.4}$$

where ${}^{(3)}D_j$ is the spinor covariant derivative on a unit three-sphere. The generalization of (5.2.4) to the case of the spin-$\frac{3}{2}$ field is then straightforward. Its consequence for flat backgrounds is that, after taking the complex conjugate of I_L defined in (5.1.2), only the effects of ${}^{(3)}D_j$ survive in the action (5.1.1). Hence we find (cf. appendix A) :

$$I_3^{(\rho,\tilde{\rho})} = \widetilde{I}_3^{(\rho,\tilde{\rho})} + I_c^{(\rho,\tilde{\rho})} \quad , \tag{5.2.5}$$

$$I_4^{(\rho,\tilde{\rho})} = \widetilde{I}_4^{(\rho,\tilde{\rho})} + I_d^{(\rho,\tilde{\rho})} \quad , \tag{5.2.6}$$

where, denoting by k_1 and k_2 two real constants whose value does not affect our results :

$$k_1 \widetilde{I}_3^{(\rho\tilde{\rho})} = \int dt \; e^{-\alpha} \sum_{npq} (\alpha_n^{pq})^2 \bar{u}_{np} \left[\frac{2}{3} ni \; m_{n-1,p}^{(\alpha)} - \left(\frac{n}{2} + \frac{3}{4} \right) m_{np}^{(\gamma)} \right] \quad , \tag{5.2.7}$$

$$k_1 \widetilde{I}_4^{(\rho\bar{\rho})} = i \int dt\ e^{-\alpha} \sum_{npq} (\alpha_n^{pq})^2 \overline{m}_{np}^{(\alpha)} u_{n+1,p} \left(\frac{2}{3}n + \frac{14}{3}\right)$$

$$+ \int dt\ e^{-\alpha} \sum_{npq} (\alpha_n^{pq})^2 \overline{m}_{np}^{(\gamma)} u_{np} \left(\frac{n}{2} + \frac{3}{4}\right) \quad , \tag{5.2.8}$$

$$[I_c - I_d]^{(\rho,\bar{\rho})} = k_2 \int dt\ e^{-\alpha} \sum_{npq} \sum_{n'p'q'} \alpha_n^{pq} \alpha_{n'}^{p'q'} \left(n' + \frac{3}{2}\right) \delta^{nn'} C_n^{qq'} \left(\overline{u}_{np} \overline{s}_{n'p'} - s_{n'p'} u_{np}\right). \tag{5.2.9}$$

In light of (5.2.9), $[I_c - I_d]^{(\rho,\bar{\rho})} + H.C.[I_c - I_d]^{(\rho,\bar{\rho})} = 0$, so that the requirement $I_B + H.C.(I_B) = 0$ is found to imply the following relation :

$$\int dt\ e^{-\alpha} \sum_{n=1}^{\infty} \sum_{pq} (\alpha_n^{pq})^2 \overline{u}_{np} \left[\left(\frac{n}{2} + \frac{3}{4}\right) m_{np}^{(\gamma)} - \left(\frac{2n}{3} + 2\right) i m_{n-1,p}^{(\alpha)}\right] = 0 \ . \tag{5.2.10}$$

Therefore, for each $n \geq 1$:

$$m_{np}^{(\gamma)} = \frac{4}{3} i \frac{(2n+6)}{(2n+3)} m_{n-1,p}^{(\alpha)} \quad , \tag{5.2.11}$$

whereas $m_{0,p}^{(\gamma)}$ has to vanish. Entirely analogous formulae hold for the coefficients $r_{np}^{(\nu)}$ and $r_{np}^{(\lambda)}$ appearing in (5.1.7).

5.3 Gauge Condition

No further progress can be made in our calculations without choosing a gauge condition. For this purpose, we now require that (cf. section 2 of D'Eath 1986) :

$$e_{AA'}{}^{j} \psi_j^A = 0 \quad . \tag{5.3.1}$$

Note that here we are studying the Lorentzian form of the action of the spin-$\frac{3}{2}$ field. The expansion in modes of the PDF form of the Euclidean action will be given in section 5.5

(cf. (5.5.1)), as the analytic continuation of the corresponding Lorentzian action. This is done to obtain the correct result in the simplest possible way. But if one performs the whole PDF calculation in the Euclidean regime, condition (5.3.1) should be supplemented by the requirement $e_{AA'}{}^{j}\,\widetilde{\psi}_j^{A'}=0$.

In light of (5.1.7), our gauge choice implies that :

$$0=\sum_{npq}\alpha_n^{pq}\Big[m_{np}^{(\alpha)}\alpha^{nqACC'}e_{CC'j}e_{AA'}{}^{j}+m_{np}^{(\beta)}\beta^{nqACC'}e_{CC'j}e_{AA'}{}^{j}$$

$$+m_{np}^{(\gamma)}\gamma^{nqACC'}e_{CC'j}e_{AA'}{}^{j}+\bar{s}_{np}\bar{\rho}^{nqC'}\epsilon^{AC}e_{CC'j}e_{AA'}{}^{j}\Big]\quad . \tag{5.3.2}$$

Let us now point out that :

$$\alpha^{nqACC'}e_{CC'j}e_{AA'}{}^{j}=-\alpha^{nqACC'}\epsilon_C{}^{D}\epsilon_{DA}\epsilon_{C'}{}^{D'}\epsilon_{D'A'}+\alpha^{nqACC'}n_{CC'}\epsilon_{DA}\epsilon_{D'A'}n^{DD'}=0, \tag{5.3.3}$$

because $\alpha^{nqACC'}n_{CC'}=0$, and $\alpha^{nqACC'}\epsilon_C{}^{D}\epsilon_{DA}$ is vanishing as well in view of the symmetry of $\alpha^{nqACC'}$ in the first two spinor indices. For the same reasons we find that also $\beta^{nqACC'}e_{CC'j}e_{AA'}{}^{j}$ is vanishing, whereas (cf. (5.1.10)) :

$$\gamma^{nqACC'}e_{CC'j}e_{AA'}{}^{j}=-\gamma^{nqACC'}\epsilon_C{}^{D}\epsilon_{DA}\epsilon_{C'}{}^{D'}\epsilon_{D'A'}+\gamma^{nqACC'}n_{CC'}\epsilon_{DA}\epsilon_{D'A'}n^{DD'}$$

$$=\frac{3}{4}\rho^{nqA}n_{AA'}\quad , \tag{5.3.4}$$

$$\bar{\rho}^{nqC'}\epsilon^{AC}e_{CC'j}e_{AA'}{}^{j}=-\bar{\rho}^{nqC'}\epsilon^{AD}\epsilon_{C'A'}\epsilon_{DA}+\bar{\rho}^{nqC'}n_{AA'}n^{A}{}_{C'}=\frac{3}{2}\bar{\rho}_{A'}^{nq}\quad . \tag{5.3.5}$$

Taking into account (5.3.3-5) we find that, in order to satisfy (5.3.2), the coefficients $m_{np}^{(\gamma)}$ and s_{np} should vanish. In view of (5.2.11), this implies that also $m_{np}^{(\alpha)}$ should vanish. The same holds for $r_{np}^{(\lambda)}$, $r_{np}^{(\nu)}$ and t_{np} appearing in (5.1.7). In other words, the imposition of the supersymmetry constraints and the choice of a gauge condition have greatly simplified the general expansion (5.1.7) of the spin-$\frac{3}{2}$ field, which finally assumes the form :

$$\psi_j^{A}=\frac{e^{-\frac{3\alpha}{2}}}{2\pi}\sum_{npq}\alpha_n^{pq}\left(m_{np}^{(\beta)}\beta^{nqABB'}+\bar{r}_{np}^{(\mu)}\bar{\mu}^{nqABB'}\right)e_{BB'j}\quad . \tag{5.3.6}$$

It is also very useful to write in a more synthetic way the form assumed by ψ_j^A before the imposition of the supersymmetry constraints and of the gauge condition :

$$\psi_j^A = \left[\alpha^{(AC)C'}(\rho) + \beta^{(AC)C'}(\rho) + \delta^{ACC'}(\rho)\right] e_{CC'j}$$
$$+ \left[\bar{\lambda}^{(AC)C'}(\sigma) + \bar{\mu}^{(AC)C'}(\sigma) + \bar{\delta}^{ACC'}(\sigma)\right] e_{CC'j} \quad , \tag{5.3.7}$$

where we have set (analogous formulae holding for the σ harmonics) :

$$\alpha^{(AC)C'}(\rho) = \frac{e^{-\frac{3\alpha}{2}}}{2\pi} \sum_{npq} \alpha_n^{pq} m_{np}^{(\alpha)} \alpha^{nqACC'} \quad , \tag{5.3.8}$$

$$\beta^{(AC)C'}(\rho) = \frac{e^{-\frac{3\alpha}{2}}}{2\pi} \sum_{npq} \alpha_n^{pq} m_{np}^{(\beta)} \beta^{nqACC'} \quad , \tag{5.3.9}$$

whereas $\delta^{ACC'}$ contains both a symmetric and an antisymmetric part :

$$\delta^{ACC'}(\rho) = \delta_1^{(AC)C'}(\rho) + \delta_2^{[AC]C'}(\rho) \quad . \tag{5.3.10}$$

Indeed one has :

$$\delta_1^{(AC)C'} e_{CC'j} = \frac{e^{-\frac{3\alpha}{2}}}{2\pi} \sum_{npq} \alpha_n^{pq} m_{np}^{(\gamma)} \gamma^{nqACC'} e_{CC'j}$$
$$= \frac{e^{-\frac{3\alpha}{2}}}{2\pi} \sum_{npq} \alpha_n^{pq} m_{np}^{(\gamma)} \frac{\rho^{nqA}}{2} n^{CC'} e_{CC'j}$$
$$+ \frac{e^{-\frac{3\alpha}{2}}}{2\pi} \sum_{npq} \alpha_n^{pq} m_{np}^{(\gamma)} \frac{\rho^{nqC}}{2} n^{AC'} e_{CC'j} \quad , \tag{5.3.11}$$

where the first term on the right-hand side of (5.3.11) vanishes in view of the relation : $n^{CC'} e_{CC'j} = 0$. Thus we can define :

$$\delta_1^{ACC'}(\rho) \equiv \frac{e^{-\frac{3\alpha}{2}}}{2\pi} \sum_{npq} \alpha_n^{pq} \frac{m_{np}^{(\gamma)}}{2} \rho^{nqC} n^{AC'} \quad , \tag{5.3.12}$$

$$\delta_2^{[AC]C'}(\rho) \equiv \frac{e^{-\frac{3\alpha}{2}}}{2\pi} \sum_{npq} \alpha_n^{pq} \bar{s}_{np} \bar{\rho}^{nqC'} \epsilon^{AC} \quad . \tag{5.3.13}$$

5.4 Final Form of the Action

We are now in a position to expand to quadratic order the action (5.1.1). For this purpose we must use the formulae (5.1.3), (5.1.17), (5.1.19), (5.2.4) and (5.3.6). This leads to :

$$I = \int dt\, N \sum_{np} \left[\frac{i}{2N} \left(\overline{m}_{np}^{(\beta)} \dot{m}_{np}^{(\beta)} + m_{np}^{(\beta)} \dot{\overline{m}}_{np}^{(\beta)} \right) + e^{-\alpha} \left(n + \frac{5}{2} \right) \overline{m}_{np}^{(\beta)} m_{np}^{(\beta)} \right]$$
$$+ \int dt\, N \sum_{np} \left[\frac{i}{2N} \left(\overline{r}_{np}^{(\mu)} \dot{r}_{np}^{(\mu)} + r_{np}^{(\mu)} \dot{\overline{r}}_{np}^{(\mu)} \right) + e^{-\alpha} \left(n + \frac{5}{2} \right) \overline{r}_{np}^{(\mu)} r_{np}^{(\mu)} \right] \quad . \tag{5.4.1}$$

The full action should contain the boundary term (D'Eath 1984) :

$$\frac{1}{2} \int_{\partial M} d^3x\, \epsilon^{ijk}\, \psi_i^A\, e_{AA'j}\, \overline{\psi}_k^{A'} \quad , \tag{5.4.2}$$

as mentioned in section 1.3. However, the Euclidean version of (5.4.2) vanishes if the whole regular part of the spin-$\frac{3}{2}$ field is set equal to zero on S^3.

5.5 PDF Prefactor of the Semiclassical Wave Function with Global Boundary Conditions

So far we have generalized what has been done by Schleich for pure gravity, i.e. we have found the physical degrees of freedom using the Hamiltonian formulation of the theory with an explicit choice of gauge. Therefore, using this approach the PDF contribution to the prefactor due to the spin-$\frac{3}{2}$ field is expressed by a path integral whose measure only involves the physical degrees of freedom.

In order to do this calculation, we must perform an analytic continuation from the Lorentzian to the Euclidean regime. Denoting with a dot the time derivatives with respect to the Euclidean time τ, the action (5.4.1) becomes :

$$\widetilde{I} = \int d\tau \; N \sum_{np} \left[\frac{1}{2N} \left(\widetilde{m}_{np}^{(\beta)} \dot{m}_{np}^{(\beta)} + m_{np}^{(\beta)} \dot{\widetilde{m}}_{np}^{(\beta)} \right) - \hat{\nu} \widetilde{m}_{np}^{(\beta)} m_{np}^{(\beta)} \right]$$

$$+ \int d\tau \; N \sum_{np} \left[\frac{1}{2N} \left(\widetilde{r}_{np}^{(\mu)} \dot{r}_{np}^{(\mu)} + r_{np}^{(\mu)} \dot{\widetilde{r}}_{np}^{(\mu)} \right) - \hat{\nu} \widetilde{r}_{np}^{(\mu)} r_{np}^{(\mu)} \right] \quad , \tag{5.5.1}$$

where we have defined : $\hat{\nu} \equiv e^{-\alpha}(n + \frac{5}{2}) = \frac{(n+\frac{5}{2})}{\tau}$. It is worth emphasizing that in (5.5.1), $\widetilde{m}_{np}^{(\beta)}$ and $\widetilde{r}_{np}^{(\mu)}$ are not the complex conjugates of $m_{np}^{(\beta)}$ and $r_{np}^{(\mu)}$. Hence in our case, if $N = 1$, the evolution of the modes for a flat Euclidean background is given by :

$$\widetilde{m}_{np}^{(\beta)} = \widetilde{A}_{np} \left(\frac{\tau}{\tau_0} \right)^{-(n+\frac{5}{2})} \quad , \quad m_{np}^{(\beta)} = \widetilde{B}_{np} \left(\frac{\tau}{\tau_0} \right)^{(n+\frac{5}{2})} \quad , \tag{5.5.2}$$

$$\widetilde{r}_{np}^{(\mu)} = \widetilde{C}_{np} \left(\frac{\tau}{\tau_0} \right)^{-(n+\frac{5}{2})} \quad , \quad r_{np}^{(\mu)} = \widetilde{D}_{np} \left(\frac{\tau}{\tau_0} \right)^{(n+\frac{5}{2})} \quad . \tag{5.5.3}$$

The important result is thus that in the classical theory the modes regular at $\tau = 0$ are just $m_{np}^{(\beta)}$ and $r_{np}^{(\mu)}$. In the quantum theory, we set these untwiddled modes equal to zero on S^3 (but all modes are regular if we study the eigenvalue equations). Now, defining :

$$\widetilde{I}^{(\beta)} \equiv \int d\tau \sum_{np} \left[\frac{1}{2} \left(\widetilde{m}_{np}^{(\beta)} \dot{m}_{np}^{(\beta)} + m_{np}^{(\beta)} \dot{\widetilde{m}}_{np}^{(\beta)} \right) - \hat{\nu} \widetilde{m}_{np}^{(\beta)} m_{np}^{(\beta)} \right] \quad , \tag{5.5.4}$$

$$\widetilde{I}^{(\mu)} \equiv \int d\tau \sum_{np} \left[\frac{1}{2} \left(\widetilde{r}_{np}^{(\mu)} \dot{r}_{np}^{(\mu)} + r_{np}^{(\mu)} \dot{\widetilde{r}}_{np}^{(\mu)} \right) - \hat{\nu} \widetilde{r}_{np}^{(\mu)} r_{np}^{(\mu)} \right] \quad , \tag{5.5.5}$$

the PDF contribution to the prefactor from the spin-$\frac{3}{2}$ field is given by :

$$P = \int d[m^{(\beta)}] d[\widetilde{m}^{(\beta)}] e^{-\widetilde{I}^{(\beta)}} \int d[r^{(\mu)}] d[\widetilde{r}^{(\mu)}] e^{-\widetilde{I}^{(\mu)}} = P_1 P_2 \quad . \tag{5.5.6}$$

In other words, we consider flat Euclidean space with a three-sphere boundary, and we study the PDF contribution of the spin-$\frac{3}{2}$ field to one-loop calculations about this flat space. Thus we have to evaluate P_1 and P_2 using the technique described in sections 4.3-4. For example we may focus our attention on P_1, because $P_2 = P_1$ in light of (5.5.4-6). The Lagrangian which defines $\widetilde{I}_{np}^{(\beta)}$ has the form :

$$L_{np}^{(\beta)} = \frac{1}{2}\left(\widetilde{m}_{np}^{(\beta)}\dot{m}_{np}^{(\beta)} + m_{np}^{(\beta)}\dot{\widetilde{m}}_{np}^{(\beta)}\right) - \hat{\nu}\widetilde{m}_{np}^{(\beta)}m_{np}^{(\beta)} \quad . \tag{5.5.7}$$

Following the technique of section 4.3 (cf. (4.3.7-9)), we therefore study the eigenvalue equation :

$$\left(-\frac{d}{d\tau} - \hat{\nu}\right)\left(\frac{d}{d\tau} - \hat{\nu}\right) m_{np}^{(\beta)} = \left(-\frac{d^2}{d\tau^2} + \frac{k(k-1)}{\tau^2}\right) m_{np}^{(\beta)} = E^2 m_{np}^{(\beta)} \quad , \tag{5.5.8}$$

where $k = (n + \frac{5}{2})$. For the operator on the left-hand side of (5.5.8), setting $\tau = t$, the heat equation is :

$$\left[\frac{\partial}{\partial\widetilde{\tau}} - \frac{\partial^2}{\partial t^2} + \frac{\left((n+2)^2 - \frac{1}{4}\right)}{t^2}\right] G_n(t,t',\widetilde{\tau}) = \delta(t-t')\delta(\widetilde{\tau}) \quad . \tag{5.5.9}$$

Taking the Laplace transform of (5.5.9) and denoting by σ^2 the parameter of the transform, we obtain the equation :

$$\left[\frac{\partial^2}{\partial t^2} - \sigma^2 - \frac{\left((n+2)^2 - \frac{1}{4}\right)}{t^2}\right] G_n(t,t',\sigma^2) = -\delta(t-t') \quad . \tag{5.5.10}$$

As already done for spin $\frac{1}{2}$ and 2, we now require that G_n should be regular at the origin : t or $t' = 0$, and should vanish on the boundary : t or $t' = a$, so that for a single set of modes we have (using the same notation as in (4.1.10-12)) :

$$G(t,t',\sigma^2) = \sum_{n=0}^{\infty}(n+4)(n+1)G_n(t,t',\sigma^2) = \sum_{n=2}^{\infty}(n+2)(n-1)G_{n-2}(t,t',\sigma^2)$$

$$= \sum_{n=2}^{\infty}(n+2)(n-1)\sqrt{t_<}\sqrt{t_>}\frac{I_n(\sigma t_<)}{I_n(\sigma a)}$$

$$\left[I_n(\sigma a)K_n(\sigma t_>) - I_n(\sigma t_>)K_n(\sigma a)\right] \quad , \tag{5.5.11}$$

because the harmonics which define $\beta^{nqACC'}$ have degeneracy $d(n) = (n+4)(n+1)$ if $n \geq 0$. Equation (5.5.11) implies that in the formula :

$$G(\sigma^2) = G^F(\sigma^2) + G^{int}(\sigma^2) \quad , \tag{5.5.12}$$

the free part $G^F(\sigma^2)$ does not contribute to $\zeta(0)$ (cf. appendix C). We may therefore focus our attention on that part of $G^{int}(\sigma^2)$ which gives rise to a constant after having taken the inverse Laplace transform L_I. This is the part involving just $\frac{1}{\sigma^2}$, and we denote its contribution to $\zeta(0)$ by means of the symbol $L_I^C G^{int}(\sigma^2)$, where :

$$G^{int}(\sigma^2) = -\frac{a^2}{2}\sum_{n=2}^{\infty}(n^2-2)f(n;\sigma a) - \frac{a^2}{2}\sum_{n=2}^{\infty} n f(n;\sigma a) \ . \tag{5.5.13}$$

Let us now point out that :

$$L_I^C\left[-\frac{a^2}{2}\sum_{n=2}^{\infty}(n^2-2)f(n;\sigma a)\right] = -\frac{31}{180} + C \quad , \tag{5.5.14}$$

where C denotes the contribution from the poles at $0, \pm 1$. In fact, from the study of pure gravity (Schleich 1985) we already know that :

$$-\int_0^{\infty} \nu^2 f(\nu,t)\, d\nu \sim -\frac{\sqrt{\pi}}{8}t^{-\frac{3}{2}} + \frac{t^{-1}}{4} - \frac{11\sqrt{\pi}}{256}t^{-\frac{1}{2}} - \frac{1}{90} + O(\sqrt{t}) \ , \tag{5.5.15}$$

$$2\int_0^{\infty} f(\nu,t)\, d\nu \sim \frac{\sqrt{\pi}}{2}t^{-\frac{1}{2}} - \frac{1}{3} + O(\sqrt{t}) \ . \tag{5.5.16}$$

These formulae are needed because the Watson transform W of the first sum in (5.5.13) is defined by :

$$W\left[-\frac{a^2}{2}\sum_{n=2}^{\infty}(n^2-2)f(n;\sigma a)\right] \equiv \frac{1}{2}\left[-\frac{a^2}{4i}\int_{C'-Q}(\nu^2-2)f(\nu;\sigma a)\cot(\pi\nu)d\nu\right] \ , \tag{5.5.17}$$

where the contour of integration $C'-Q$ encloses all poles along the real axis (owing to the singular behaviour of $\cot(\pi\nu)$ at $\nu = 0, \pm1, \pm2, \ldots$) with the exception of the poles at 0 and ± 1, which are excluded because the sum over all n only starts from $n = 2$ (cf. Schleich 1985). These poles at 0 and ± 1 are thus enclosed by the contour Q, with anti-clockwise orientation. This is why the constant contribution to (5.5.17) arising from the poles is obtained by means of :

$$W^C_{poles} = \frac{1}{2}\left(-\frac{a^2}{4i}\right)(-2\pi i)\left[\lim_{\nu\to 0}\nu(\nu^2-2)f^c(0;\sigma a)i\frac{(e^{i\pi\nu}+e^{-i\pi\nu})}{(e^{i\pi\nu}-e^{-i\pi\nu})}\right]$$
$$+\frac{1}{2}\left(-\frac{a^2}{4i}\right)(-2\pi i)2\left[\lim_{\nu\to 1}(\nu-1)(\nu^2-2)f^c(1;\sigma a)i\frac{(e^{i\pi\nu}+e^{-i\pi\nu})}{(e^{i\pi\nu}-e^{-i\pi\nu})}\right]$$
$$= -\frac{1}{2\sigma^2} \quad . \tag{5.5.18}$$

Hence we find (cf. (5.5.14-16)) :

$$-\frac{31}{180} = \frac{1}{2}\left(-\frac{1}{90}-\frac{1}{3}\right) \quad , \quad C = L_I(W^C_{poles}) = -\frac{1}{2} \quad . \tag{5.5.19}$$

Finally, in light of sections 4.3-5 one has that :

$$L^C_I\left[-\frac{a^2}{2}\sum_{n=2}^{\infty} nf(n;\sigma a)\right] = -\frac{a^2}{2}L^C_I\left[\sum_{n=1}^{\infty} nf(n;\sigma a) - f(1;\sigma a)\right] = \frac{1}{48}+\frac{1}{4} = \frac{13}{48}. \tag{5.5.20}$$

Thus, taking the L^C_I of both members of (5.5.13), and using (5.5.14) and (5.5.19-20) we find :

$$L^C_I G^{int}(\sigma^2) = -\frac{289}{720} = \zeta_{el}(0) \quad , \tag{5.5.21}$$

which implies :

$$P_1 = D_1 a^{-\zeta_{el}(0)} = D_1 a^{\frac{289}{720}} \quad , \tag{5.5.22}$$

where D_1 is a constant. Therefore the total PDF contribution to the prefactor which is due to the spin-$\frac{3}{2}$ field is given by :

$$P = P_1 P_2 = (P_1)^2 = D a^{-\zeta_{tot}(0)} = D a^{\frac{289}{360}} \quad . \tag{5.5.23}$$

Indeed, this value does not cancel the PDF contribution due to the gravitational field, which, as seen in section 4.1, is given by : $P = Da^{-\frac{278}{45}}$. However, we might think this happens because, in computing $\zeta(0)$, we have used boundary conditions (setting all untwiddled modes equal to zero on S^3) which do not respect supersymmetry. Our next task is therefore to look for local boundary conditions which involve the supersymmetric properties of the theory. Such a generalization is better understood by discussing at first some basic results of twistor theory in flat space.

5.6 Basic Results of Twistor Theory in Flat Space

There are two main ways (Penrose 1986) of defining twistors in flat space, and it is indeed worth mentioning both of them.

5.6.1.1 Twistors via α-planes

α-planes are complex two-surfaces, all of whose tangent vectors at any point have the form $\lambda^A \pi^{A'}$ for some $\pi^{A'}$, and whose tangent spaces go into themselves under parallel propagation in tangent directions. This last condition can be written as :

$$\lambda^A \pi^{A'} \nabla_{AA'} (\mu_B \pi_{B'}) = \nu_B \pi_{B'} \quad , \tag{5.6.1}$$

for some ν_B and for each λ^A, μ_B. In other words, so as to define α-planes we need the complexification of Minkowski space and its conformal compactification (Penrose and Rindler 1986). An α-plane is then a complex two-plane which is totally null, in that the complexification of the space-time metric g vanishes identically over the entire plane. Indeed there are two disjoint families of such totally null planes, called α-planes and β-planes. Any complex null vector has the spinor form : $\lambda^A \pi^{A'}$, and the tangents to an α-plane are the vectors of this form for which $\pi^{A'}$ is held fixed and λ^A varies. By contrast, for the tangents to β-planes λ^A is held fixed and $\pi^{A'}$ is varying.

A nonzero twistor is, by definition, an α-plane with constant $\pi_{A'}$ associated with it. The choice of scaling for $\pi_{A'}$ fixes the scaling for the twistor. A remarkable result (Penrose

1976, Ward and Wells 1990) states that a complex space-time admits α-planes if and only if it has an anti-self-dual Weyl tensor.

5.6.1.2 Twistors via the Twistor Equation

The general solution of the twistor equation in flat space :

$$\nabla_{A'}^{(A}\omega^{B)} = 0 \quad , \tag{5.6.2}$$

is of the form (Penrose and Rindler 1986):

$$\omega^A = (\omega^o)^A - ix^{AA'}\pi_{A'}^o \quad , \tag{5.6.3}$$

$$\pi_{A'} = \pi_{A'}^o \quad , \tag{5.6.4}$$

ω_A^o and $\pi_{A'}^o$ being arbitrary constant spinors. A twistor is then represented by the pair of spinor fields : $(\omega^A, \pi_{A'}) \leftrightarrow Z^\alpha$. A very important property of (5.6.2) is its conformal invariance. In fact, putting : $\hat{g}_{ab} = \Omega^2 g_{ab}$, $\hat{\epsilon}_{AB} = \Omega\epsilon_{AB}$, $\hat{\epsilon}^{AB} = \Omega^{-1}\epsilon^{AB}$, $T_a = \nabla_a \log\Omega$, and choosing $\hat{\omega}^B = \omega^B$, one finds (Penrose and Rindler 1986) :

$$\hat{\nabla}_{AA'}\hat{\omega}^B = \nabla_{AA'}\omega^B + \epsilon_A^{\ B}T_{CA'}\omega^C \quad , \tag{5.6.5}$$

which implies :

$$\hat{\nabla}_{A'}^{(A}\hat{\omega}^{B)} = \Omega^{-1}\nabla_{A'}^{(A}\omega^{B)} \quad . \tag{5.6.6}$$

The solutions ω^A of the twistor equation are fully determined by the four complex components at O of ω^A and $\pi_{A'}$ in a spin-frame at O. These solutions ω^A constitute therefore a four-dimensional vector space over the complex numbers called twistor space.

It is important to remark that it is the twistor concept associated with a β-plane which is dual to that associated with a solution of the twistor equation (Penrose 1986). In fact the tangent vectors to a β-plane have the form $\lambda^A\xi^{A'}$, for varying $\xi^{A'}$. Requiring that λ_A be constant over the β-planes implies that : $\lambda^A\xi^{A'}\nabla_{AA'}\lambda_B = 0$, for each $\xi^{A'}$, i.e. $\lambda^A\nabla_{AA'}\lambda_B = 0$. In addition, a scalar product can be defined between the ω^A field and the λ_A-scaled β-plane : $\omega^A\lambda_A$. Its constancy over the β-plane implies that :

$$\lambda^A\xi^{A'}\nabla_{AA'}\left(\omega^B\lambda_B\right) = 0 \quad , \tag{5.6.7}$$

$\forall \xi^{A'}$, which finally implies :

$$\lambda_A \lambda_B \left(\nabla_{A'}^{(A} \omega^{B)} \right) = 0 \quad , \tag{5.6.8}$$

for each β-plane and thus for each λ_A. Thus (5.6.8) becomes just the twistor equation (5.6.2). After having shown this important link between the two ways of defining twistors, we can focus our attention on the approach 5.6.1.2, which leads more directly to the links with theoretical physics. In particular, we are interested in the following properties :

(a) Any solution ω^A of the twistor equation acts as a spin-lowering operator for massless fields. In other words, if $\phi_{AB...L} = \phi_{(AB...L)}$ satisfies the massless free-field equation

$$\nabla_{A'}^{A} \phi_{AB...L} = 0 \quad , \tag{5.6.9}$$

then so does $\omega^L \phi_{AB...L}$, since :

$$\nabla_{A'}^{A} (\omega^L \phi_{AB...L}) = \phi_{AB...L} \nabla_{A'}^{(A} \omega^{L)} = 0 \quad . \tag{5.6.10}$$

(b) ω^A can also be used as a spin-raising operator since, by a direct calculation : $\overline{\omega}^{K'} \nabla_{K'(K} \phi_{AB...L)} - (s + \frac{1}{2}) \phi_{(AB...L} \nabla_{K)}^{K'} \overline{\omega}_{K'}$ also satisfies the massless free-field equation (Hawking 1983, Penrose and Rindler 1986).

5.7 Local Boundary Conditions and Spin-Lowering Operators

Let us consider a complex space-time given by a four-dimensional complex manifold with holomorphic metric (Ward 1980), so that the spinorial expression of the Weyl tensor C_{abcd} is :

$$C_{abcd} = \left[\phi_{ABCD}\, \epsilon_{A'B'}\, \epsilon_{C'D'} + \widetilde{\phi}_{A'B'C'D'}\, \epsilon_{AB}\, \epsilon_{CD} \right] \sigma_a^{AA'} \sigma_b^{BB'} \sigma_c^{CC'} \sigma_d^{DD'} \quad . \tag{5.7.1}$$

Therefore, the Weyl spinors ϕ_{ABCD} and $\widetilde{\phi}_{A'B'C'D'}$ are completely independent (there is no complex conjugation which relates them). They are linearized spin-2 field strengths which satisfy the massless free-field equation (cf. (5.6.9)). Similarly, ϕ_{ABC} and $\widetilde{\phi}_{A'B'C'}$

denote completely independent linearized spin-$\frac{3}{2}$ field strengths obeying a massless free-field equation. We are only interested in linearized field strengths because we are studying perturbation theory about a flat Euclidean background, but of course (5.7.1) also holds in the fully curved case. Moreover, let us recall that in a Hamiltonian formalism the lapse function must be rotated : $N \rightarrow -iN$, so that we obtain the Euclidean normal (D'Eath and Halliwell 1987) :

$$\left({}_e n^{AA'}\right) \equiv -i n^{AA'} \quad . \tag{5.7.2}$$

We set $\epsilon = \pm 1$, $\gamma = \pm 1$. The following result can now be proved :

Theorem 5.7.1 Let ω^D be a solution of the twistor equation in flat Euclidean space with a three-sphere boundary, and let $\widetilde{\omega}^{D'}$ be its independent primed counterpart in this space (see now appendix D). Then a form exists of this spin-lowering operator which preserves the local boundary conditions on S^3 :

$$4\left({}_e n^{AA'}\right)\left({}_e n^{BB'}\right)\left({}_e n^{CC'}\right)\left({}_e n^{DD'}\right)\phi_{ABCD} = \epsilon\, \widetilde{\phi}^{A'B'C'D'} \quad , \tag{5.7.3}$$

$$2^{\frac{3}{2}}\left({}_e n^{AA'}\right)\left({}_e n^{BB'}\right)\left({}_e n^{CC'}\right)\phi_{ABC} = \gamma\, \widetilde{\phi}^{A'B'C'} \quad . \tag{5.7.4}$$

Proof. At first let us choose $\epsilon = \gamma$ on the right-hand sides of (5.7.3-4) (finally we shall also deal with the case when $\epsilon = -\gamma$). Multiplying both sides of (5.7.3) by $({}_e n_{FD'})$ we get, in view of (5.7.2) :

$$-2\left({}_e n^{AA'}\right)\left({}_e n^{BB'}\right)\left({}_e n^{CC'}\right)\phi_{ABCF} = \epsilon\, \widetilde{\phi}^{A'B'C'D'}\left({}_e n_{FD'}\right) \; . \tag{5.7.5}$$

Taking into account the total symmetry of the Weyl spinor, putting F=D and multiplying both sides of (5.7.5) by $\sqrt{2}\,\omega^D$ we finally get :

$$-2^{\frac{3}{2}}\left({}_e n^{AA'}\right)\left({}_e n^{BB'}\right)\left({}_e n^{CC'}\right)\phi_{ABCD}\,\omega^D = \epsilon\,\sqrt{2}\,\widetilde{\phi}^{A'B'C'D'}\left({}_e n_{DD'}\right)\omega^D \quad , \tag{5.7.6}$$

$$2^{\frac{3}{2}}\left({}_e n^{AA'}\right)\left({}_e n^{BB'}\right)\left({}_e n^{CC'}\right)\phi_{ABCD}\,\omega^D = \epsilon\, \widetilde{\phi}^{A'B'C'D'}\,\widetilde{\omega}_{D'} \quad , \tag{5.7.7}$$

where (5.7.7) is obtained inserting into (5.7.4) the definition of the spin-lowering operator. The comparison of (5.7.6) and (5.7.7) yields the preservation condition :

$$\sqrt{2}\,({}_e n_{DA'})\,\omega^D = -\widetilde{\omega}_{A'} \quad . \tag{5.7.8}$$

We also know that (cf. (5.6.3)) :

$$\omega^D = (\omega^o)^D - i\left({}_e x^{DD'}\right)\pi^o_{D'} \quad , \tag{5.7.9}$$

$$\widetilde{\omega}^{D'} = (\widetilde{\omega}^o)^{D'} - i\left({}_e x^{DD'}\right)\widetilde{\pi}^o_D \quad , \tag{5.7.10}$$

where the spinors on the right-hand sides of (5.7.9-10) are constant, with the obvious exception of $\left({}_e x^{DD'}\right)$. Thus (5.7.8) is found to imply :

$$\sqrt{2}\,({}_e n_{DA'})\,(\omega^o)^D - i\sqrt{2}\,({}_e n_{DA'})\left({}_e x^{DD'}\right)\pi^o_{D'} = -\widetilde{\omega}^o_{A'} - i\,({}_e x_{DA'})\,(\widetilde{\pi}^o)^D \quad . \tag{5.7.11}$$

Requiring that (5.7.11) should be identically satisfied, and using the identity : $\left({}_e n^{AA'}\right) = \frac{1}{r}\left({}_e x^{AA'}\right)$ on a three-sphere of radius r, we find :

$$\widetilde{\omega}^o_{A'} = i\sqrt{2}\,r\,({}_e n_{DA'})\left({}_e n^{DD'}\right)\pi^o_{D'} = -\frac{ir}{\sqrt{2}}\pi^o_{A'} \quad , \tag{5.7.12}$$

$$-\sqrt{2}\,({}_e n_{DA'})\,(\omega^o)^D = ir\,({}_e n_{DA'})\,(\widetilde{\pi}^o)^D \quad . \tag{5.7.13}$$

Multiplying both sides of (5.7.13) by $\left({}_e n^{BA'}\right)$, and then acting with ϵ_{BA} on both sides of the resulting relation we get :

$$\omega^o_A = -\frac{ir}{\sqrt{2}}\widetilde{\pi}^o_A \quad . \tag{5.7.14}$$

We can now write these formulae in a more general way, taking into account that one can also have $\epsilon = -\gamma$ in (5.7.3-4). Thus one finds :

$$\sqrt{2}\,({}_e n_{DA'})\,\omega^D = \mp\widetilde{\omega}_{A'} \quad , \tag{5.7.15}$$

$$\omega^{o}_{A} = \mp \frac{ir}{\sqrt{2}} \widetilde{\pi}^{o}_{A} \quad , \tag{5.7.16}$$

$$\widetilde{\omega}^{o}_{A'} = \mp \frac{ir}{\sqrt{2}} \pi^{o}_{A'} \quad . \tag{5.7.17}$$

The relations (5.7.15-17) completely solve the problem of finding a spin-lowering operator which preserves (5.7.3-4). Q. E. D.

Three important remarks are now in order :

(a) Our theorem only proves that a form of the spin-lowering operator exists which preserves the local boundary conditions (5.7.3-4) on S^3. However, we still have to check whether also the spin-raising operator preserves (5.7.3-4). This problem can be more easily studied starting from lower spins. Thus we now focus on the comparison spin 0 vs spin $\frac{1}{2}$. Denoting by ϕ a complex scalar field whose complex conjugate is $\overline{\phi}$ (cf. problem 5.1 after chapter eleven), the spin-raising operator is such that :

$$\phi_A = \widetilde{\omega}^{K'} (\nabla_{AK'} \phi) - \frac{\phi}{2} \nabla_{A}^{\ K'} \widetilde{\omega}_{K'} \quad . \tag{5.7.18}$$

The fields ϕ_A and ϕ are assumed to satisfy the Weyl equation and the Laplace equation respectively inside S^3. Moreover, we require the following boundary conditions on S^3 (Hawking 1983) :

$$\sqrt{2}\, {}_{e}n^{AA'} \phi_A = \epsilon\, \widetilde{\phi}^{A'} \quad , \tag{5.7.19}$$

$$\phi = \epsilon\, \overline{\phi} \quad , \tag{5.7.20}$$

$${}_{e}n^{AA'} \nabla_{AA'} \phi = -\epsilon\, {}_{e}n^{BB'} \nabla_{BB'} \overline{\phi} \quad , \tag{5.7.21}$$

where $\epsilon = \pm 1$. We must now insert (5.7.18) into (5.7.19), and check whether it leads to a condition in agreement with (5.7.20-21). Indeed, the relations (5.7.18-20) lead to the following boundary conditions on S^3 :

$$\sqrt{2}\, {}_{e}n^{AA'} \widetilde{\omega}^{K'} (\nabla_{AK'} \phi) - \epsilon\, \omega^{K} \nabla_{K}^{\ A'} \overline{\phi} + \epsilon \overline{\phi} \left[\frac{1}{2} \left(\nabla^{KA'} \omega_K \right) - \frac{{}_{e}n^{AA'}}{\sqrt{2}} \nabla_{A}^{\ K'} \widetilde{\omega}_{K'} \right] = 0. \tag{5.7.22}$$

Using (5.7.9-10), the boundary conditions (5.7.22) can be finally cast in the form :

$$\sqrt{2}\, {}_{e}n^{AA'} \left[\widetilde{\omega}^{K'} (\nabla_{AK'} \phi) - i\epsilon\, \widetilde{\phi}\, \widetilde{\pi}^{o}_{A} \right] - \epsilon \left[\omega^{K} \left(\nabla_{K}^{\ A'} \widetilde{\phi} \right) - i \widetilde{\phi} (\pi^{o})^{A'} \right] = 0 \ . \qquad (5.7.23)$$

We have not been able to prove that (5.7.23) is identically satisfied in view of (5.7.20-21), and going to higher spins the consistency condition becomes even more involved (this difficulty had not been realized in the original proof of our theorem at first appearing in Esposito 1989a). Thus we leave this question for future investigation.

(b) Let us require local boundary conditions on S^3 involving field strengths and normals also for lower spins (i.e. spin $\frac{3}{2}$ vs spin 1, spin 1 vs spin $\frac{1}{2}$, spin $\frac{1}{2}$ vs spin 0). Then, using the same technique of the theorem just proved, we find that the preservation condition to be obeyed by the spin-lowering operator is still expressed by (5.7.15-17). In this calculation, we are interested in field strengths which solve the massless free-field equations.

(c) In our theorem we just apply to the case of a compact boundary the boundary conditions studied in Breitenlohner and Freedman 1982 and Hawking 1983 for the case of gauged supergravity. These boundary conditions also have another remarkable property, in that they lead to a well-posed classical boundary-value problem. In fact, if we study linearized gravity using (5.7.3), we find a unique regular solution for ϕ_{ABCD} and $\widetilde{\phi}_{A'B'C'D'}$ inside S^3 if $\epsilon = 1$ (cf. sections 7.2-3). Moreover, the theorem now proved strengthens the evidence in favour of the complex or Euclidean regime being a natural framework for problems of interest in fundamental physics. In other words, only when ω^{o}_{A} and $\widetilde{\omega}^{o}_{A'}$ (as well as $\pi^{o}_{A'}$ and $\widetilde{\pi}^{o}_{A}$) are independent spinors, there is agreement between (5.7.16) and (5.7.17). Otherwise, $\widetilde{\omega}^{o}_{A'}$ should be replaced by $\overline{\omega}^{o}_{A'}$ (which is the complex conjugate of ω^{o}_{A}), and the sign on the right-hand side of (5.7.17) would not be compatible with (5.7.16) and with the definition of $\overline{\omega}^{o}_{A'}$. However, we cannot yet claim to have solved our problem. In fact, we have to check by analytic calculation whether local boundary conditions yield one-loop finiteness of amplitudes in quantum cosmology. This is what we will do from now on.

5.8 Application of Local Boundary Conditions to the Spin-$\frac{1}{2}$ Field

We now study in detail local boundary conditions briefly described in section 4.6 in the case of the massless Majorana spin-$\frac{1}{2}$ field on a three-sphere of radius a, bounding a region of Euclidean four-space centred on the origin. The field $\left(\psi^A, \widetilde{\psi}^{A'}\right)$ may be expanded in terms of harmonics on the family of three-spheres centred on the origin, as :

$$\psi^A = \frac{t^{-\frac{3}{2}}}{2\pi} \sum_{npq} \alpha_n^{pq} \left(m_{np}(t) \rho^{nqA} + \widetilde{r}_{np}(t) \overline{\sigma}^{nqA} \right) \quad , \tag{5.8.1}$$

$$\widetilde{\psi}^{A'} = \frac{t^{-\frac{3}{2}}}{2\pi} \sum_{npq} \alpha_n^{pq} \left(\widetilde{m}_{np}(t) \overline{\rho}^{nqA'} + r_{np}(t) \sigma^{nqA'} \right) \quad . \tag{5.8.2}$$

Here t is the radius of a three-sphere, and the notation is the one described after (4.2.13). The *twiddle* symbol for $\widetilde{\psi}^{A'}$ and for the coefficients $\widetilde{m}_{np}$ and $\widetilde{r}_{np}$ does not denote any conjugation operation. In fact, on analytic continuation to a smooth and positive-definite metric, unprimed and primed spinors transform under independent groups $SU(2)$ and $\widetilde{SU(2)}$. Moreover, setting $\epsilon = \pm 1$, we know that Hawking's local boundary conditions may be written in the form :

$$\sqrt{2}\, {}_e n_A^{\ A'} \psi^A = \epsilon\, \widetilde{\psi}^{A'} \ \text{ on } \ S^3 \quad , \tag{5.8.3}$$

and we also know that the harmonics $\overline{\rho}^{nqA'}$ and $\overline{\sigma}^{nqA}$ may be re-expressed in terms of the harmonics $n_A^{\ A'} \rho^{npA}$ and $n^A_{\ A'} \sigma^{npA'}$ using the relations (D'Eath and Halliwell 1987) :

$$\overline{\rho}^{nqA'} = 2n_A^{\ A'} \sum_d \rho^{ndA} (A_n^{-1} H_n)^{dq} \quad , \tag{5.8.4}$$

$$\overline{\sigma}^{nqA} = 2n^A_{\ A'} \sum_d \sigma^{ndA'} (A_n^{-1} H_n)^{dq} \quad . \tag{5.8.5}$$

Boundary conditions of this kind have also been studied in Luckock 1991, where (though using a different formalism) the more general possibility of an ϵ which is a complex-valued function of position on the boundary has been considered. We shall see later that self-adjointness of the boundary-value problem implies reality of ϵ.

Let us now recall that (cf. appendix A) : $\alpha_n = \begin{pmatrix} 1 & 1 \\ 1 & -1 \end{pmatrix}$, $A_n = \sqrt{2}\begin{pmatrix} 0 & 1 \\ -1 & 0 \end{pmatrix}$, $A_n^{-1} = \frac{1}{\sqrt{2}}\begin{pmatrix} 0 & -1 \\ 1 & 0 \end{pmatrix} = A_n^{-1} H_n$. For simplicity, consider first that part of (5.8.3) which involves the ρ harmonics. The relations (5.8.1-4) yield therefore for each n :

$$-i\sum_{pq}\begin{pmatrix} 1 & 1 \\ 1 & -1 \end{pmatrix}^{pq} m_{np}(a)\rho^{nqA} = \epsilon\sum_{pq}\begin{pmatrix} 1 & 1 \\ 1 & -1 \end{pmatrix}^{pq} \widetilde{m}_{np}(a)\sum_d \rho^{ndA}\begin{pmatrix} 0 & -1 \\ 1 & 0 \end{pmatrix}^{dq} . \tag{5.8.6}$$

In the typical case of the indices $p, q = 1, 2$, this implies :

$$-im_{n1}(a)\Big(\rho^{n1A} + \rho^{n2A}\Big) - im_{n2}(a)\Big(\rho^{n1A} - \rho^{n2A}\Big) = \epsilon\widetilde{m}_{n1}(a)\Big(\rho^{n2A} - \rho^{n1A}\Big) + \epsilon\widetilde{m}_{n2}(a)\Big(\rho^{n2A} + \rho^{n1A}\Big) . \tag{5.8.7}$$

Similar equations hold for adjacent indices $p, q = 2k+1, 2k+2$ $(k = 0, 1, ..., \frac{n}{2}(n+3))$. Since the harmonics ρ^{n1A} and ρ^{n2A} on the bounding three-sphere are linearly independent, we have the following system :

$$-i\Big[m_{n1}(a) + m_{n2}(a)\Big] = \epsilon\Big[\widetilde{m}_{n2}(a) - \widetilde{m}_{n1}(a)\Big] \quad , \tag{5.8.8}$$

$$-i\Big[m_{n1}(a) - m_{n2}(a)\Big] = \epsilon\Big[\widetilde{m}_{n2}(a) + \widetilde{m}_{n1}(a)\Big] \quad , \tag{5.8.9}$$

whose solution is :

$$-im_{n1}(a) = \epsilon\, \widetilde{m}_{n2}(a) \quad , \tag{5.8.10}$$

$$im_{n2}(a) = \epsilon\, \widetilde{m}_{n1}(a) \quad . \tag{5.8.11}$$

In the same way, the part of (5.8.3) involving the σ harmonics leads to :

$$\epsilon\sum_{pq}\begin{pmatrix} 1 & 1 \\ 1 & -1 \end{pmatrix}^{pq} r_{np}(a)\sigma^{nqA'} = i\sum_{pq}\begin{pmatrix} 1 & 1 \\ 1 & -1 \end{pmatrix}^{pq} \widetilde{r}_{np}(a)\sum_d \sigma^{ndA'}\begin{pmatrix} 0 & -1 \\ 1 & 0 \end{pmatrix}^{dq} , \tag{5.8.12}$$

which implies, for example :

$$\epsilon\Big[r_{n1}(a)+r_{n2}(a)\Big]\sigma^{n1A'}+\epsilon\Big[r_{n1}(a)-r_{n2}(a)\Big]\sigma^{n2A'}=-i\Big[\widetilde{r}_{n1}(a)-\widetilde{r}_{n2}(a)\Big]\sigma^{n1A'}$$
$$+i\Big[\widetilde{r}_{n1}(a)+\widetilde{r}_{n2}(a)\Big]\sigma^{n2A'} \quad (5.8.13)$$

so that we finally get :

$$-i\widetilde{r}_{n1}(a)=\epsilon\, r_{n2}(a) \quad , \quad (5.8.14)$$

$$i\widetilde{r}_{n2}(a)=\epsilon\, r_{n1}(a) \quad , \quad (5.8.15)$$

and similar equations for other adjacent indices p,q. Thus, defining :

$$x\equiv m_{n1} \quad , \quad X\equiv m_{n2} \quad , \quad (5.8.16)$$

$$\widetilde{x}\equiv \widetilde{m}_{n1} \quad , \quad \widetilde{X}\equiv \widetilde{m}_{n2} \quad , \quad (5.8.17)$$

$$y\equiv r_{n1} \quad , \quad Y\equiv r_{n2} \quad , \quad (5.8.18)$$

$$\widetilde{y}\equiv \widetilde{r}_{n1} \quad , \quad \widetilde{Y}\equiv \widetilde{r}_{n2} \quad , \quad (5.8.19)$$

we may cast (5.8.10-11) and (5.8.14-15) in the form :

$$-ix(a)=\epsilon\,\widetilde{X}(a) \quad , \quad iX(a)=\epsilon\,\widetilde{x}(a) \quad , \quad (5.8.20)$$

$$-i\widetilde{y}(a)=\epsilon\,Y(a) \quad , \quad i\widetilde{Y}(a)=\epsilon\,y(a) \quad . \quad (5.8.21)$$

Again, similar equations hold relating $m_{np},\widetilde{m}_{np},r_{np}$ and $\widetilde{r}_{np}$ at the boundary for adjacent indices $p=2k+1,2k+2$. The task now remains to work out the eigenvalue condition for this problem in the massless case. Indeed, we know that the Euclidean action I_E for a massless Majorana field is a sum of terms of the kind :

$$I_{n,E}(x,\widetilde{x},y,\widetilde{y})=\int_0^a d\tau\left[\frac{1}{2}\Big(\widetilde{x}\dot{x}+x\dot{\widetilde{x}}+\widetilde{y}\dot{y}+y\dot{\widetilde{y}}\Big)-\frac{1}{\tau}\left(n+\frac{3}{2}\right)(\widetilde{x}x+\widetilde{y}y)\right] \quad . \quad (5.8.22)$$

Thus, writing $\kappa_n=n+\frac{3}{2}$ and introducing, $\forall n\geq 0$, the operators :

$$L_n\equiv\frac{d}{d\tau}-\frac{\kappa_n}{\tau} \quad , \quad M_n\equiv\frac{d}{d\tau}+\frac{\kappa_n}{\tau} \quad , \quad (5.8.23)$$

the eigenvalue equations are found to be (cf. section 4.3) :

$$L_n x = E\widetilde{x} \quad , \quad M_n \widetilde{x} = -Ex \quad , \tag{5.8.24}$$

$$L_n y = E\widetilde{y} \quad , \quad M_n \widetilde{y} = -Ey \quad , \tag{5.8.25}$$

$$L_n X = E\widetilde{X} \quad , \quad M_n \widetilde{X} = -EX \quad , \tag{5.8.26}$$

$$L_n Y = E\widetilde{Y} \quad , \quad M_n \widetilde{Y} = -EY \quad . \tag{5.8.27}$$

We now define $\forall n \geq 0$ the differential operators :

$$P_n \equiv \frac{d^2}{d\tau^2} + \left[E^2 - \frac{((n+2)^2 - \frac{1}{4})}{\tau^2}\right] \quad , \tag{5.8.28}$$

$$Q_n \equiv \frac{d^2}{d\tau^2} + \left[E^2 - \frac{((n+1)^2 - \frac{1}{4})}{\tau^2}\right] \quad . \tag{5.8.29}$$

Equations (5.8.24-27) lead straightforwardly to the following second-order equations, $\forall n \geq 0$:

$$P_n \widetilde{x} = P_n \widetilde{X} = P_n \widetilde{y} = P_n \widetilde{Y} = 0 \quad , \tag{5.8.30}$$

$$Q_n y = Q_n Y = Q_n x = Q_n X = 0 \quad . \tag{5.8.31}$$

The solutions of (5.8.30-31) which are regular at the origin are :

$$\widetilde{x} = C_1 \sqrt{\tau} J_{n+2}(E\tau) \quad , \quad \widetilde{X} = C_2 \sqrt{\tau} J_{n+2}(E\tau) \quad , \tag{5.8.32}$$

$$x = C_3 \sqrt{\tau} J_{n+1}(E\tau) \quad , \quad X = C_4 \sqrt{\tau} J_{n+1}(E\tau) \quad , \tag{5.8.33}$$

$$\widetilde{y} = C_5 \sqrt{\tau} J_{n+2}(E\tau) \quad , \quad \widetilde{Y} = C_6 \sqrt{\tau} J_{n+2}(E\tau) \quad , \tag{5.8.34}$$

$$y = C_7 \sqrt{\tau} J_{n+1}(E\tau) \quad , \quad Y = C_8 \sqrt{\tau} J_{n+1}(E\tau) \quad . \tag{5.8.35}$$

In order to find the condition obeyed by E, we must now insert (5.8.32-35) into the boundary conditions (5.8.20-21), taking into account also the first-order system given by (5.8.24-27). This gives the following eight equations :

$$-iC_3 J_{n+1}(Ea) = \epsilon \, C_2 J_{n+2}(Ea) \quad , \tag{5.8.36}$$

$$iC_4 J_{n+1}(Ea) = \epsilon\, C_1 J_{n+2}(Ea) \quad , \tag{5.8.37}$$

$$-iC_5 J_{n+2}(Ea) = \epsilon\, C_8 J_{n+1}(Ea) \quad , \tag{5.8.38}$$

$$iC_6 J_{n+2}(Ea) = \epsilon\, C_7 J_{n+1}(Ea) \quad , \tag{5.8.39}$$

$$C_1 = -\frac{EC_3 J_{n+1}(Ea)}{\left[E\dot{J}_{n+2}(Ea) + (n+2)\frac{J_{n+2}(Ea)}{a}\right]} \quad , \tag{5.8.40}$$

$$C_2 = -\frac{EC_4 J_{n+1}(Ea)}{\left[E\dot{J}_{n+2}(Ea) + (n+2)\frac{J_{n+2}(Ea)}{a}\right]} \quad , \tag{5.8.41}$$

$$C_7 = \frac{EC_5 J_{n+2}(Ea)}{\left[E\dot{J}_{n+1}(Ea) - (n+1)\frac{J_{n+1}(Ea)}{a}\right]} \quad , \tag{5.8.42}$$

$$C_8 = \frac{EC_6 J_{n+2}(Ea)}{\left[E\dot{J}_{n+1}(Ea) - (n+1)\frac{J_{n+1}(Ea)}{a}\right]} \quad . \tag{5.8.43}$$

Note that these give separate relations among the constants C_1, C_2, C_3, C_4 and among C_5, C_6, C_7, C_8. For example, eliminating C_1, C_2, C_3, C_4, using (5.8.36-37), (5.8.40-41) and the useful identities (Gradshteyn and Ryzhik 1965) :

$$Ea\dot{J}_{n+1}(Ea) - (n+1)J_{n+1}(Ea) = -EaJ_{n+2}(Ea) \quad , \tag{5.8.44}$$

$$Ea\dot{J}_{n+2}(Ea) + (n+2)J_{n+2}(Ea) = EaJ_{n+1}(Ea) \quad , \tag{5.8.45}$$

one finds :

$$-i\epsilon\frac{J_{n+1}(Ea)}{J_{n+2}(Ea)} = \epsilon^2\frac{C_2}{C_3} = \epsilon^2\frac{C_4}{C_1} = -i\epsilon^3\frac{J_{n+2}(Ea)}{J_{n+1}(Ea)} \quad , \tag{5.8.46}$$

which implies (since $\epsilon = \pm 1$) :

$$\left[J_{n+1}(Ea)\right]^2 - \left[J_{n+2}(Ea)\right]^2 = 0 \quad , \quad \forall n \geq 0 \quad . \tag{5.8.47}$$

The desired set of eigenvalue conditions can also be written in the form :

$$J_{n+1}(Ea) = \pm J_{n+2}(Ea) \quad , \quad \forall n \geq 0 \quad . \tag{5.8.48}$$

Exactly the same set of eigenvalue conditions arises from eliminating C_5, C_6, C_7, C_8. The limiting behaviour of the eigenvalues can be found under certain approximations. For example, if n is fixed and $| z | \rightarrow \infty$, one has the standard asymptotic expansion (Abramowitz and Stegun 1964) :

$$J_n(z) \sim \sqrt{\frac{2}{\pi z}} \cos\left(z - \frac{n\pi}{2} - \frac{\pi}{4}\right) + \mathrm{O}\left(z^{-\frac{3}{2}}\right) \quad . \tag{5.8.49}$$

Thus, writing (5.8.48) in the form : $J_{n+1}(E) = \hat{\kappa} J_{n+2}(E)$, where $\hat{\kappa} = \pm 1$ and we set $a = 1$ for simplicity, two asymptotic sets of eigenvalues result :

$$E^{+} \sim \pi\left(\frac{n}{2} + L\right) \text{ if } \hat{\kappa} = 1 \quad , \tag{5.8.50}$$

$$E^{-} \sim \pi\left(\frac{n}{2} + M + \frac{1}{2}\right) \text{ if } \hat{\kappa} = -1 \quad , \tag{5.8.51}$$

where L and M are large integers (both positive and negative). This result will be very useful in section 8.1. One can also obtain an estimate of the smallest eigenvalues, for a given large n. The asymptotic expansions in section 9.3 of Abramowitz and Stegun 1964 show that these eigenvalues have the asymptotic form :

$$| E | \sim \Big[n + \mathrm{o}(n)\Big] \quad . \tag{5.8.52}$$

In general it is very difficult to solve numerically (5.8.48), because the recurrence relations which enable one to compute Bessel functions starting from J_0 and J_1 are a source of large errors when the argument is comparable with the order. Alternatively, one can easily work out a fourth-order differential equation whose solution is of the form : $\sqrt{x}\Big(J_{n+1}(x) \mp J_{n+2}(x)\Big)$. The coefficients of this equation do not depend on the Bessel functions, but unfortunately they involve the eigenvalues E. This is why we have not been able to compute numerically the eigenvalues. However, we should say that a more successful numerical study has been carried out in Berry and Mondragon 1987. In that paper, the authors study eigenvalues of the Dirac operator with local boundary conditions, in the case of neutrino billiards. This corresponds to massless spin-$\frac{1}{2}$ particles moving under the action of a potential describing a hard wall bounding a finite domain. The authors end up with an eigenvalue condition of the kind : $J_l(k_{nl}) = J_{l+1}(k_{nl})$, and compute the lowest 2600 positive eigenvalues k_{nl}.

We now study the Hartle-Hawking path integral (D'Eath and Halliwell 1987) :

$$K_{HH} = \int e^{-I_E} \, D\psi^A \, D\widetilde{\psi}^{A'} \quad , \tag{5.8.53}$$

taken over the class of massless spin-$\frac{1}{2}$ fields $\left(\psi^A, \, \widetilde{\psi}^{A'}\right)$ which obey (5.8.3) on the bounding S^3. Here, with our conventions (see below) :

$$I_E \equiv \frac{i}{2} \int d^4x \sqrt{g} \left[\widetilde{\psi}^{A'} \left(\nabla_{AA'} \psi^A \right) - \left(\nabla_{AA'} \widetilde{\psi}^{A'} \right) \psi^A \right] \tag{5.8.54}$$

is the Euclidean action for a massless spin-$\frac{1}{2}$ field $\left(\psi^A, \, \widetilde{\psi}^{A'}\right)$. The fermionic fields are taken to be anti-commuting, and Berezin integration is being used (D'Eath and Halliwell 1987 and Eq. (3.1.15)) :

$$\int dy = 0 \quad , \quad \int y \, dy = 1 \quad . \tag{5.8.55}$$

Let us now assume provisionally that ψ^A and $\widetilde{\psi}^{A'}$ can be expanded in a complete set of eigenfunctions $\left\{ \psi_n^A, \, \widetilde{\psi}_n^{A'} \right\}$ obeying the eigenvalue equations which arise from variation of the action (5.8.54) :

$$\nabla_{AA'} \psi_n^A = \lambda_n \widetilde{\psi}_{nA'} \quad , \tag{5.8.56}$$

$$\nabla_{AA'} \widetilde{\psi}_n^{A'} = \lambda_n \psi_{nA} \quad , \tag{5.8.57}$$

and the boundary condition : $\sqrt{2} \, {}_e n_{AA'} \psi_n^A = \epsilon \, \widetilde{\psi}_{nA'}$ on S^3 (cf. (5.8.3)). As with a bosonic one-loop path integral, one would like to be able to express the fermionic Hartle-Hawking path integral K_{HH} in terms of a suitable product of eigenvalues. One then needs the cross-terms in I_E to vanish. Indeed, the typical cross-term Σ appearing in I_E is :

$$\Sigma = \frac{i}{2} \int d^4x \sqrt{g} \left[\lambda_n \widetilde{\psi}_m^{A'} \widetilde{\psi}_{nA'} + \lambda_m \psi_n^A \psi_{mA} + \lambda_m \widetilde{\psi}_m^{A'} \widetilde{\psi}_{nA'} + \lambda_n \psi_n^A \psi_{mA} \right]$$

$$= \frac{i}{2} \int d^4x \sqrt{g} \left[(\lambda_n + \lambda_m) \widetilde{\psi}_m^{A'} \widetilde{\psi}_{nA'} + (\lambda_n + \lambda_m) \psi_n^A \psi_{mA} \right] \quad . \tag{5.8.58}$$

In deriving (5.8.58), where $n \neq m$, we have raised and lowered spinor indices, and used the anticommutation relations obeyed by our spinor fields. However, commutation can be assumed when the path integral is not involved, as we shall do later. The use of the eigenvalue equations (5.8.56-57) shows now that the square bracket in (5.8.58) may be cast in the form : $a\nabla_{AA'}\left(\psi_m^A \widetilde{\psi}_n^{A'}\right) + b\nabla_{AA'}\left(\psi_n^A \widetilde{\psi}_m^{A'}\right)$, with $a = -b = \frac{(\lambda_n+\lambda_m)}{(\lambda_n-\lambda_m)}$, provided $\lambda_n \neq \lambda_m$. Thus Σ becomes :

$$\Sigma = \frac{i}{2}\frac{(\lambda_n + \lambda_m)}{(\lambda_n - \lambda_m)}\left[\int_{\partial M} d^3x \sqrt{h}\ {}_e n_{AA'}\psi_m^A \widetilde{\psi}_n^{A'} - \int_{\partial M} d^3x \sqrt{h}\ {}_e n_{AA'}\psi_n^A \widetilde{\psi}_m^{A'}\right] = 0 \ , \quad (5.8.59)$$

by virtue of the local boundary conditions (5.8.3). In the degenerate case : $\lambda_n = \lambda_m$, a linear transformation within the degenerate eigenspace can be found such that the cross-terms again vanish. It is indeed well-known that every Hermitian matrix can be cast in diagonal form with real eigenvalues. Thus the asymptotic calculations (5.8.50-52), and the property that I_E can be written as a diagonal expression in terms of a sum over eigenfunctions suggest that the Dirac action used here, subject to local boundary conditions, can be expressed in terms of a self-adjoint differential operator acting on fields $\left(\psi^A,\ \widetilde{\psi}^{A'}\right)$. We shall now prove that this is indeed the case.

So far we have seen that the framework for the formulation of local boundary conditions involving normals and field strengths or fields is the Euclidean regime, where one deals with Riemannian metrics. Thus, we will pay special attention to the conjugation of $SU(2)$ spinors in Euclidean four-space. In fact such a conjugation will play a key role in proving self-adjointness. For this purpose, it can be useful to recall at first some basic results about $SU(2)$ spinors on an abstract Riemannian three-manifold (Σ, h). In that case, one considers (Ashtekar 1988) a bundle over the three-manifold, each fibre of which is isomorphic to a two-dimensional complex vector space W. It is then possible to define a nowhere vanishing antisymmetric ϵ_{AB} (the usual one of appendix A) so as to raise and lower internal indices, and a positive-definite Hermitian inner product on each fibre : $(\psi, \phi) = \overline{\psi}^{A'} G_{A'A}\phi^A$. The requirements of Hermiticity and positivity imply respectively that : $\overline{G}_{A'A} = G_{A'A}$, $\overline{\psi}^{A'} G_{A'A}\psi^A > 0, \forall\ \psi^A \neq 0$. This $G_{A'A}$ converts primed indices to unprimed ones, and it is given by $i\sqrt{2}\, n_{AA'}$. Given the space H of all objects $\alpha^{A}{}_{B}$ such that : $\alpha^{A}{}_{A}=0$ and $\left(\alpha^{\dagger}\right)^{A}{}_{B}=-\alpha^{A}{}_{B}$, one finds there always exists a $SU(2)$ soldering form $\sigma^{a}{}_{A}{}^{B}$ (i.e. a global isomorphism) between H and the tangent space on (Σ, h) such that : $h^{ab}=-\sigma^{a}{}_{A}{}^{B}\sigma^{b}{}_{B}{}^{A}$. Therefore one also finds : $\sigma^{a}{}_{A}{}^{A}=0$ and $\left(\sigma^{a}{}_{A}{}^{B}\right)^{\dagger}=-\sigma^{a}{}_{A}{}^{B}$. One then defines ψ^{A} an $SU(2)$ spinor on (Σ, h). A basic remark is that $SU(2)$ transformations are those $SL(2,C)$ transformations which preserve $n^{AA'}=n^{a}\sigma_{a}{}^{AA'}$, where $n^{a}=(1,0,0,0)$ is the normal to Σ. The Euclidean conjugation used here (not to be confused with complex conjugation in Minkowski space) is such that (see now (A.34-36)) :

$$\left(\psi_{A}+\lambda\phi_{A}\right)^{\dagger}={\psi_{A}}^{\dagger}+\lambda^{*}{\phi_{A}}^{\dagger} \quad , \quad \left({\psi_{A}}^{\dagger}\right)^{\dagger}=-\psi_{A} \quad , \tag{5.8.60}$$

$${\epsilon_{AB}}^{\dagger}=\epsilon_{AB} \quad , \quad \left(\psi_{A}\phi_{B}\right)^{\dagger}={\psi_{A}}^{\dagger}{\phi_{B}}^{\dagger} \quad , \tag{5.8.61}$$

$$\left(\psi_{A}\right)^{\dagger}\psi^{A}>0 \quad , \quad \forall\, \psi_{A}\neq 0 \quad . \tag{5.8.62}$$

In (5.8.60) and in the following pages, the symbol $*$ denotes complex conjugation of scalars. How to generalize this picture to the Euclidean four-space ? For this purpose, let us now focus our attention on states that are pairs of spinor fields, defining :

$$w\equiv\left(\psi^{A}, \widetilde{\psi}^{A'}\right) \quad , \quad z\equiv\left(\phi^{A}, \widetilde{\phi}^{A'}\right) \quad , \tag{5.8.63}$$

on the ball of radius a in Euclidean four-space, subject always to the boundary conditions (5.8.3). In other words, w and z are Majorana spinors in two-component language in the sense of section 1.4 (Hawking and Pope 1978), subject also to suitable differentiability conditions, to be specified later. Let us also define the operator C :

$$C:\left(\psi^{A}, \widetilde{\psi}^{A'}\right)\rightarrow\left(\nabla^{A}{}_{B'}\widetilde{\psi}^{B'}, \nabla_{B}{}^{A'}\psi^{B}\right) \quad , \tag{5.8.64}$$

and the *dagger* operation :

$$\left(\psi^{A}\right)^{\dagger}\equiv\epsilon^{AB}\delta_{BA'}\overline{\psi}^{A'} \quad , \quad \left(\widetilde{\psi}^{A'}\right)^{\dagger}\equiv\epsilon^{A'B'}\delta_{B'A}\overline{\widetilde{\psi}}^{A} \quad . \tag{5.8.65}$$

The consideration of C is suggested of course by the action (5.8.54) and by the eigenvalue equations (5.8.56-57). In (5.8.65), $\delta_{BA'}$ is an identity matrix playing the same role of $G_{AA'}$ for $SU(2)$ spinors on (Σ, h), so that $\delta_{BA'}$ is preserved by $SU(2)$ transformations. Moreover, the *bar* symbol $\overline{\psi^A} = \overline{\psi}^{A'}$ denotes the usual complex conjugation of $SL(2,C)$ spinors. Hence we find :

$$\left((\psi^A)^\dagger\right)^\dagger = \epsilon^{AC}\delta_{CB'}\overline{(\psi^{B\dagger})'} = \epsilon^{AC}\delta_{CB'}\epsilon^{B'D'}\delta_{D'F}\psi^F = -\psi^A \quad , \tag{5.8.66}$$

in view of the definition of ϵ^{AB} (cf. appendix A). Thus the *dagger* operation defined in (5.8.65) is anti-involutory, because, when applied twice to ψ^A, it yields $-\psi^A$.

As anticipated after (5.8.58), from now on we study commuting spinors, for simplicity of exposition of the self-adjointness. It is easy to check that the *dagger*, also called in the literature Euclidean conjugation (Woodhouse 1985), satisfies all properties (5.8.60-62). We suggest to read now the last part of appendix A, where this matter is discussed in more detail. We can now define the scalar product :

$$(w,z) \equiv \int_M \left[\psi^\dagger_A \phi^A + \widetilde{\psi}^\dagger_{A'}\widetilde{\phi}^{A'}\right]\sqrt{g}\, d^4x \quad . \tag{5.8.67}$$

This is indeed a scalar product, because it satisfies all following properties for all vectors u, v, w and $\forall \lambda \in C$:

$$(u,u) > 0 \quad , \quad \forall u \neq 0 \quad , \tag{5.8.68}$$

$$(u, v+w) = (u,v) + (u,w) \quad , \tag{5.8.69}$$

$$(u, \lambda v) = \lambda (u,v) \quad , \quad (\lambda u, v) = \lambda^*(u,v) \quad , \tag{5.8.70}$$

$$(v,u) = (u,v)^* \quad . \tag{5.8.71}$$

We are now aiming to check that C or iC is a symmetric operator, i.e. that $(Cz,w) = (z,Cw)$ or : $(iCz,w) = (z,iCw), \forall z,w$. This will be used in the course of proving further that the symmetric operator has self-adjoint extensions. In order to prove this result it is clear, in view of (5.8.67), we need to know the properties of the spinor covariant derivative acting on $SU(2)$ spinors. In the case of $SL(2,C)$ spinors (Penrose and Rindler 1984,

Stewart 1990) it is known this derivative is a linear, torsion-free map $\nabla_{AA'}$ which satisfies the Leibniz rule, annihilates ϵ_{AB} and is real (i.e. $\psi = \nabla_{AA'}\theta \Rightarrow \overline{\psi} = \nabla_{AA'}\overline{\theta}$). Moreover, we know that :

$$\nabla^{AA'} = e^{AA'}{}_{\mu}\nabla^{\mu} = e^{a}{}_{\mu}\,\sigma_{a}{}^{AA'}\nabla^{\mu} \quad . \tag{5.8.72}$$

In Euclidean four-space, we use both (5.8.72) and the relation :

$$\sigma_{\mu AC'}\,\sigma_{\nu B}{}^{C'} + \sigma_{\nu BC'}\,\sigma_{\mu A}{}^{C'} = \delta_{\mu\nu}\epsilon_{AB} \quad , \tag{5.8.73}$$

where $\delta_{\mu\nu}$ has signature $(+,+,+,+)$. This implies that : $\sigma_0 = -\frac{i}{\sqrt{2}}I$, $\sigma_i = \frac{\Sigma_i}{\sqrt{2}}$, $\forall i = 1,2,3$, where Σ_i are the Pauli matrices. Now, in view of (5.8.64) and (5.8.67) one finds :

$$(Cz,w) = \int_M \left(\nabla_{AB'}\phi^A\right)^{\dagger}\widetilde{\psi}^{B'}\sqrt{g}\,d^4x + \int_M \left(\nabla_{BA'}\widetilde{\phi}^{A'}\right)^{\dagger}\psi^B\sqrt{g}\,d^4x \;, \tag{5.8.74}$$

whereas, using the Leibniz rule in computing $\nabla^{A}{}_{B'}\left(\phi^{\dagger}_{A}\widetilde{\psi}^{B'}\right)$ and $\nabla_{B}{}^{A'}\left(\left(\widetilde{\phi}_{A'}\right)^{\dagger}\psi^B\right)$, and integrating by parts, one finds :

$$\begin{aligned}(z,Cw) = &\int_M \left(\nabla_{AB'}\phi^{A\dagger}\right)\widetilde{\psi}^{B'}\sqrt{g}\,d^4x + \int_M \left(\nabla_{BA'}\left(\widetilde{\phi}^{A'}\right)^{\dagger}\right)\psi^B\sqrt{g}\,d^4x \\ &- \int_{\partial M}({}_e n_{AB'})\phi^{A\dagger}\widetilde{\psi}^{B'}\sqrt{h}\,d^3x - \int_{\partial M}({}_e n_{BA'})\left(\widetilde{\phi}^{A'}\right)^{\dagger}\psi^B\sqrt{h}\,d^3x. \end{aligned} \tag{5.8.75}$$

Now we use (5.8.65), the identity :

$$\left({}_e n^{AA'}\phi_A\right)^{\dagger} = \epsilon^{A'B'}\delta_{B'C}\,\overline{{}_e n^{DC'}}\,\overline{\phi_D} = -\epsilon^{A'B'}\delta_{B'C}\left({}_e n^{CD'}\right)\overline{\phi}_{D'} \quad , \tag{5.8.76}$$

the relation (A.9) and the boundary conditions on S^3 : $\sqrt{2}\,{}_e n^{CB'}\psi_C = \widetilde{\psi}^{B'}$, $\sqrt{2}\,{}_e n^{AA'}\phi_A = \widetilde{\phi}^{A'}$. In so doing, the sum of the boundary terms in (5.8.75) is found to vanish. This implies in turn that equality of the volume integrands is sufficient to show that (Cz,w) and (z,Cw) are equal. For example, one finds in flat space, using also (5.8.65) : $\left(\nabla_{BA'}\widetilde{\phi}^{A'}\right)^{\dagger} =$

$\delta_{BF'}\overline{\sigma}^{F'}{}_{C}{}^{a}\partial_a\left(\widetilde{\overline{\phi}}^{C}\right)$, whereas : $\left(\nabla_{BA'}\left(\widetilde{\phi}^{A'}\right)^{\dagger}\right) = -\delta_{CF'}\sigma_{B}{}^{F'a}\partial_a\left(\widetilde{\overline{\phi}}^{C}\right)$. In other words, we are led to study the condition :

$$\delta_{BF'}\,\overline{\sigma}^{F'}{}_{C}{}^{a} = \pm\delta_{BF'}\,\sigma_{C}{}^{F'a} \quad , \tag{5.8.77}$$

$\forall\ a = 0,1,2,3$. Now, using the relations :

$$\sqrt{2}\,\sigma_{AA'}{}^{0} = \begin{pmatrix} -i & 0 \\ 0 & -i \end{pmatrix} \quad , \quad \sqrt{2}\,\sigma_{AA'}{}^{1} = \begin{pmatrix} 0 & 1 \\ 1 & 0 \end{pmatrix} \quad , \tag{5.8.78}$$

$$\sqrt{2}\,\sigma_{AA'}{}^{2} = \begin{pmatrix} 0 & -i \\ i & 0 \end{pmatrix} \quad , \quad \sqrt{2}\,\sigma_{AA'}{}^{3} = \begin{pmatrix} 1 & 0 \\ 0 & -1 \end{pmatrix} \quad , \tag{5.8.79}$$

$$\sigma^{A}{}_{A'}{}^{a} = \epsilon^{AB}\sigma_{BA'}{}^{a} \quad , \quad \sigma_{A}{}^{A'a} = -\sigma_{AB'}{}^{a}\,\epsilon^{B'A'} \quad , \tag{5.8.80}$$

one finds that the complex conjugate of $\sigma^{A}{}_{A'}{}^{a}$ is always equal to $\sigma_{A}{}^{A'a}$, which is not in agreement with the choice of the $(-)$ sign on the right-hand side of (5.8.77). This implies in turn that the symmetric operator we are looking for is iC, where C has been defined in (5.8.64). The generalization to a curved four-dimensional Riemannian space is obtained via the relation : $e^{AA'}{}_{\mu} = e^{a}{}_{\mu}\,\sigma_a{}^{AA'}$.

Now, it is known that every symmetric operator has a closure, and the operator and its closure have the same closed extensions. Moreover, a closed symmetric operator on a Hilbert space is self-adjoint if and only if its spectrum is a subset of the real axis (Reed and Simon 1975, page 136). To prove self-adjointness for our boundary-value problem, we may recall an important result due to von Neumann (Reed and Simon 1975, page 143). This theorem states that, given a symmetric operator A with domain $D(A)$, if a map $F : D(A) \rightarrow D(A)$ exists such that :

$$F(\alpha w + \beta z) = \alpha^{*}F(w) + \beta^{*}F(z) \quad , \tag{5.8.81}$$

$$(w,w) = (Fw,Fw) \quad , \tag{5.8.82}$$

$$F^{2} = \pm I \quad , \tag{5.8.83}$$

$$FA = AF \quad , \tag{5.8.84}$$

then A has self-adjoint extensions. In our case, denoting by D the operator (cf. (5.8.65))

$$D : \left(\psi^A, \widetilde{\psi}^{A'}\right) \rightarrow \left(\left(\psi^A\right)^\dagger, \left(\widetilde{\psi}^{A'}\right)^\dagger\right) \quad , \tag{5.8.85}$$

let us focus our attention on the operators : $F = iD$ and $A = iC$. The operator F maps indeed $D(A)$ to $D(A)$. In fact, bearing in mind the definitions :

$$G \equiv \left\{\varphi = \left(\phi^A, \widetilde{\phi}^{A'}\right) : \varphi \text{ is at least } C^1\right\} \quad , \tag{5.8.86}$$

$$D(A) \equiv \left\{\varphi \in G : \sqrt{2}\, {}_e n^{AA'} \phi_A = \epsilon\, \widetilde{\phi}^{A'} \text{ on } S^3\right\} \quad , \tag{5.8.87}$$

one finds that F maps $\left(\phi^A, \widetilde{\phi}^{A'}\right)$ to $\left(\beta^A, \widetilde{\beta}^{A'}\right) = \left(i\left(\phi^A\right)^\dagger, i\left(\widetilde{\phi}^{A'}\right)^\dagger\right)$ such that :

$$\sqrt{2}\, {}_e n^{AA'} \beta_A = \gamma\, \widetilde{\beta}^{A'} \text{ on } S^3 \quad , \tag{5.8.88}$$

where $\gamma = \epsilon^*$. The boundary condition (5.8.88) is clearly of the type which occurs in (5.8.87) provided ϵ is real, and the differentiability of $\left(\beta^A, \widetilde{\beta}^{A'}\right)$ is not affected by the action of F (cf. (5.8.85)). In deriving (5.8.88), we have used the result for $\left({}_e n^{AA'} \phi_A\right)^\dagger$ obtained in (5.8.76). It is worth emphasizing that the requirement of self-adjointness enforces the choice of a real function ϵ, which is a constant in our case. Moreover, in view of (5.8.66), one immediately sees that (5.8.81) and (5.8.83) hold when $F = iD$, provided we write (5.8.83) as $F^2 = -I$. This is indeed a crucial point which deserves special attention. Condition (5.8.83) is written in Reed and Simon 1975 as $F^2 = I$, and examples are later given (see page 144 therein) where F is complex conjugation. But we are formulating our problem in the Euclidean regime, where we have seen that the only possible conjugation is the *dagger* operation, which is anti-involutory on spinors with an odd number of indices. Thus we are here generalizing von Neumann's theorem in the following way. If F is a map $D(A) \rightarrow D(A)$ which satisfies (5.8.81-84), then the same is clearly true of $\widetilde{F} = -iD = -F$. Hence :

$$-F\, D(A) \subseteq D(A) \quad , \tag{5.8.89}$$

$$F\ D(A) \subseteq D(A) \quad . \tag{5.8.90}$$

Acting with F on both sides of (5.8.89), we find :

$$D(A) \subseteq F\ D(A) \quad , \tag{5.8.91}$$

using the property : $F^2 = -I$. But then the relations (5.8.90-91) imply that $F\ D(A) = D(A)$, so that F takes $D(A)$ onto $D(A)$ also in the case of the anti-involutory Euclidean conjugation that we called *dagger*. Comparison with the proof presented at the beginning of page 144 in Reed and Simon 1975 shows that this is all what we need so as to generalize von Neumann's theorem to the Dirac operator acting on $SU(2)$ spinors in Euclidean four-space (one later uses properties (5.8.84), (5.8.81) and (5.8.82) as well in completing the proof).

It remains to verify conditions (5.8.82) and (5.8.84). First, note that :

$$\begin{aligned}(Fw, Fw) &= (iDw, iDw)\\ &= \int_M \left(i\psi_A^{\dagger}\right)^{\dagger} i\left(\psi^A\right)^{\dagger} \sqrt{g}\ d^4x + \int_M \left(i\tilde{\psi}_{A'}^{\dagger}\right)^{\dagger} i\left(\tilde{\psi}^{A'}\right)^{\dagger} \sqrt{g}\ d^4x\\ &= (w, w) \quad ,\end{aligned} \tag{5.8.92}$$

where we have used (5.8.66-67) and the commutation property of our spinors. Second, one finds :

$$FAw = (iD)(iC)\,w = i\left[i\left(\nabla^A{}_{B'}\tilde{\psi}^{B'}, \nabla_B{}^{A'}\psi^B\right)\right]^{\dagger} = \left(\nabla^A{}_{B'}\tilde{\psi}^{B'}, \nabla_B{}^{A'}\psi^B\right)^{\dagger} , \tag{5.8.93}$$

$$AFw = (iC)(iD)\,w = iCi\left(\psi^{A\dagger}, \left(\tilde{\psi}^{A'}\right)^{\dagger}\right) = -\left(\nabla^A{}_{B'}\left(\tilde{\psi}^{B'}\right)^{\dagger}, \nabla_B{}^{A'}\psi^{B\dagger}\right) , \tag{5.8.94}$$

which in turn implies that also (5.8.84) holds in view of what we found just before (5.8.77) and after (5.8.80). To sum up, we have proved that the operator iC arising in our boundary-value problem is symmetric and has self-adjoint extensions (a problem for future research is the proof of essential self-adjointness). Hence the eigenvalues of iC are real, and the eigenvalues λ_n of C are purely imaginary. This is in agreement with (5.8.48), because there $E = i\lambda_n = -Im(\lambda_n)$. Further, they occur in equal and opposite pairs in our example of flat Euclidean four-space bounded by a three-sphere. Hence the fermionic one-loop path integral K_{HH} is formally given by the dimensionless product (Allen 1983a and our section 4.3) :

$$K_{HH} = \prod_{n} \left(\frac{|\lambda_n|}{\widetilde{\mu}} \right) \quad , \tag{5.8.95}$$

where $\widetilde{\mu}$ is a normalization constant with dimensions of mass, and the right-hand side of (5.8.95) should be interpreted as explained following (4.3.10). This one-loop K_{HH} can be regularized using the zeta-function :

$$\zeta_M(s) \equiv \sum_{n,k} d_k(n) \Big(|\lambda_{n,k}|^2 \Big)^{-s} \quad , \tag{5.8.96}$$

as explained in Hawking 1977, Allen 1983a. With this more accurate notation, we write $d_k(n)$ for the degeneracy of the eigenvalues $|\lambda_{n,k}|$ (see comments after (4.5.4)). Two indices are used because, for each value of the integer n, there are infinitely many eigenvalues labeled by k. This $\zeta_M(s)$ has the properties described at the beginning of appendix B.

Our next task seems to be the calculation of $\zeta(0)$ (cf. chapter eight), but this is more easily obtained after completing a series of $\zeta(0)$ calculations for bosonic fields. Thus this chapter contains some mathematical results on local boundary conditions, whereas the corresponding $\zeta(0)$ calculations are studied and (when possible) performed in chapters six, seven, eight and nine, after the preliminary results of section 5.9. Also the comparison between different approaches to zeta-function calculations for fermionic fields is postponed until the end of chapter eight.

5.9 Spin-1 and Spin-0 Fields

We now begin the analysis of local boundary conditions for the boundary-value problems involving the spin-1 and spin-0 fields, so as to compare the values found for $\zeta(0)$. This is an important check needed so as to understand whether or not local boundary conditions involving field strengths and normals yield one-loop finiteness in quantum cosmology.

Let us begin by recalling that local boundary conditions in the spin-1 case are :

$$2\, n^{AA'} n^{BB'} \phi_{AB} = \epsilon\, \widetilde{\phi}^{A'B'} \ \text{ on } \ S^3 \quad , \tag{5.9.1}$$

where $\epsilon = \pm 1$. In fact, the product of two Lorentzian normals is minus the product of two Euclidean normals, so that the only difference due to a sign can be absorbed into the definition of ϵ. Moreover, we can expand ϕ_{AB} and $\widetilde{\phi}^{A'B'}$ in spinor harmonics on S^3, according to the relations :

$$\phi_{AB} = \frac{t^{-\frac{3}{2}}}{2\pi} \sum_{npq} \alpha_n^{pq} \Bigl(m_{np}(t) \rho_{AB}^{nq} + \widetilde{r}_{np}(t) \overline{\sigma}_{AB}^{nq} \Bigr) \quad , \tag{5.9.2}$$

$$\widetilde{\phi}^{A'B'} = \frac{t^{-\frac{3}{2}}}{2\pi} \sum_{npq} \alpha_n^{pq} \Bigl(\widetilde{m}_{np}(t) \overline{\rho}^{nqA'B'} + r_{np}(t) \sigma^{nqA'B'} \Bigr) \quad . \tag{5.9.3}$$

The basic identities we need are :

$$\int d\mu\, \rho_{AB}^{np}\, n^{AA'} n^{BB'} \overline{\rho}_{A'B'}^{mq} = \delta^{nm} H_n^{pq} \quad , \tag{5.9.4}$$

$$\int d\mu\, \rho_{AB}^{np}\, \epsilon^{AC} \epsilon^{BD} \rho_{CD}^{mq} = \delta^{nm} A_n^{pq} \quad . \tag{5.9.5}$$

From (5.9.4-5) we can now derive the relation between the ρ and $\overline{\rho}$ harmonics, and the consistency condition obeyed by A_n and H_n. In fact, writing $\phi_{AB}(x)$ in the form (equivalent to (5.9.2)) :

$$\phi_{AB}(x) = \sum_{np} \Bigl(a_{np} \rho_{AB}^{np}(x) + b_{np} \overline{\sigma}_{AB}^{np}(x) \Bigr) \quad , \tag{5.9.6}$$

we derive from (5.9.4) and (5.9.5) respectively that :

$$a_{np} = \int d\mu\, \phi_{AB}(x) n^{AA'} n^{BB'} \sum_q \overline{\rho}_{A'B'}^{nq} \left(H_n^{-1}\right)^{qp} \quad , \tag{5.9.7}$$

$$a_{np} = \int d\mu\, \phi_{AB}(x) \epsilon^{AC} \epsilon^{BD} \sum_q \rho_{CD}^{nq} \left(A_n^{-1}\right)^{qp} \quad . \tag{5.9.8}$$

Thus, equality of the integrands in (5.9.7-8) is sufficient, which finally implies :

$$\tilde{\rho}^{npA'B'} = 4n^{CA'}n^{DB'} \sum_{q} \rho_{CD}^{nq} \left(A_n^{-1} H_n\right)^{qp} \quad . \tag{5.9.9}$$

In addition, taking the complex conjugate of (5.9.9), inserting it into the right-hand side of (5.9.9) and using the relations : $n_{CC'}n^{CA'} = \frac{1}{2}\epsilon_{C'}^{\ \ A'}$, $n_{DD'}n^{DB'} = \frac{1}{2}\epsilon_{D'}^{\ \ B'}$, we find the consistency condition :

$$A_n^{-1} H_n A_n^{-1} H_n = \frac{1}{4} 1_n \quad . \tag{5.9.10}$$

The relation (5.9.10) has a (+) sign on the right-hand side instead of a (−) sign appearing in the spin-$\frac{1}{2}$ case (D'Eath and Halliwell 1987), because two normals are involved in the calculation (cf. (5.9.9)). Thus the matrices A_n and H_n can be chosen to be :

$$A_n = 2\begin{pmatrix} 0 & 1 \\ 1 & 0 \end{pmatrix} \quad , \quad H_n = \begin{pmatrix} 1 & 0 \\ 0 & 1 \end{pmatrix} \quad . \tag{5.9.11}$$

The insertion of (5.9.2-3), (5.9.9) and (5.9.11) into (5.9.1) yields therefore :

$$\sum_{pq} \begin{pmatrix} 1 & 1 \\ 1 & -1 \end{pmatrix}^{pq} m_{np}(a) \rho_{AB}^{nq} = \epsilon \sum_{pq} \begin{pmatrix} 1 & 1 \\ 1 & -1 \end{pmatrix}^{pq} \widetilde{m}_{np}(a) \sum_{d} \rho_{AB}^{nd} \begin{pmatrix} 0 & 1 \\ 1 & 0 \end{pmatrix}^{dq} , \tag{5.9.12}$$

which implies for example :

$$\rho_{AB}^{n1}(m_{n1} + m_{n2})(a) + \rho_{AB}^{n2}(m_{n1} - m_{n2})(a) = \epsilon \rho_{AB}^{n1}(\widetilde{m}_{n1} - \widetilde{m}_{n2})(a) + \epsilon \rho_{AB}^{n2}(\widetilde{m}_{n1} + \widetilde{m}_{n2})(a) \, . \tag{5.9.13}$$

The solution of (5.9.13) is :

$$m_{n1}(a) = \epsilon \, \widetilde{m}_{n1}(a) \quad , \tag{5.9.14}$$

$$m_{n2}(a) = -\epsilon \, \widetilde{m}_{n2}(a) \quad . \tag{5.9.15}$$

Thus, the difference with respect to the spin-$\frac{1}{2}$ case is that m and $\widetilde{m}$ modes with the same degeneracy label are related by the boundary conditions (cf. (5.8.10-11)). This holds by virtue of (5.9.10-11). Studying the σ and $\tilde{\sigma}$ harmonics, two more boundary conditions are obtained :

$$\widetilde{r}_{n1}(a) = \epsilon\, r_{n1}(a) \quad , \tag{5.9.16}$$

$$\widetilde{r}_{n2}(a) = -\epsilon\, r_{n2}(a) \quad . \tag{5.9.17}$$

The relations (5.9.14-17) are important in that they yield the boundary conditions obeyed by the modes, to be compared with (5.8.10-11) and (5.8.14-15) so as to understand what happens for various values of the spin.

However, if we want to compute the PDF $\zeta(0)$ for spin 1, there is a more useful way of doing this calculation. In fact, we know that the components of the electric field $\vec{E}$ and of the magnetic field $\vec{B}$ are given by :

$$E_j + iB_j = n^A{}_{C'} e^{BC'}{}_j \, \phi_{AB} \quad , \tag{5.9.18}$$

$$E_j - iB_j = n_C{}^{A'} e^{CB'}{}_j \, \widetilde{\phi}_{A'B'} \quad . \tag{5.9.19}$$

Thus, if we require on S^3 a local boundary condition of the type :

$$\widetilde{\phi}_{A'B'} = -2n^C{}_{A'} n^D{}_{B'} \phi_{CD} \quad , \tag{5.9.20}$$

the insertion of (5.9.20) into (5.9.19) yields :

$$E_j(a) + iB_j(a) = E_j(a) - iB_j(a) \quad , \tag{5.9.21}$$

in view of the identity : $n_{EA'} n^{CA'} = \frac{1}{2}\epsilon_E{}^C$. Condition (5.9.21) implies in turn that $B_j(a) = 0$, $\forall j = 1,2,3$, i.e. the vanishing of F_{ij} on S^3, $\forall i,j = 1,2,3$, where $F_{\mu\nu}$ is the electromagnetic field tensor. In addition, we know that : $F_{ij} = \partial_i A_j - \partial_j A_i$, where A_μ is the vector potential. Its spatial components can be expanded according to the relations (Louko 1988b) :

$$A_k(x,t) = \sum_{n=2}^{\infty} \Big[f_n(t) S_k^{(n)}(x) + g_n(t) P_k^{(n)}(x) \Big] \quad . \tag{5.9.22}$$

In (5.9.22), we have omitted for simplicity the full set of degeneracy labels. The $S_k^{(n)}$ and $P_k^{(n)}$ are the transverse and longitudinal eigenfunctions respectively of the vector Laplacian.

Choosing the Coulomb gauge (cf. section 5.10) : $A_0 \approx 0$, ${A_k}^{|k} \approx 0$, the Lorentzian action can be finally cast in the form (Louko 1988b) :

$$S_{em} = -\frac{1}{4}\int \sqrt{-g}F_{\mu\nu}F^{\mu\nu}d^4x = \sum_{n=2}^{\infty}\int \left[\frac{a}{2N}\dot{f}_n^2 - \frac{N}{2a}n^2 f_n^2\right] dt \quad . \tag{5.9.23}$$

Making the analytic continuation to the Euclidean regime, and applying exactly the same technique we have used so far, Louko was able to show that the prefactor of the semiclassical approximation of the wave function behaves as $a^{\zeta(0)}$, where $\zeta(0) = -\frac{77}{180}$. The modes $f_n(\tau)$ were required to vanish both at the origin (regularity requirement) and on the three-sphere boundary, $\forall n \geq 2$.

Now, using the Coulomb gauge, F_{ij} becomes :

$$F_{ij}(x,\tau) = \sum_{n=2}^{\infty} f_n(\tau)\left[\partial_i S_j^{(n)}(x) - \partial_j S_i^{(n)}(x)\right] \quad , \tag{5.9.24}$$

and we already know that the local boundary condition (5.9.20) implies the vanishing of F_{ij} on S^3. In view of the properties of the transverse vector harmonics (Gerlach and Sengupta 1978), this is finally found to imply that $f_n(\tau = a) = 0$, $\forall n \geq 2$. Therefore we end up with the same boundary conditions studied by Louko, and we obtain his same PDF value for $\zeta(0)$. However, choosing the (+) sign on the right-hand side of (5.9.20), we get a different PDF result. In fact, in such a case, (5.9.21) is replaced by : $E_j(a) + iB_j(a) = -E_j(a) + iB_j(a)$, which implies : $E_j(a) = 0$, $\forall j = 1,2,3$. But we know that E_j is obtained by means of :

$$F_{0j}(x,\tau) = \sum_{n=2}^{\infty} \dot{f}_n(\tau) S_j^{(n)}(x) \quad . \tag{5.9.25}$$

Thus, the vanishing of $E_j(a)$, $\forall j = 1,2,3$, implies the vanishing of $\dot{f}_n(a)$, $\forall n \geq 2$. In other words, setting $\tau = t$, we have to study the generalized zeta-function for the eigenvalues of the operator (Louko 1988b) :

$$D_n \equiv -\frac{1}{t}\frac{d}{dt}\left(t\frac{d}{dt}\right) + \frac{n^2}{t^2} \quad , \tag{5.9.26}$$

whose eigenfunctions obey the boundary conditions : $u_n(0) = 0, \dot{u}_n(a) = 0, \ \forall n \geq 2$. Indeed, a comment is now in order. Given the differential equation :

$$x'' + \frac{(1-2\alpha)}{z}x' + \left(\beta^2 + \frac{(\alpha^2 - \nu^2)}{z^2}\right) x = 0 \quad , \tag{5.9.27}$$

the solutions regular at $z = 0$ are of the form : $x = z^{\alpha} J_{\nu}(\beta z)$ (Gradshteyn and Ryzhik 1965). For the operator appearing in (5.9.26), one has : $\alpha = 0, \beta = \sqrt{E}, \nu = n$, so that its regular eigenfunctions (omitting a constant) are : $u_n(t) = J_n(\sqrt{E}t)$, where $E > 0$ are the eigenvalues. We now recall that the eigenfunctions of D_n are required to be the modes for the expansion of the field strength F_{ij} : $u_n(t) = f_n(t)$. This finally leads to the boundary conditions : $u_n(0) = 0, \dot{u}_n(a) = 0, \ \forall n \geq 2$, so that the eigenvalue condition is $\dot{J}_n(\sqrt{E}a) = 0, \ \forall n \geq 2$. Moreover, we know that for spin $\frac{1}{2}$ both the eigenvalue equations and the boundary conditions involve the field (section 5.8). But for spin > 1, Hawking's local boundary conditions involve the field strengths, whereas the action integral is not expressed in terms of the field strengths. Thus, the relation between eigenfunctions of differential operators and perturbative modes for field strengths should be derived in a case-by-case analysis.

When we take the Laplace transform of the heat equation, we know that $J_n(\sqrt{E}t)$ is replaced by the linear combination $AI_n(\sigma t) + BK_n(\sigma t)$. The ratio $\frac{A}{B}$ is found by requiring that :

$$\frac{d}{dt}\Big(AI_n(\sigma t) + BK_n(\sigma t)\Big)(a) = 0 \quad , \quad \forall n \geq 2 \quad , \tag{5.9.28}$$

which takes into account the eigenvalue condition $\dot{J}_n(\sqrt{E}a) = 0$. The linear combinations of modified Bessel functions are multiplied by t, where this factor is determined from the integration in the heat kernel, as explained at the end of section 4.3. Thus the Laplace transform of the kernel of the heat equation for spin 1 when $\dot{u}_n(a) = 0, \ \forall n \geq 2$, is an infinite sum of products G_n (cf. (4.3.13)) of functions $\widetilde{G}_n$ of the type : $\widetilde{G}_n = T\Big(AI_n(\sigma T) + BK_n(\sigma T)\Big)$. More precisely, using again the notation : $t_< = min(t,t')$, $t_> = max(t,t')$, we find that : $G_n(t,t',\sigma^2) = \widetilde{G}_n(t_<,\sigma^2)\widetilde{G}_n(t_>,\sigma^2)$, where :

$$\widetilde{G}_n(t_<,\sigma^2) = t_< I_n(\sigma t_<) \quad , \tag{5.9.29}$$

$$\widetilde{G}_n(t_>,\sigma^2) = t_> \left(K_n(\sigma t_>) - \frac{K'_n(\sigma a)}{I'_n(\sigma a)} I_n(\sigma t_>)\right) \quad . \tag{5.9.30}$$

This implies that the free part of the heat kernel is equal to the one found in Louko 1988b, and hence does not contribute to $\zeta(0)$. We have therefore to study the interacting part :

$$G^{int}(\sigma^2) = -\sum_{n=2}^{\infty}\left(n^2-1\right)\frac{K'_n(\sigma a)}{I'_n(\sigma a)}\left[\left(1+\frac{n^2}{\sigma^2 a^2}\right)I_n^2(\sigma a)\right.$$
$$\left. -I'_n(\sigma a)K'_n(\sigma a)\frac{I_n(\sigma a)}{K_n(\sigma a)} - \frac{I'_n(\sigma a)}{\sigma a K_n(\sigma a)}\right]$$
$$= -\sum_{n=2}^{\infty}\left(n^2-1\right)\left(\frac{K'_n(\sigma a)}{K_n(\sigma a)}\frac{I_n(\sigma a)}{I'_n(\sigma a)}\right) f(n;\sigma a) \quad , \tag{5.9.31}$$

where $f(n;\sigma a)$ is the function at first defined in (4.1.18). We have thus to work out the uniform asymptotic expansions of the various terms on the right-hand side of (5.9.31) according to the relations (4.4.13-22). In so doing, defining again : $\tau \equiv \frac{n}{\sqrt{n^2+\sigma^2a^2}}$, and setting $a=1$ for simplicity, we find :

$$\frac{K'_n(\sigma)}{K_n(\sigma)}\frac{I_n(\sigma)}{I'_n(\sigma)} \sim -\left(1+\frac{\gamma_1(\tau)}{n}+\frac{\gamma_2(\tau)}{n^2}+\frac{\gamma_3(\tau)}{n^3}+\ldots\right)$$
$$\left(1+\frac{\delta_1(\tau)}{n}+\frac{\delta_2(\tau)}{n^2}+\frac{\delta_3(\tau)}{n^3}+\ldots\right) \quad , \tag{5.9.32}$$

where :

$$\gamma_1(\tau) = \frac{\tau}{2}\left(1-\tau^2\right) \quad , \quad \gamma_2(\tau) = -\frac{\tau^2}{8}\left(1-\tau^2\right)\left(1-5\tau^2\right) \quad , \tag{5.9.33}$$

$$\gamma_3(\tau) = \frac{\tau^3}{8}\left(1-\tau^2\right)\left(1-12\tau^2+15\tau^4\right) \quad , \tag{5.9.34}$$

$$\delta_1(\tau) = \frac{\tau}{2}\left(1-\tau^2\right) \quad , \quad \delta_2(\tau) = \frac{\tau^2}{8}\left(1-\tau^2\right)\left(3-7\tau^2\right) \quad , \tag{5.9.35}$$

$$\delta_3(\tau) = \frac{\tau^3}{8}\left(1-\tau^2\right)\left(3-20\tau^2+21\tau^4\right) \quad . \tag{5.9.36}$$

In deriving (5.9.33-36), we have also used the well-known relations for the inverse of an asymptotic series (Abramowitz and Stegun 1964, page 15). Thus we find :

$$\frac{K_n'(\sigma)}{K_n(\sigma)}\frac{I_n(\sigma)}{I_n'(\sigma)} \sim \left[A_0(\tau)+\frac{A_1(\tau)}{n}+\frac{A_2(\tau)}{n^2}+\frac{A_3(\tau)}{n^3}+...\right] \quad , \tag{5.9.37}$$

where :

$$A_0(\tau)=-1 \quad , \quad A_1(\tau)=-\delta_1-\gamma_1=-\tau\Bigr(1-\tau^2\Bigr) \quad , \tag{5.9.38}$$

$$A_2(\tau)=-\delta_2-\gamma_1\delta_1-\gamma_2=-\frac{\tau^2}{2}\Bigr(1-\tau^2\Bigr)^2 \quad , \tag{5.9.39}$$

$$A_3(\tau)=-\delta_3-\gamma_1\delta_2-\gamma_2\delta_1-\gamma_3=-\frac{\tau^3}{16}\Bigr(1-\tau^2\Bigr)\Bigr(10-68\tau^2+74\tau^4\Bigr) \ . \tag{5.9.40}$$

Therefore, since : $f(n;\sigma) \sim \frac{\sqrt{n^2+\sigma^2}}{\sigma^2}\Bigr(\frac{B_1(\tau)}{n}+\frac{B_2(\tau)}{n^2}+\frac{B_3(\tau)}{n^3}+\frac{B_4(\tau)}{n^4}+...\Bigr)$, in light of the relations (4.4.12) and (5.9.37-40) we find :

$$\frac{K_n'(\sigma)}{K_n(\sigma)}\frac{I_n(\sigma)}{I_n'(\sigma)}f(n;\sigma) \sim \frac{\sqrt{n^2+\sigma^2}}{\sigma^2}\left(\frac{C_1(\tau)}{n}+\frac{C_2(\tau)}{n^2}+\frac{C_3(\tau)}{n^3}+\frac{C_4(\tau)}{n^4}+...\right) \quad , \tag{5.9.41}$$

where :

$$C_1(\tau)=A_0B_1=-\frac{\tau}{2}\Bigr(1-\tau^2\Bigr) \quad , \tag{5.9.42}$$

$$C_2(\tau)=A_0B_2+A_1B_1=\frac{\tau^2}{2}\Bigr(1-\tau^2\Bigr)\Bigr(2\tau^2-1\Bigr) \quad , \tag{5.9.43}$$

$$C_3(\tau)=A_0B_3+A_1B_2+A_2B_1=-\frac{\tau^3}{8}\Bigr(1-\tau^2\Bigr)\Bigr(3-20\tau^2+21\tau^4\Bigr) \quad , \tag{5.9.44}$$

$$C_4(\tau)=A_0B_4+A_1B_3+A_2B_2+A_3B_1=-\frac{\tau^4}{16}\Bigr(1-\tau^2\Bigr)\Bigr(9-122\tau^2+301\tau^4-196\tau^6\Bigr). \tag{5.9.45}$$

Note that we have not computed additional terms in our series because they give a contribution equal to $O(\sqrt{t})$, which thus does not affect the $\zeta(0)$ value. From (5.9.31) and

(5.9.41-45), taking the inverse Laplace transforms, we find (from now on, t is not the variable appearing in (5.9.26)) :

$$G^{int}(t) \sim -\sum_{n=2}^{\infty}\left(n^2-1\right)\sum_{i=1}^{4}\widetilde{f_i}(n,t)+O(\sqrt{t}) \quad , \tag{5.9.46}$$

where :

$$\widetilde{f_1}(n,t) = -\frac{1}{2}e^{-n^2t} \quad , \tag{5.9.47}$$

$$\widetilde{f_2}(n,t) = \frac{4}{3}\frac{t^{\frac{3}{2}}}{\sqrt{\pi}}n^2e^{-n^2t} - \sqrt{\frac{t}{\pi}}e^{-n^2t} \quad , \tag{5.9.48}$$

$$\widetilde{f_3}(n,t) = -\frac{3}{8}te^{-n^2t} + \frac{5}{4}t^2n^2e^{-n^2t} - \frac{7}{16}t^3n^4e^{-n^2t} \quad , \tag{5.9.49}$$

$$\widetilde{f_4}(n,t) = -\frac{3}{4}\frac{t^{\frac{3}{2}}}{\sqrt{\pi}}e^{-n^2t} + \frac{61}{15}\frac{t^{\frac{5}{2}}}{\sqrt{\pi}}n^2e^{-n^2t} - \frac{301}{105}\frac{t^{\frac{7}{2}}}{\sqrt{\pi}}n^4e^{-n^2t} + \frac{392}{945}\frac{t^{\frac{9}{2}}}{\sqrt{\pi}}n^6e^{-n^2t} . \tag{5.9.50}$$

The interacting part $G^{int}(t)$ is an even function of n, and we can thus compute its contribution to $\zeta(0)$ using the Watson transform (cf. (5.5.17)). Again, the poles of the integrand at 0 and ± 1 are excluded because the sum over all n in (5.9.46) only starts from $n=2$. The poles at ± 1 do not contribute because the integrand of the Watson transform has zeros at ± 1. Doing the calculation as in (5.5.18), we find in our case that the constant contribution C arising from the poles is given by :

$$C = L_I^C\left[-\frac{1}{2}\widetilde{f}(0,\sigma)\right] = \frac{1}{4} \quad , \tag{5.9.51}$$

so that (5.9.46-51) yield :

$$G^{int}(t) \sim \left[\frac{1}{4}+O(\sqrt{t})-\int_0^{\infty}(\nu^2-1)\sum_{i=1}^{4}\widetilde{f_i}(\nu,t)d\nu\right]$$

$$\sim \frac{\sqrt{\pi}}{8}t^{-\frac{3}{2}} - \frac{1}{4}t^{-1} - \frac{55}{256}\sqrt{\pi}t^{-\frac{1}{2}} - \frac{1}{6} - \frac{1}{90} + \frac{1}{4} + O(\sqrt{t}) \quad , \tag{5.9.52}$$

where the contributions $-\frac{1}{6}$ and $-\frac{1}{90}$ are due to (5.9.48) and (5.9.50) respectively. We thus find the PDF value :

$$\zeta_E(0) = \frac{13}{180} \quad . \tag{5.9.53}$$

Finally, in the case of a complex scalar field ϕ, local boundary conditions (Hawking 1983) require that on S^3 :

$$\phi = \pm \overline{\phi} \quad , \tag{5.9.54}$$

$$_e n^{AA'} \nabla_{AA'} \phi = \mp \, _e n^{BB'} \nabla_{BB'} \overline{\phi} \quad . \tag{5.9.55}$$

Thus, writing $\phi = \phi_1 + i\phi_2$, where ϕ_1 and ϕ_2 are real, and choosing the (+) sign in (5.9.54) and the (−) sign in (5.9.55), we obtain the boundary conditions : $\phi_2 = 0$, $\frac{\partial \phi_1}{\partial n} = 0$ on S^3. The converse holds if we make the alternative choice of sign. In other words, the imaginary part obeys Dirichlet boundary conditions, and the real part obeys Neumann boundary conditions, or viceversa. In the case of Dirichlet boundary conditions, we already know that (Kennedy 1978) : $\zeta(0) = -\frac{1}{180}$. In the case of Neumann boundary conditions, the derivatives of the modes $p_n(t)$ appearing in the expansion :

$$\phi(x,t) = \sum_{n=1}^{\infty} p_n(t) Q^{(n)}(x) \quad , \tag{5.9.56}$$

vanish on S^3. Now, using the form of the Laplacian for scalars (Schleich 1985) and denoting by E its eigenvalues, one finds the following differential equation :

$$\left[\frac{d^2}{dt^2} + \frac{3}{t}\frac{d}{dt} + \left(E + \frac{(1-n^2)}{t^2} \right) \right] p_n(t) = 0 \quad , \tag{5.9.57}$$

whose solutions, using (5.9.27), are found to be : $p_n(t) = t^{-1} J_n(\sqrt{E}t)$. Thus the eigenvalue condition is :

$$-J_n(\sqrt{E}a) + \sqrt{E}a \dot{J}_n(\sqrt{E}a) = 0 \quad , \quad \forall n \geq 1 \quad . \tag{5.9.58}$$

This eigenvalue condition has been studied in detail in Moss 1989 (see also our section 7.3), which deals with the more general case of Robin boundary condition (our Neumann boundary condition can be seen as a special case). Thus, the result for $\zeta(0)$ can be derived from Moss 1989 setting $\psi = 0$ in the four-dimensional formula (36) of that paper, which implies (see the end of section 7.3) :

$$\zeta^N(0) = -\frac{1}{180} + \frac{1}{6} = \frac{29}{180} \quad , \tag{5.9.59}$$

where N means obviously Neumann. The full $\zeta(0)$ for spin 0 is then given by :

$$\zeta(0) = \zeta^D(0) + \zeta^N(0) = -\frac{1}{180} + \frac{29}{180} = \frac{7}{45} \quad . \tag{5.9.60}$$

In fact, writing $\phi = Re\ \phi + iIm\ \phi = \phi_1 + i\phi_2$, the full path integral is given by the product of the path integral involving ϕ_1 and the one involving ϕ_2. The result (5.9.60) should be compared with our new PDF result (but see now section 6.5) :

$$\zeta_B(0) = -\frac{77}{180} \quad , \quad \zeta_E(0) = \frac{13}{180} \text{ for spin } 1 \quad . \tag{5.9.61}$$

Now, had supersymmetry been *working*, the value of $\zeta(0)$ should be the same for all values of spin. Thus in particular $\zeta(0)$ for spin 0 should be equal to $\zeta(0)$ for spin 1, whereas it turns out this is not true. This in turn seems to be a preliminary indication that there is not one-loop finiteness for the amplitudes of our problem in the presence of a S^3 boundary, even using local boundary conditions involving field strengths and normals. An important question to be addressed is therefore : why spin-lowering and spin-raising operators (cf. sections (5.6-7)) do not lead to a matching of the $\zeta(0)$ values ? We have not a complete and rigorous answer to this question. However, there is an important calculation which can be done. Let us compare the scalar field and the spin-$\frac{1}{2}$ field. The massless free-field equation for ϕ_A and $\widetilde{\phi}_{A'}$ is :

$$\nabla^{AA'}\phi_A = 0 \quad , \quad \nabla^{AA'}\widetilde{\phi}_{A'} = 0 \quad . \tag{5.9.62}$$

We know that theorem 5.7.1 can be shown to hold also for the comparison spin 0 vs spin $\frac{1}{2}$, and the preservation conditions are again given by (5.7.15-17). However, in computing

$\zeta(0)$ for spin $\frac{1}{2}$ we are no more studying (5.9.62), but the system of eigenvalue equations (cf. (5.8.56-57)) :

$$\nabla^{AA'}\phi_A = \lambda\widetilde{\phi}^{A'} \quad , \quad \nabla^{AA'}\widetilde{\phi}_{A'} = \lambda\phi^A \quad . \tag{5.9.63}$$

Let us denote by $\square$ the operator : $\nabla^{AA'}\nabla_{AA'}$, and by ω^L a solution of the twistor equation in flat Euclidean space. We now ask the question : given ϕ_L obeying (5.9.63), which is the equation obeyed by $\square\left(\omega^L\phi_L\right)$? This is important in understanding what happens to the eigenvalues for the problem involving the scalar field $\omega^L\phi_L$. Are they obtained just squaring up λ appearing in (5.9.63) ? Indeed, one finds :

$$\square\left(\omega^L\phi_L\right) = 2\left(\nabla_{AA'}\omega^L\right)\left(\nabla^{AA'}\phi_L\right) + \left(\nabla^{AA'}\nabla_{AA'}\omega^L\right)\phi_L$$
$$+\,\omega^L\left(\nabla^{AA'}\nabla_{AA'}\phi_L\right) \quad . \tag{5.9.64}$$

In view of (5.9.63), we expect the first and third term on the right-hand side of (5.9.64) to modify the result. In fact, using (5.7.9) we find :

$$\square\phi(x) = \lambda^2\phi(x) + \lambda G(x) \quad , \tag{5.9.65}$$

where :

$$\phi(x) \equiv \omega^L\phi_L \quad , \quad G(x) \equiv -2i\widetilde{\phi}^{A'}\pi^o_{A'} \quad . \tag{5.9.66}$$

Using the same technique, we also find the equation :

$$\square\widetilde{\phi}(x) = \lambda^2\widetilde{\phi}(x) + \lambda H(x) \quad , \tag{5.9.67}$$

where :

$$\widetilde{\phi}(x) \equiv \widetilde{\omega}^{L'}\widetilde{\phi}_{L'} \quad , \quad H(x) \equiv -2i\phi^A\widetilde{\pi}^o_A \quad . \tag{5.9.68}$$

Therefore the eigenvalues for spin 0 and spin $\frac{1}{2}$ are not the same. Our argument does not yield the value of $\zeta(0)$ for spin $\frac{1}{2}$, but at least it proves why the eigenvalues are not the same even though ω^D is available. The basic reason is that (5.9.62) is replaced by (5.9.63), and ω^D and $\widetilde{\omega}^{D'}$ are given by (5.7.9-10). Going to higher spins, a similar comparison becomes much harder, but maybe more work could be attempted along the lines of section 2 of Hawking and Pope 1978. In general, one has to construct covariantly constant spinors after a careful definition of a suitable derivative operator, and then use these spinors to realize a mapping between eigenfunctions for bosonic and fermionic fields whose spins differ by $\frac{1}{2}$ (we are grateful to Dr. G. Gibbons for helping us in understanding this point).

5.10 Preservation in Time of the Gauge Constraint

We now clarify a point which will play a role also in sections 7.2-3. In the previous section, following Louko 1988b, we wrote the gauge constraint in the form :

$$A_0 \approx 0 \quad , \quad {A_k}^{|k} \approx 0 \quad .$$

More precisely, what we have done is to impose the Coulomb gauge constraint :

$${A_k}^{|k} \approx 0 \quad . \tag{5.10.1}$$

Requiring the preservation in time of (5.10.1) we finally obtain :

$$A_0 \approx 0 \quad . \tag{5.10.2}$$

In fact, the expansion of A_0 on the three-sphere is (Louko 1988b) :

$$A_0(x,t) = \sum_{n=1}^{\infty} r_n(t) Q^{(n)}(x) \quad , \tag{5.10.3}$$

where $Q^{(n)}(x)$ are the scalar harmonics. The expansion of A_k has already been written in (5.9.22). Thus, the expansion of the Lorentzian action I yields (Louko 1988b) :

$$I = \sum_{n=2}^{\infty} \int L_n dt \quad , \tag{5.10.4}$$

$$L_n = \frac{a}{2N} \left(\dot{f}_n^2 + \frac{\dot{g}_n^2}{(n^2-1)} + (n^2-1) r_n^2 - 2 r_n \dot{g}_n \right) - \frac{N}{2a} n^2 f_n^2 \quad . \tag{5.10.5}$$

From (5.10.5) we have the primary constraint : $\pi_{r_n} \approx 0, \forall n \geq 2$. Moreover, the gauge constraint (5.10.1) can be written in the form : $g_n \approx 0, \forall n \geq 2$. For our present purposes, we can treat g_n on the same footing of π_{r_n}, adding both of them to the Legendre transform of L_n with some coefficients to be determined (whose values are not strictly needed here). The effective Hamiltonian $\widetilde{H}_n$ is thus given by :

$$\widetilde{H}_n \equiv H_n + \lambda_1 \phi_n^1 + \lambda_2 \phi_n^2 \quad , \tag{5.10.6}$$

where :

$$H_n \equiv \frac{N}{2a}\left[(n^2-1)\pi_{g_n}^2 + \left(\pi_{f_n}^2 + n^2 f_n^2\right)\right] + (2n^2-3) r_n \pi_{g_n} \quad , \tag{5.10.7}$$

$$\phi_n^1 \equiv \pi_{r_n} \quad , \quad \phi_n^2 \equiv g_n \quad . \tag{5.10.8}$$

Requiring the preservation in time of ϕ_n^1 and ϕ_n^2 (Dirac 1964) we obtain the secondary constraints :

$$\phi_n^3 \equiv \left\{\phi_n^1, \widetilde{H}_n\right\} = (3-2n^2)\pi_{g_n} \quad , \tag{5.10.9}$$

$$\phi_n^4 \equiv \left\{\phi_n^2, \widetilde{H}_n\right\} = \frac{N}{a}(n^2-1)\pi_{g_n} + (2n^2-3) r_n \quad . \tag{5.10.10}$$

Thus, when all constraints are preserved in time one finds :

$$\pi_{g_n} \approx 0 \quad \forall n \geq 2 \quad , \quad r_n \approx 0 \quad \forall n \geq 2 \quad , \tag{5.10.11}$$

in addition to : $\pi_{r_n} \approx 0, \forall n \geq 2$, and $g_n \approx 0, \forall n \geq 2$. The insertion of (5.10.11) into (5.10.3) finally yields (5.10.2) as we claimed (note that r_1 does not affect the action I of (5.10.4), since $\frac{a}{2N}(n^2-1)$ vanishes at $n=1$ (cf. (5.10.5))).

CHAPTER SIX

GHOST FIELDS AND GAUGE MODES IN ONE-LOOP QUANTUM COSMOLOGY

Abstract. At first we briefly review the geometric formulation of gauge theories and the problems one faces in trying to quantize them. We then apply the extended phase-space method of Batalin, Fradkin and Vilkovisky to the spin-1 field, which is described by a constrained Hamiltonian system with first-class constraints. The charge Q and the gauge-fixed action are derived. The Lorentzian path integral is restricted to the trajectories of the extended phase space which satisfy the boundary conditions which project on the original phase space. This path integral is independent of gauge fixing provided the charge Q remains nilpotent on quantization, and the Faddeev-Popov formula is formally derived from the path integral involving the gauge-fixed action.

We then study the effect of ghost fields and gauge-fixing terms in the path integral for linearized gravity. The quantum theory is only well-defined when expressed in terms of its physical degrees of freedom, the transverse-traceless modes. One can formally show that a suitable measure exists such that the gauge-invariant form of the path integral for the ground-state wave function is equal to the one expressed in terms of the physical degrees of freedom only. However, it remains to be seen whether this result can be generalized to the boundary-value problems which occur in quantum cosmology. In fact different quantization techniques for gauge fields have led to discrepancies in the literature. As an example, we show how to obtain the Moss and Poletti result, according to which magnetic and electric boundary conditions for electromagnetic fields lead to the same result for $\zeta(0)$, thus correcting the PDF values of chapter five. This is obtained by using an *indirect* technique, where $\zeta(0)$ is expressed in terms of functions of the geometrical objects of the problem, multiplied by coefficients which take into account the contribution of all degrees of freedom (PDF and gauge) and of the ghost field.

We finally study the problem of the corresponding *direct* $\zeta(0)$ calculation for spin 1 using the Faddeev-Popov formula for perturbation theory. In the Lagrangian of the Lorentzian action, the gauge-averaging term $-\frac{\Phi^2}{2\alpha}$ is added, where Φ is a linear function of the covariant derivatives of (A_0, A_k), α is a constant, and at the end the whole action is analytically continued to the Euclidean time variable $\tau = it$. The PDF contribution decouples from the ghost and gauge contribution. The radial component A_0 of the vector potential is expanded on S^3 in terms of R_n-modes and scalar harmonics, and the spatial components A_k in terms of f_n-modes (the PDF multiplying transverse vector harmonics) and g_n-modes (multiplying longitudinal vector harmonics). When the ghost field obeys Dirichlet boundary conditions, remaining boundary conditions compatible with BRST invariance are that f_n-modes vanish on S^3 (magnetic case), together with g_n-modes and first derivatives of R_n-modes (vanishing on S^3 of A_k and gauge-fixing term). Viceversa, if the ghost field obeys Neumann boundary conditions, we should require the vanishing on S^3 of $\dot{f}_n$, together with $\dot{g}_n$ and R_n-modes separately. The final result is independent of α, and the calculation is easier in the $\alpha = 1$ gauge. In that case, two second-order differential operators appear in the Euclidean action for gauge modes, whose eigenfunctions are of the kind $J_{\sqrt{n^2-1}}(a\tau)$ and $\tau^{-1}J_n(b\tau)$ respectively. However, g_n-modes and R_n-modes obey a coupled system of two second-order ordinary differential equations. This coupling seems to be the effect of the 3 + 1 split of the vector potential. In the $\alpha = 1$ gauge it is easy to obtain the decoupled fourth-order differential equations obeyed by g_n-modes and R_n-modes. The boundary conditions previously mentioned and regularity at the origin lead in principle to the complete knowledge of the solution, which can be written in the form : $g_n(\tau) = \tau^{\mu} \sum_{k=0}^{\infty} a_{n,k}\tau^k$, $\forall n \geq 2$. In fact μ obeys a fourth-order algebraic equation, and the $a_{n,k}$ coefficients obey very involved recurrence relations, which show that only half of them are nonvanishing. However, the $3+1$ split of the vector potential, which is needed for a *direct* $\zeta(0)$ calculation, does not seem to lead to a formula for $g_n(\tau)$ and $R_n(\tau)$ expressible in terms of special functions. Thus the full $\zeta(0)$ in the presence of boundaries, corrected for the effect of ghost fields and gauge-averaging terms, has not yet been computed using *direct* techniques.

6.1 Main Ideas About Gauge Theories and Their Quantization

Gauge theories are at the core of modern classical and quantum field theory. We now summarize some relevant geometrical (Poénaru and Tanasi 1986, Atiyah 1988b) and quantum aspects of these theories.

6.1.1.1 Geometry

The fundamental elements of a gauge theory are the total space of a fibre bundle, a structure group, a connection form, a curvature form and Bianchi identities. The correspondence between these mathematical concepts and the familiar physical language is as follows :

Total Space of a Fibre Bundle → Phase-Factor Space,

Structure Group → Gauge Group G ,

Connection Form → Gauge Potential A_μ ,

Curvature Form → Gauge Field $F_{\mu\nu}$,

Bianchi Identities → Faraday's Law.

In simpler terms, we consider a compact Lie group G with Lie algebra $L(G)$. A gauge potential is a set of $L(G)$-valued functions $A_\mu(x)$, and the gauge field $F_{\mu\nu}$ is initially introduced as the commutator of covariant derivatives ∇_μ defined as follows :

$$\nabla_\mu \equiv \partial_\mu + A_\mu \quad , \tag{6.1.1}$$

so that :

$$F_{\mu\nu} \equiv 2\partial_{[\mu} A_{\nu]} + [A_\mu, A_\nu] \quad . \tag{6.1.2}$$

As far as the general theory is concerned, we can write all relations in terms of the connection A and of the curvature $F_A = F$, as one does in modern differential geometry. The theory is invariant under the following gauge transformations g :

$$A' = g^{-1} \nabla g \quad , \tag{6.1.3}$$

$$F' = g^{-1} F g \quad , \tag{6.1.4}$$

where g is a G-valued function. We say that (6.1.3-4) express the affine nature of A_μ and the tensorial nature of $F_{\mu\nu}$ respectively.

6.1.1.2 Quantum Theory

In a relativistic quantum field theory, the basic objects are Green's functions. From the mathematical point of view, we know (Streater and Wightman 1964) that Green's functions are boundary values of holomorphic functions, and we also know under which conditions a set of tempered distributions can be seen as the Green's functions of the theory (but the suitable generalization of Wightman's axioms to gauge theories is a hard task). In modern terms (we are here interested in Minkowski space), one uses the path-integral formalism (though affected by divergence problems) in defining the generating functional $W[J]$ of Green's functions and the corresponding effective action.

In the case of gauge theories, the measure problem in the functional integral is even more dramatic because of the invariance under (6.1.3-4). In fact, integrating over all field configurations, one integrates infinitely many times over the volume of the gauge group, whereas we need to concentrate the measure over a subset of configurations containing a single point for each orbit of the gauge group. This is achieved using the Faddeev-Popov method, which leads to the introduction of anticommuting ghost fields $c(x)$ and $c^*(x)$. The geometrical interpretation of ghost fields can be obtained as follows (Bonora and Cotta-Ramusino 1983). We denote by E the Lie algebra of the group G of gauge transformations. The anticommuting nature of ghosts shows that they are E-valued differential forms. Moreover, the BRST transformation :

$$c^a(x) \rightarrow c^a(x) - \frac{\lambda}{2} f^{abd} c_b(x) c_d(x) \quad ,$$

is a relation of the Maurer-Cartan type. This leads to the identification of ghost fields with the Maurer-Cartan form ω on the group G, i.e. the left-invariant E-valued one-form on G such that : $\omega(\xi) = \xi^*$. Here, ξ is a left-invariant vector field on G, and ξ^* is the element of the Lie algebra E corresponding to ξ.

For example, when the relativistic gauge condition is imposed :

$$\partial \cdot A \equiv \partial_\mu A^\mu \approx 0 \quad , \tag{6.1.5}$$

and defining :

$$L_{eff} \equiv -\frac{F^2}{4} - \frac{1}{2\alpha}(\partial \cdot A)^2 + (\partial c^* \cdot \widetilde{D} c) \quad , \tag{6.1.6}$$

the generating functional $W[J]$ assumes the form :

$$W[J] = \lim_{\alpha \to 0} \int DA \, Dc \, Dc^* e^{\left[i \int (L_{eff} + J \cdot A) d^4x\right]} \quad . \tag{6.1.7}$$

In (6.1.6), $\widetilde{D}$ is the covariant derivative acting on ghosts, and $F^2 = F_{\mu\nu}F^{\mu\nu}$. The formalism must be independent of the gauge condition chosen, and requiring (6.1.5) this is only achieved using the Hamiltonian theory of Batalin, Fradkin and Vilkovisky (hereafter referred to as BFV). Now, denoting by λ the Lagrange multiplier, (6.1.5) can be seen to be of the form : $\dot{\lambda} \approx \partial_i A^i$. In the BFV theory, the Lagrange multipliers of Dirac's theory are regarded as dynamical degrees of freedom, enlarging the phase space with n new couples of coordinates, i.e. the Lagrange multipliers and their conjugate momenta. The advantages of this formalism are :

(1) the imposition of the constraints is reduced to a single condition;

(2) the Hamiltonian nature of the formalism ensures a priori the unitarity of the S matrix;

(3) the functional $W[J]$ is independent of gauge conditions (thus including relativistic gauges).

Recent work on the geometrical derivation of the Faddeev-Popov ansatz can be found in Ellicott et al. 1989. We now study the extended phase space for electromagnetism.

6.2 Extended Phase Space for the Spin-1 Field

In view of (6.1.6-7), the formula for the prefactor for spin 1 can be obtained by analytic continuation to the Euclidean time variable $\tau = it$ of the following path-integral formula :

$$\lim_{\alpha \to 0} \int DA \, Dc \, Dc^* e^{\left[i \int \left(-\frac{F^2}{4} - \frac{1}{2\alpha}\left(A_{\mu}^{\ |\mu}\right)^2 + \partial_\mu c^* \partial^\mu c\right) d^4x\right]} \quad , \tag{6.2.1}$$

where $\int -\frac{F^2}{4} d^4x$ is given in (5.10.4-5). In studying this problem in a way as general as possible, it is useful to apply the Hamiltonian BFV technique. In fact, we know (Hanson et al. 1976) that the electromagnetic field can be seen as a constrained Hamiltonian system with first-class constraints ψ^a. Setting $L \equiv -\frac{F^2}{4}$, $G^\mu \equiv \frac{\delta L}{\delta \dot{A}_\mu}$, one finds :

$$\psi^1 = G^0 \quad , \qquad \psi^2 = div\, \vec{G} = -div\, \vec{E} \quad , \tag{6.2.2}$$

so that the extended Hamiltonian is (cf. (2.1.16)) :

$$H = \int \left(\frac{E^2}{2} + \frac{B^2}{2} - \dot{A}^0 G^0 - A^0 div\, \vec{E} \right) d^3x \quad . \tag{6.2.3}$$

6.2.1.1 Classical Theory

We can now write H as the Legendre transform H_c of L, plus a linear combination $l_a \psi^a$ of all first-class constraints. Thus the Hamiltonian form I_H of the action is :

$$I_H = \int_{t'}^{t''} \left[p_j \dot{q}^j - H_c(p,q) - l_a \psi^a(p,q) \right] dt \quad , \tag{6.2.4}$$

where $a = 1, 2$. According to the BFV method (Bimonte 1988), we now enlarge the phase space with 2 couples of new canonical coordinates, with the same parity of the constraints (L^a, l_a), and 4 canonical couples (M^a, m_a) and (R^a, r_a), the ghost fields, having parity of the opposite sign. The only nonvanishing Poisson brackets among these fields are :

$$\left\{ l_a, L^b \right\} = \left\{ m_a, M^b \right\} = \left\{ r_a, R^b \right\} = \delta_a^{\ b} \quad , \qquad \left\{ q^i, p_j \right\} = \delta^i_{\ j} \; . \tag{6.2.5}$$

The central object of the theory is the charge Q :

$$Q \equiv \psi^a m_a + L^a r_a \quad . \tag{6.2.6}$$

Note that (6.2.6) only holds for $U(1)$ gauge theory. In the nonAbelian case, the charge Q also contains some nonvanishing structure constants (Henneaux 1985). If m_a and r_a are given a ghost number $= 1$, M^a and R^a a ghost number $= -1$, and q^j, p_j, l_a and L^a a vanishing ghost number, then Q is determined (up to a canonical transformation) by the relations :

$$gh(Q) = 1 \quad , \qquad Q^* = Q \quad , \qquad \{Q, Q\} = 0 \quad . \tag{6.2.7}$$

The gauge-fixed action I_{GF} in the extended phase space is :

$$I_{GF} = \int_{t'}^{t''} \left[p_j \dot{q}^j + L^a \dot{l}_a + M^a \dot{m}_a + R^a \dot{r}_a - H_{GF} \right] dt \quad , \tag{6.2.8}$$

where H_{GF} is the gauge-fixed Hamiltonian :

$$H_{GF} = H_c + l_a \psi^a + \{\beta, Q\} \quad , \tag{6.2.9}$$

where β is an arbitrary fermionic function of canonical coordinates, such that the gauge is suitably fixed. It has the property : $\{\{\beta, Q\}, Q\} = 0$. The charge Q is classically a constant of motion, because it has vanishing Poisson bracket with H_{GF}. The solutions to the original equations of motion coincide with the solutions in the extended phase space, provided : $Q(t') = Q(t'') = 0$.

6.2.1.2 Quantum Theory

On quantization, we consider the path integral :

$$P = \int D\mu \; e^{iI_{GF}} \quad , \tag{6.2.10}$$

where $D\mu = [Dq]\,[Dp]\,[Dl]\,[DL]\,[Dm]\,[DM]\,[Dr]\,[DR]$ is the measure invariant under canonical transformations (Liouville's measure). The canonical coordinates are subject to the boundary conditions : $l_a(t') = l_a(t'') = m_a(t') = m_a(t'') = r_a(t') = r_a(t'') = 0$, $q^j(t') = q^j_{t'}$, $q^j(t'') = q^j_{t''}$, which are needed so as to project on the original phase space, the one without ghosts. Thus (6.2.10) is restricted to the trajectories of the extended phase space which satisfy these boundary conditions.

The path integral (6.2.10) is independent of β provided Q remains nilpotent at the quantum level, so that the anticommutator of the operator version $\hat{Q}$ of Q with itself vanishes. Setting : $\beta = R^a \omega_a(p, q, l, L, m, M, r, R)$, the equations of motion for the Lagrange multipliers are the most general relativistic gauge conditions : $\dot{l}_a = \omega_a(p, q, l, L, m, M, r, R)$.

6.3 Formal Equivalence to the Faddeev-Popov Result

If we make the rescalings : $\omega^a \rightarrow \epsilon^{-1}\omega^a$, $L^a \rightarrow \epsilon L^a$, $R^a \rightarrow \epsilon R^a$, $M^a \rightarrow \epsilon M^a$, we find that (Bimonte 1988) :

$$\lim_{\epsilon \to 0} I_{GF}(\epsilon) = \int_{t'}^{t''} \left[p_j \dot{q}^j + \omega^a L_a - R_a \left\{ \omega^a, \psi^b \right\} m_b - H_c - l_a \psi^a \right] dt \quad . \tag{6.3.1}$$

In so doing, we are essentially using the argument on page 46 of Henneaux 1985, where a change of variables is made whose super-Jacobian is one, and such that the transition amplitude does not depend on the parameter involved in the transformation law in light of the BFV theory. The introduction of Dirac's delta, though not being a completely rigorous mathematical technique in the form here used, becomes now very useful. In fact, inserting (6.3.1) into (6.2.10) and using the formal identities (Bimonte 1988) :

$$\int DL \, e^{i \int \omega^a L_a \, dt} = \delta \left[\omega^a \right] \quad , \tag{6.3.2}$$

$$\int Dl \, e^{-i \int l_a \psi^a \, dt} = \delta \left[\psi^a \right] \quad , \tag{6.3.3}$$

$$\int DR \, Dm \, e^{-i \int R_a \left\{ \omega^a, \psi^b \right\} m_b \, dt} = \left| \det \left\{ \omega^a, \psi^b \right\} \right| \quad , \tag{6.3.4}$$

where ω^a is now assumed to be independent of L, R and m, we find (omitting the infinite constants due to the M and r integrations) :

$$P \cong \int Dp \, Dq \, \delta \left[\psi^a \right] \delta \left[\omega^a \right] \left| \det \left\{ \omega^a, \psi^b \right\} \right| e^{i \int \left(p_j \dot{q}^j - H_c \right) dt} \quad . \tag{6.3.5}$$

In (6.3.5), the symbol $\cong$ means that we have omitted the infinite constants and used (6.3.2-4). The formula (6.3.5) finally leads to (6.2.1) as it is well-known (Frampton 1987).

However, it should be emphasized that the outlined proofs of section 6.2 and of (6.3.5) are only formal (Halliwell 1988). In fact, we have performed a generalized canonical transformation on the variables of integration, but without reference to a particular definition of the path integral. It is indeed very difficult to implement canonical transformations in phase-space path integrals, and it is not yet clear whether these results still hold if we use a skeletonized version of the path integral (Halliwell 1988).

6.4 Physical Degrees of Freedom and Redundant Variables

In Schleich 1985 a crucial point was only briefly mentioned, i.e. the possibility of computing the prefactor P using the Hamiltonian formulation of the theory with an explicit choice of gauge. This technique has been used so far in chapters four and five. We now clarify its motivation. The known Hamiltonian methods for quantization of physical theories are (Sundermeyer 1982, Dayi 1989, Dunne et al. 1989, Romano and Tate 1989, Gitman and Tyutin 1990, Schleich 1990, Govaerts and Troost 1991, Govaerts 1991, Henneaux and Teitelboim 1992) :

(1) reduced phase-space approach;

(2) Dirac's method described in sections 2.1-3;

(3) the BFV theory outlined in section 6.2.

It is worth recalling that in a canonical theory the important distinction is not the one between primary and secondary constraints, but the one between first-class and second-class constraints (chapter two). By taking linear combinations of second-class constraints one can bring a number of them into the first-class set. In Dirac's theory, the remaining second-class constraints become equations between operators in the quantum theory, and the first-class constraints are supplementary conditions on the wave functions (chapter two). The Dirac brackets (Dirac 1964) of the classical theory are the tool needed to treat the surviving second-class constraints as strong (however, as explained in Hanson et al. 1976, a suitable choice of gauge constraints and invariant relations leads in the end to second-class constraints only).

Now, as explained in Henneaux 1985, the main formal proofs in the Hamiltonian theory of the path integral are :

(a) the BFV theory is *formally* equivalent to Dirac's theory;

(b) Dirac's theory is *formally* equivalent to the reduced phase-space method;

(c) the Faddeev-Popov determinant can be derived from the Hamiltonian path integral of the BFV theory (cf. section 6.3);

(d) the generating functional of Green's functions is gauge-invariant.

The results (a) and (b) imply the formal equivalence of the extended phase-space method to the reduced phase-space approach. As we already said in section 1.4 and chapter five, we denote by PDF the method based on the reduction of the theory to its physical degrees of freedom. In light of (a-d), one could be naively tempted to justify the PDF method, its validity arising as a corollary. Actually, what one has to do in general is to prove that quantum effects due to ghosts and to pure gauge degrees of freedom (both of them hereafter referred to as redundant variables) mutually cancel (Henneaux 1985). Following Hartle and Schleich 1987, we now describe under which conditions this result is proved for linearized gravity. The conceptual steps are as follows (Hartle and Schleich 1987).

(1) The relation between Euclidean functional integrals in the PDF method and those over redundant variables can be explicitly derived for linearized real Euclidean general relativity. In fact the physical degrees of freedom of the theory are the two transverse-traceless components of the perturbed three-metric, and the action is quadratic in the perturbative modes.

(2) Although the Euclidean action for linearized gravity is not positive-definite, a consistent quantum theory does exist. In fact the Hamiltonian expressed in terms of the physical degrees of freedom is positive, so that sum over histories and ground-state wave function are meaningful concepts. This is an example of the general result according to which the physical degrees of freedom play a key role in deciding whether the quantization of a theory can be implemented.

(3) The effect of the nonpositivity of the Euclidean action is that in the Euclidean functional integrals the action is not local in the metric perturbations (though it is local in terms of some redundant variables).

(4) The Euclidean functional integral for linearized gravity over the conformally rotated action is best understood as arising from the quantization of a theory which possesses gauge and reparametrization invariance. In other words (Hartle and Schleich 1987) :

(4a) At first we express the theory in terms of its physical degrees of freedom;

(4b) As a second step we formulate the sum over histories in terms of these degrees of freedom;

(4c) We finally add back the integration over redundant variables, so as to preserve the invariance of the theory.

The step (4c) is made possible by the existence of a suitable measure which ensures the validity of some identities involving gauge-fixing conditions and the Faddeev-Popov determinant. Thus, we decompose the metric h into conformal equivalence classes as :

$$h_{\alpha\beta} = t_{\alpha\beta} + l_{\alpha\beta} + \gamma^T_{\alpha\beta} + \gamma^L_{\alpha\beta} + 2\chi\delta_{\alpha\beta} \quad . \tag{6.4.1}$$

The decomposition (6.4.1) is uniquely fixed by requiring that the linearized scalar curvature from $(h_{\alpha\beta} - 2\chi\delta_{\alpha\beta})$ should vanish :

$$R = \nabla_\alpha \nabla_\beta \Bigl(h^{\alpha\beta} - 2\chi\delta^{\alpha\beta}\Bigr) - \nabla^2 \Bigl(h^\alpha_\alpha - 2\chi\delta^\alpha_\alpha\Bigr) = 0 \quad . \tag{6.4.2}$$

Moreover, we can choose for example the following gauge-fixing conditions :

$$C_\alpha \equiv \nabla^\beta \left[h_{\alpha\beta} - \frac{1}{2}\delta_{\alpha\beta}(\det h)\right] = 0 \quad . \tag{6.4.3}$$

Given any $a > 0$, and defining :

$$16\pi G I^g_2 \equiv \frac{1}{4}\int \Bigl[\bigl(\nabla_\alpha l_{\beta\rho} \nabla^\alpha l^{\beta\rho}\bigr) + a\bigl(\nabla_\alpha \chi \nabla^\alpha \chi\bigr)\Bigr] d^4x \quad , \tag{6.4.4}$$

the following theorem can be proved (see appendix of Hartle and Schleich 1987) :

Theorem 6.4.1 Suitable measures $\mu\Bigl[dl, d\gamma^L, d\chi\Bigr]$ and $\mu\Bigl[d\gamma^T\Bigr]$ exist such that the following identities hold :

$$1 = \int \mu\Bigl[dl, d\gamma^L, d\chi\Bigr] \delta[C^\alpha] \det\left[\frac{\delta C^\alpha}{\delta\xi^\beta}\right] e^{-I^g_2[l,\chi]} \quad , \tag{6.4.5}$$

$$1 = \int \mu\left[d\gamma^T\right]\delta[R]\det\left[\frac{\delta R}{\delta\omega}\right] \quad . \tag{6.4.6}$$

This is why the ground-state wave function for linearized gravity is given by (Hartle and Schleich 1987) :

$$\psi_0 = \int \mu\left[dh^{TT}\right]e^{-I_2[h^{TT}]} \quad , \tag{6.4.7a}$$

or equivalently :

$$\psi_0 = \int \mu\left[d\gamma, d\chi\right]\delta[C^\alpha]\delta[R]\det\left[\frac{\delta C^\alpha}{\delta\xi^\beta}\right]\det\left[\frac{\delta R}{\delta\omega}\right]e^{-\widetilde{I}_2[\gamma,\chi]} \quad , \tag{6.4.7b}$$

where :

$$16\pi G\widetilde{I}_2[\gamma,\chi] \equiv 16\pi G I_2^g + \frac{1}{4}\int \left(\nabla_\alpha t_{\beta\rho}\nabla^\alpha t^{\beta\rho}\right)d^4x \quad . \tag{6.4.8}$$

In other words, ψ_0 is given by the gauge-invariant formula (6.4.7b), which is equal to the PDF form (6.4.7a) in light of (6.4.5-6). The choice of the measures appearing in (6.4.5-6) and (6.4.7b) is enforced by the requirement of having a well-defined quantum theory of the linearized gravitational field (Hartle and Schleich 1987).

Thus, in computing semiclassical wave functions we may try to start (reversing the argument in Hartle and Schleich 1987) with gauge-invariant formulae of the kind (6.4.7b), and check whether they can be finally reduced to the PDF form (6.4.7a) using the same technique which leads to (6.4.5-6). However, it should be emphasized that it is not yet clear whether theorem 6.4.1 can be generalized to quantum cosmology. In fact, as we anticipated at the end of section 4.1, the work in Moss and Poletti 1990a and Poletti 1990 disagrees with the $\zeta(0)$ value obtained in Schleich 1985. In Moss and Poletti 1990a the following BRST transformations for the metric perturbations γ_{ab} and the ghost field ϵ_a are studied :

$$\delta\gamma_{ab} = 2\nabla_{(a}\epsilon_{b)} \quad , \quad \delta\epsilon_a = 0 \quad , \quad \delta\bar{\epsilon}_a = \nabla^b\left[\gamma_{ab} - \frac{\delta_{ab}}{2}(\det\gamma)\right] \quad . \tag{6.4.9}$$

The authors point out that Dirichlet boundary conditions on the ghost field, together with (6.4.9), imply that the normal component of the gauge-fixing term vanishes. They then derive the boundary conditions on γ_{ab} which are consistent with this requirement and with self-adjointness. These enable them to write three relations. Requiring BRST invariance of the third formula : ${P_a}^c \, n^d \gamma_{cd} = 0$, they derive the following boundary conditions on the ghost field :

$$n \cdot \epsilon = 0 \quad , \qquad \Bigl(-{k_a}^b + n \cdot \nabla\Bigr){P_a}^b \, \epsilon_b = 0 \quad . \tag{6.4.10}$$

In (6.4.10), k is the intrinsic curvature of the boundary, and ${P_a}^b$ is the tangential projection operator : ${P_a}^b \equiv {\delta_a}^b - n_a n^b$. The authors then remark that the boundary conditions (6.4.10) are not BRST-invariant because surface components change under a BRST transformation, but they say this does not affect the gauge invariance of the transition amplitude. In Moss and Poletti 1990a and Poletti 1990, the authors do not discuss the problem of whether a measure exists such that (6.4.5-6) hold in the presence of S^3 boundaries. Thus, it is not easy to understand whether the present discrepancies in the $\zeta(0)$ values can be eliminated (for the problem studied in Schleich 1985, Poletti finds $\zeta(0) = -\frac{803}{45}$ rather than $\zeta(0) = -\frac{278}{45}$). If a rigorous mathematical analysis shows that theorem 6.4.1 in Hartle and Schleich 1987 cannot be generalized to quantum cosmology in the presence of S^3 boundaries, the $\zeta(0)$ values we have derived for spin 1, $\frac{3}{2}$ and 2 and summarized in section 11.2, should be seen just as PDF parts of the full $\zeta(0)$ values. As an example, we now examine the modifications of the $\zeta(0)$ calculation for spin 1 introduced by ghost fields and gauge modes.

6.5 A More Careful Study of the Spin-1 Problem

The Fradkin-Vilkovisky theorem states that, provided half of the canonical variables are subject to suitable BRST-invariant boundary conditions, the path integral (6.2.10) is independent of gauge fixing (cf. Henneaux 1985, pp 38-47). We chose these boundary conditions to be of the Dirichlet type on ghost fields and Lagrange multipliers. Moreover, in section 6.3 we proved under which conditions the BFV formula (6.2.10) can be formally reduced to the Faddeev-Popov formula. Thus it appears possible to derive the ghost contribution to the $\zeta(0)$ calculation at least in the spin-1 case. The related work in Moss and Poletti 1990b relies on previous work for scalar fields in Melmed 1988 and Moss 1989, which was later corrected in Moss and Dowker 1989. In four dimensions, the form of $\zeta(0)$ is :

$$16\pi^2\zeta(0) = \int_M b_4 \, d\mu + \int_{\partial M} c_4 \, d\mu \quad . \tag{6.5.1}$$

For example, for a real scalar field ϕ obeying the equation : $\Big(\nabla_\mu \nabla^\mu - X\Big)\phi = 0$, in light of the work in Moss 1989 and Moss and Dowker 1989 it is known that for Dirichlet boundary conditions one has :

$$c_4^D = \alpha_D q + \beta_D g + \gamma_D \left(X - \frac{R}{6}\right) K + \epsilon_D n \cdot \nabla \left(X - \frac{R}{6}\right) \quad . \tag{6.5.2}$$

In (6.5.2), $\alpha_D, \beta_D, \gamma_D$ and ϵ_D are constant coefficients, K is the extrinsic curvature of the boundary, whereas :

$$q = -8det(K) + 4K\widetilde{r} - 8Tr(K\widetilde{r}) \quad , \tag{6.5.3}$$

$$g = Tr\Big(K^3\Big) - (TrK)Tr\Big(K^2\Big) + \frac{2}{9}(TrK)^3 \quad , \tag{6.5.4}$$

where $\widetilde{r}$ is the intrinsic curvature of the boundary. Hence c_4 contains invariants of dimension $(\text{length})^{-3}$. The generalization of this technique to the case of a spin-1 field subject to magnetic boundary conditions ($B_i = 0$ on S^3, $\forall i = 1,2,3$), for a flat Euclidean background, leads to (Moss and Poletti 1990b) :

$$16\pi^2\zeta(0) = \int_{S^3} \Big[A(TrK)^3 + B(TrK)K_{ab}K^{ab} + C K_a^{\ b} K_b^{\ c} K_c^{\ a}\Big] \, d\mu \quad . \tag{6.5.5}$$

This is why, using the relation $K_{ab} = \frac{h_{ab}}{r}$ for a three-sphere of radius r, one finds :

$$\zeta(0) = \frac{1}{8}\Big[27A + 9B + 3C\Big] \quad . \tag{6.5.6}$$

In (6.5.6), the coefficients A, B and C contain the effects of ghost fields plus physical and unphysical (gauge) degrees of freedom, and are given by (Moss and Poletti 1990b) :

$$A = -\frac{338}{945} \quad , \qquad B = \frac{16}{63} \quad , \qquad C = \frac{194}{945} \quad . \tag{6.5.7}$$

The form (6.5.7) holds for both magnetic and electric boundary conditions, and leads to :

$$\zeta_B(0) = \zeta_E(0) = \zeta(0) = -\frac{38}{45} \quad , \tag{6.5.8}$$

which substantially differs from our PDF result (5.9.61). The formula (6.5.8) shows that different quantization techniques for gauge fields can lead to substantially different $\zeta(0)$ values. Interestingly, the full $\zeta(0)$ values with magnetic and electric boundary conditions are equal. Thus it seems that the symmetries of the classical theory are also respected by the quantum amplitudes. In Moss and Poletti 1990b, this result has also been derived using cohomology theory.

In a *direct* $\zeta(0)$ calculation, we must separately compute the contributions to the full $\zeta(0)$ due to the ghost field, PDF and gauge degrees of freedom, after the analytic continuation to the Euclidean time variable $\tau = it$ of (6.2.1), where d^4x should be understood as $\sqrt{-g}d^4x$ (Louko 1988b). The Euclidean action is then obtained multiplying by $-i$ the Lorentzian action, setting $t = -i\tau$, and bearing in mind that $(A_0)_L dt = (A_0)_E d\tau$, so that the r_n-modes appearing in the Lorentzian formula (5.10.3) are related to the Euclidean R_n-modes by : $r_n = iR_n$. Moreover, for flat Euclidean backgrounds bounded by a three-sphere, we choose the gauge-averaging term $\frac{\Phi^2}{2\alpha}$, where Φ is defined as follows :

$$\Phi \equiv \frac{\partial A_0}{\partial \tau} + \tau^{-2} A_l{}^{|l} = \sum_{n=1}^{\infty} \dot{R}_n(\tau) Q^{(n)}(x) - \tau^{-2} \sum_{n=2}^{\infty} g_n(\tau) Q^{(n)}(x) \quad . \tag{6.5.9}$$

Note that in (6.5.9) the vertical stroke $\mid l$ denotes three-dimensional covariant differentiation with respect to the unit three-sphere metric. This is why the τ^{-2} factor must be included. With this choice of gauge-averaging term, the corresponding differential operator acting on R_n-modes will turn out to be the one-dimensional Laplace operator for scalars if $\alpha = 1$, as we would expect in light of the expansion (5.10.3). Thus, the part $I_E(g, R)$ of the Euclidean action quadratic in gauge modes is (see also (5.10.5)) :

$$I_E(g,R) - \frac{1}{2\alpha} \int_0^1 \tau^3 (\dot{R}_1)^2 \, d\tau = \sum_{n=2}^{\infty} \int_0^1 \left[\frac{\tau}{2(n^2-1)} \Big(\dot{g}_n - (n^2-1) R_n \Big)^2 \right.$$

$$\left. + \frac{\tau}{2\alpha} \Big(-\frac{g_n}{\tau} + \tau \dot{R}_n \Big)^2 \right] d\tau \quad , \tag{6.5.10}$$

where we have finally inserted the flat-background hypothesis : $N = 1$, $a(\tau) = \tau$. The PDF and the ghost field decouple from (6.5.10). Because we are here dealing with all degrees of freedom, we need further boundary conditions on the modes for A_0 and the whole of A_k. For example, we may set to zero on S^3 the whole of A_k : $f_n(1) = g_n(1) = 0$, $\forall n \geq 2$. This of course implies the vanishing on S^3 of the magnetic field **B**, whereas the converse does not hold, because **B** only depends on the f_n-modes. As explained in Moss and Poletti 1990a, in this case the gauge-fixing term has to vanish as well on S^3. In light of (6.5.9), this implies that $\dot{R}_n(1) = 0$, $\forall n \geq 1$. The ghost operator is then self-adjoint only if Dirichlet boundary conditions are imposed. Viceversa, if Neumann boundary conditions are chosen for the ghost field, remaining boundary conditions compatible with BRST invariance are : $\dot{f}_n(1) = 0$ and $\dot{g}_n(1) = 0$, $\forall n \geq 2$; $R_n(1) = 0$, $\forall n \geq 1$. This case is then called electric.

We now integrate by parts in (6.5.10) and use the generalized magnetic or electric boundary conditions discussed above. Thus, defining the following second-order differential operators, $\forall n \geq 2$:

$$\hat{A}_n(\tau) \equiv -\frac{d^2}{d\tau^2} - \frac{1}{\tau}\frac{d}{d\tau} + \frac{(n^2-1)}{\alpha\tau^2} \quad , \tag{6.5.11}$$

$$\hat{B}_n(\tau) \equiv \frac{1}{\alpha}\left(-\frac{d^2}{d\tau^2} - \frac{3}{\tau}\frac{d}{d\tau}\right) + \frac{(n^2-1)}{\tau^2} \quad , \tag{6.5.12}$$

we find the following fundamental result, $\forall n \geq 2$:

$$I_E^{(n)}(g,R) = \frac{1}{2}\int_0^1 \frac{\tau g_n}{(n^2-1)}(\hat{A}_n g_n)\, d\tau + \frac{1}{2}\int_0^1 \tau^3 R_n(\hat{B}_n R_n)\, d\tau$$
$$+ \left(1 - \frac{1}{\alpha}\right)\int_0^1 \tau g_n \dot{R}_n\, d\tau + \int_0^1 g_n R_n\, d\tau \quad . \tag{6.5.13}$$

Even though we wrote (6.2.1) in the limit $\alpha \to 0$, we know that the final one-loop result is independent of α (Henneaux 1985, Frampton 1987). Thus we can perform the $\zeta(0)$ calculation in the $\alpha = 1$ gauge, so that the contribution of $\tau g_n \dot{R}_n$ vanishes. In light of (5.9.27), we also know that the eigenfunctions of $\hat{A}_n$ and $\hat{B}_n$ in this gauge are respectively of the kind :

$$u_n^{(1)} = J_{\sqrt{n^2-1}}(a\tau) \quad , \quad u_n^{(2)} = \tau^{-1}J_n(b\tau) \quad . \tag{6.5.14}$$

However, $\int_0^1 g_n R_n \, d\tau$ does not vanish if $g_n = u_n^{(1)}$, $R_n = u_n^{(2)}$, when generalized magnetic or electric boundary conditions are imposed. In fact, setting $\rho = \sqrt{n^2-1}$, $\sigma = n$, if $\mid \frac{b}{a} \mid < 1$, this integral becomes (Luke 1962) :

$$\int_0^1 \frac{J_\rho(at)J_\sigma(bt)}{t} \, dt = \frac{\left(\frac{a}{2}\right)^\rho \left(\frac{b}{2}\right)^\sigma}{\Gamma(\rho+1)\Gamma(\sigma+1)} \cdot \sum_{k=0}^{\infty} \left[\frac{(-1)^k \left(\frac{a}{2}\right)^{2k} {}_2F_1\left(-k; -\rho-k; \sigma+1; \frac{b^2}{a^2}\right)}{k!(\rho+\sigma+2k)(\rho+1)_k} \right] . \tag{6.5.15}$$

In the magnetic case, we would have : $J_\rho(a) = 0$, $J_\sigma(b) - bJ'_\sigma(b) = 0$, and in the electric case we would have : $J'_\rho(a) = 0$, $J_\sigma(b) = 0$. Moreover, if $\mid \frac{b}{a} \mid < 1$, we also have the useful identity (Luke 1962) :

$$J_\rho(a)J_\sigma(b) = \frac{\left(\frac{a}{2}\right)^\rho \left(\frac{b}{2}\right)^\sigma}{\Gamma(\rho+1)\Gamma(\sigma+1)} \cdot \sum_{k=0}^{\infty} \left[\frac{(-1)^k \left(\frac{a}{2}\right)^{2k} {}_2F_1\left(-k; -\rho-k; \sigma+1; \frac{b^2}{a^2}\right)}{k!(\rho+1)_k} \right] . \tag{6.5.16}$$

The infinite sum on the right-hand side of (6.5.16) should vanish both in the magnetic case (because $J_\rho(a) = 0$) and in the electric case (because $J_\sigma(b) = 0$). On the other hand, this sum differs substantially from the one appearing on the right-hand side of (6.5.15) since the various terms in the latter contain the k-dependent extra factor $(\rho + \sigma + 2k)$ in the denominator. Hence it follows that (6.5.15) does not vanish and our problem remains a coupled one, also in the $\alpha = 1$ gauge.

The eigenvalues for g_n-modes and R_n-modes can be obtained in principle from the boundary conditions and the variational principle : $\delta\Big(I + \lambda J\Big) = 0$, where the Lorentzian version of J is $\frac{1}{2}\int A_\mu A^\mu \sqrt{-g} \, d^4x$. We then make the analytic continuation to the Euclidean time variable $\tau = it$ in computing I (cf. (6.5.10)) and J, and we use (5.9.22), (5.10.3) and the well-known properties of longitudinal vector harmonics and scalar harmonics (Schleich 1985, Louko 1988b). This leads to the coupled system of two second-order ordinary differential equations for arbitrary α and $\forall n \geq 2$ (the case $n = 1$ only involves the R_1-mode, and should be treated separately) :

$$\frac{\tau}{(n^2-1)}\left[-\ddot{g}_n - \frac{\dot{g}_n}{\tau} + \frac{(n^2-1)}{\alpha\tau^2}g_n\right] + \left(1-\frac{1}{\alpha}\right)\tau\dot{R}_n + R_n = \frac{\lambda_n}{(n^2-1)}\tau g_n \ , \qquad (6.5.17)$$

$$\tau^3\left[\frac{1}{\alpha}\left(-\ddot{R}_n - \frac{3}{\tau}\dot{R}_n\right) + \frac{(n^2-1)}{\tau^2}R_n\right] - \tau\dot{g}_n\left(1-\frac{1}{\alpha}\right) + \frac{g_n}{\alpha} = \lambda_n\tau^3 R_n \ . \qquad (6.5.18)$$

Now we still choose the $\alpha = 1$ gauge, because it enables one to decouple much more easily the system (6.5.17-18). The boundary conditions are regularity at the origin :

$$g_n(0) = R_n(0) = 0 \quad , \quad \forall n \geq 2 \quad , \qquad (6.5.19)$$

and magnetic conditions on S^3 :

$$g_n(1) = \dot{R}_n(1) = 0 \quad , \quad \forall n \geq 2 \quad , \qquad (6.5.20)$$

or electric conditions on S^3 :

$$\dot{g}_n(1) = R_n(1) = 0 \quad , \quad \forall n \geq 2 \quad . \qquad (6.5.21)$$

In the $\alpha = 1$ gauge we can express R_n from (6.5.17) as :

$$R_n = \frac{\lambda_n}{(n^2-1)}\tau g_n + \frac{\tau}{(n^2-1)}\left[\ddot{g}_n + \frac{\dot{g}_n}{\tau} - \frac{(n^2-1)}{\tau^2}g_n\right] \quad , \qquad (6.5.22)$$

and its insertion into the corresponding form of (6.5.18) yields the following fourth-order equation :

$$\begin{aligned} 0 = g_n &\left[\left(2 - \frac{3}{(n^2-1)}\right)\lambda_n\tau^2 - \frac{\lambda_n^2}{(n^2-1)}\tau^4 - (n^2-1)\right] \\ &+ 2\tau\dot{g}_n\left[1 - \frac{3\lambda_n\tau^2}{(n^2-1)}\right] + \frac{2\tau^2}{(n^2-1)}\ddot{g}_n\left[n^2 - 4 - \lambda_n\tau^2\right] \\ &+ \tau^3 g_n^{\mathrm{III}}\left[-\frac{6}{(n^2-1)}\right] + \tau^4 g_n^{\mathrm{IV}}\left[-\frac{1}{(n^2-1)}\right] \quad . \end{aligned} \qquad (6.5.23)$$

Moreover, studying first the magnetic case, the relations (6.5.20) and (6.5.22) lead to :

$$\lambda_n = (n^2 - 1) - 2\frac{\ddot{g}_n(1)}{\dot{g}_n(1)} - \frac{g_n^{\rm III}(1)}{\dot{g}_n(1)} \quad , \quad \forall n \geq 2 \quad . \tag{6.5.24}$$

Of course, as shown by (6.5.17-18) and (6.5.22-23), the eigenvalues λ_n have dimension $(\text{length})^{-2}$. However, in (6.5.24), following (6.5.19-21), we have set $a = 1$ for simplicity. Hence the physical dimension does not appear explicitly. For the solutions of the equations (6.5.22-23) subject to the boundary conditions (6.5.19-20), an existence and uniqueness theorem holds. Thus, denoting by k an integer ≥ 0, in light of the form of (6.5.23) we write its solution as :

$$g_n(\tau) = \tau^{\mu} \sum_{k=0}^{\infty} a_{n,k}(n, k, \lambda_n)\tau^k \quad . \tag{6.5.25}$$

The insertion of (6.5.25) into (6.5.23-24), plus the requirement that $g_n(1) = 0, \forall n \geq 2$, leads to a problem formulated in purely algebraic terms. The results are :

(1) only half of the $a_{n,k}$ coefficients are nonvanishing;

(2) the nonvanishing $a_{n,k}$ coefficients obey very involved recurrence relations;

(3) the value of μ is obtained by solving a fourth-order algebraic equation.

In fact, defining :

$$F(k, n, \mu) \equiv 2(k + \mu)^2 - (n^2 - 1) - \frac{(k + \mu)^2 \Bigl((k + \mu)^2 - 1\Bigr)}{(n^2 - 1)} \quad , \tag{6.5.26}$$

we find :

$$a_{n,0}F\Bigl(0, n, \mu\Bigr) = a_{n,1}F\Bigl(1, n, \mu\Bigr) = 0 \quad , \tag{6.5.27}$$

$$0 = a_{n,m}F\Bigl(m, n, \mu\Bigr) + \left[\left(2 - \frac{3}{(n^2 - 1)}\right)\lambda_n - \frac{4\lambda_n}{(n^2 - 1)}\Bigl(m - 2 + \mu\Bigr) - \frac{2\lambda_n}{(n^2 - 1)}\Bigl(m - 2 + \mu\Bigr)^2\right] a_{n,m-2} \quad , \tag{6.5.28}$$

where $m = 2, 3$, whereas, $\forall k \geq 4$, we have :

$$0 = a_{n,k}F(k,n,\mu) + \left[\left(2 - \frac{3}{(n^2-1)}\right)\lambda_n - \frac{4\lambda_n}{(n^2-1)}(k+\mu-2)\right.$$

$$\left. - \frac{2\lambda_n}{(n^2-1)}(k+\mu-2)^2\right] a_{n,k-2} - \frac{\lambda_n^2}{(n^2-1)}a_{n,k-4} \quad . \tag{6.5.29}$$

In (6.5.27-29), the value of μ can be obtained from the equation $F(0,n,\mu) = 0$, and bearing in mind (6.5.19), which implies that only a $\mu > 1$ is an acceptable value, in light of (6.5.22). In other words, we study the fourth-order algebraic equation :

$$\mu^4 - (2n^2-1)\mu^2 + (n^2-1)^2 = 0 \quad . \tag{6.5.30}$$

This equation can be easily solved setting $\mu^2 = x$ and studying the corresponding second-order equation for x. One thus finds the four roots :

$$\mu_+^{(1)} = +\sqrt{n^2 - \frac{3}{4}} + \frac{1}{2} \quad , \tag{6.5.31}$$

$$\mu_+^{(2)} = +\sqrt{n^2 - \frac{3}{4}} - \frac{1}{2} \quad , \tag{6.5.32}$$

$$\mu_-^{(1)} = -\mu_+^{(1)} \quad , \tag{6.5.33}$$

$$\mu_-^{(2)} = -\mu_+^{(2)} \quad . \tag{6.5.34}$$

Interestingly, both $\mu_+^{(1)}$ and $\mu_+^{(2)}$ are > 1, $\forall n \geq 2$. They yield the desired regular solution of the system (6.5.17-18). In the magnetic case, the eigenvalues obey (6.5.24), whereas the $a_{n,k}$ coefficients obey (6.5.27-29) and the condition

$$\sum_{k=0}^{\infty} a_{n,k} = 0 \quad , \quad \forall n \geq 2 \quad , \tag{6.5.35}$$

which holds by virtue of the boundary condition $g_n(1) = 0$, $\forall n \geq 2$. Moreover, the conditions for the electric case are :

$$\sum_{k=0}^{\infty} (k+\mu)a_{n,k} = 0 \quad , \quad \forall n \geq 2 \quad , \tag{6.5.36}$$

$$\lambda_n = (n^2 - 1) - \frac{\ddot{g}_n(1)}{g_n(1)} \quad , \quad \forall n \geq 2 \quad , \tag{6.5.37}$$

in light of (6.5.21-22). Using (6.5.25) and (6.5.36), (6.5.37) leads to :

$$\lambda_n = (n^2 - 1) - \mu^2 - \frac{\sum_{k=0}^{\infty}(k^2 + 2\mu k)a_{n,k}}{\sum_{k=0}^{\infty} a_{n,k}} \quad . \tag{6.5.38}$$

This is why, using (6.5.31-32), two asymptotic sets of λ_n result :

$$\lambda_n^{(1)} \sim -(n+1) - \frac{\sum_{k=0}^{\infty}\left(k^2 + 2\mu_+^{(1)}k\right)a_{n,k}^{(1)}}{\sum_{k=0}^{\infty} a_{n,k}^{(1)}} \quad , \tag{6.5.39}$$

$$\lambda_n^{(2)} \sim (n-1) - \frac{\sum_{k=0}^{\infty}\left(k^2 + 2\mu_+^{(2)}k\right)a_{n,k}^{(2)}}{\sum_{k=0}^{\infty} a_{n,k}^{(2)}} \quad . \tag{6.5.40}$$

However, we have not been able to express our exact solution in terms of special functions.

Even though this point remains unclear, we think it can be interesting to anticipate the effect of the R_1-mode, which remains decoupled (see comment before (6.5.17)), on the $\zeta(0)$ calculation. The eigenvalue condition corresponding to the R_1-mode is :

$$J_1(\sqrt{\lambda_1}) - \sqrt{\lambda_1}\dot{J}_1(\sqrt{\lambda_1}) = 0 \quad , \tag{6.5.41}$$

in the case of magnetic boundary conditions, or $J_1(\sqrt{\lambda_1}) = 0$ in the case of electric boundary conditions. In the next three chapters, we describe and apply a technique (originally developed in Moss 1989) for the direct calculation of $\zeta(0)$, once the explicit form of the eigenvalue condition is given. Because our analysis is still preliminary, we here present only the final result of the calculation. Using the labels B and E for magnetic and electric boundary conditions respectively, one finds :

$$\hat{\zeta}_B(0) = -\frac{77}{180} - \frac{1}{4} = -\frac{61}{90} \quad , \tag{6.5.42}$$

$$\hat{\zeta}_E(0) = \frac{13}{180} - \frac{3}{4} = -\frac{61}{90} \quad . \tag{6.5.43}$$

The left-hand sides of (6.5.42-43) are the *parts* of $\zeta(0)$ which take into account the contribution of the physical degrees of freedom (cf. (5.9.61)) and of the R_1-mode respectively. The values $-\frac{1}{4}$ and $-\frac{3}{4}$ are obtained applying the technique described in the next three chapters (see in particular the end of section 7.3).

Interestingly, our *partial* values (6.5.42-43) are close to the value given in (6.5.8). Thus more work remains to be done, to check whether the coupled g_n- and R_n-modes and the ghost field yield the difference between (6.5.8) and (6.5.42-43), but there is now hope to be able to perform *direct* and gauge-invariant $\zeta(0)$ calculations in the presence of boundaries, at least for electromagnetism. Note also that the occurrence of two sets of gauge modes (i.e. g_n and R_n) for the spin-1 problem is a peculiarity of problems with boundaries. By contrast, one-loop calculations in the absence of boundaries for spin-1 fields only involve one set of gauge modes (cf. (4.16) in Allen 1983b).

Moreover, the reader should bear in mind that the gauge choice made in this section might be improved (cf. problem 6.2 in the problems' section), and that the Hamiltonian BFV path integral remains the ultimate goal. So far, its equivalence to the Faddeev-Popov result remains *formal* (cf. end of section 6.3).

ADDENDUM TO CHAPTER SIX

In section 6.5, we have chosen the gauge-averaging term $\frac{\Phi^2}{2\alpha}$, where the gauge-averaging functional $\Phi(A)$ can be written in the form (cf. Laenen and van Nieuwenhuizen 1991)

$$\Phi(A) \equiv \Phi_1(A) \equiv {}^{(4)}\nabla^{\mu} A_{\mu} - K^i_{\ i} A_0 = \frac{\partial A_0}{\partial \tau} + {}^{(3)}\nabla^i A_i \quad , \qquad (AD.1)$$

where $K^i_{\ i} = \frac{3}{\tau}$ is the trace of the extrinsic-curvature tensor of the boundary. This choice of $\Phi(A)$ leads to the familiar one-dimensional Laplace operator acting on the R_n-modes, which simplifies the $\zeta(0)$ calculation for the coupled gauge modes and for the R_1-mode, as shown in sections 6.5 and 7.3. However, since $\Phi_1(A)$ is not the Lorentz gauge-averaging functional, the corresponding ghost action does not involve the familiar Laplace operator. This is proved (Itzykson and Zuber 1985) by studying the gauge transformation

$${}^{\epsilon}A_{\mu} \equiv A_{\mu} + {}^{(4)}\nabla_{\mu}\epsilon = A_{\mu} + \partial_{\mu}\epsilon \quad , \qquad (AD.2)$$

where the scalar ϵ is expanded on a family of three-spheres centred on the origin as

$$\epsilon(x,\tau) = \sum_{n=1}^{\infty} \epsilon_n(\tau) Q^{(n)}(x) \quad . \qquad (AD.3)$$

One thus finds

$$\delta(\Phi_1(A)) \equiv \Phi_1(A) - \Phi_1({}^{\epsilon}A) = \sum_{n=1}^{\infty} Q^{(n)}(x) \left[-\frac{d^2}{d\tau^2} + \frac{(n^2-1)}{\tau^2} \right] \epsilon_n(\tau) \quad . \qquad (AD.4)$$

This implies that the eigenfunctions of the ghost operator are of the kind (Gradshteyn and Ryzhik 1965)

$$\widetilde{\epsilon}_n(\tau) = \sqrt{\tau} J_{\sqrt{n^2 - \frac{3}{4}}}(\sqrt{E}\tau) \quad . \qquad (AD.5)$$

More precisely, since the electromagnetic field is bosonic, the corresponding ghost field is fermionic (Itzykson and Zuber 1985). Its contribution to the full $\zeta(0)$ is thus obtained changing sign to the scalar-eigenfunctions contribution of (AD.5), and then multiplying the resulting number by two, since the ghost field is complex. We now have to perform a $\zeta(0)$ calculation which involves Bessel functions of non-integer order, generalizing the technique described in chapters seven, eight and nine. Here we show that, although eigenvalues and eigenfunctions are different, the $\zeta(0)$ calculation originating from (AD.5) is closely related to a standard $\zeta(0)$ calculation involving Bessel functions of integer order. For this purpose, we study the simplest case, i.e. when the ghost field obeys homogeneous Dirichlet conditions on S^3. This leads to the eigenvalue condition

$$J_{\sqrt{n^2-\frac{3}{4}}}(\sqrt{E}a) = 0 \qquad \forall n \geq 1 \quad . \tag{AD.6}$$

Following section 7.3 and (AD.6), it is now useful to define $\forall n \geq 1$ and at large x

$$\nu \equiv +\sqrt{n^2 - \frac{3}{4}} \quad , \tag{AD.7}$$

$$\alpha_\nu(x) \equiv \sqrt{\nu^2 + x^2} = \sqrt{n^2 - \frac{3}{4} + x^2} \quad , \tag{AD.8}$$

$$\alpha_n(x) \equiv \sqrt{n^2 + x^2} \quad . \tag{AD.9}$$

Since the application of the technique of section 7.3 to infinitely many perturbative modes for the ghost involves defining $\alpha_\nu(x)$, whereas we are only able to perform exact calculations using $\alpha_n(x)$, it is also useful to evaluate the ratio

$$\frac{\alpha_\nu(x)}{\alpha_n(x)} \sim \rho_n(x) \sim \left[1 - \frac{3}{8}\left(n^2 + x^2\right)^{-1} - \frac{9}{128}\left(n^2 + x^2\right)^{-2} + \mathrm{O}\left(\left(n^2 + x^2\right)^{-3}\right)\right] \quad . \tag{AD.10}$$

The asymptotic expansion (AD.10) is very useful in that it is uniform in n, i.e. it holds $\forall n \geq 1$, at large x. A careful study of section 7.3 shows that, if the eigenvalue condition (AD.6) holds (whose eigenvalues are positive $\forall n \geq 1$), the zeta-function at large x has the uniform asymptotic expansion

$$\Gamma(3)\zeta(3, x^2) \sim \left[\sigma_1 + \sigma_2\right] \quad , \tag{AD.11}$$

where

$$\sigma_1 \sim \sum_{n=0}^{\infty} n^2 \left[-\nu x^{-6} + \nu^2 x^{-6} \alpha_\nu^{-1} + \frac{\nu^2}{2} x^{-4} \alpha_\nu^{-3} + \frac{3}{8} \nu^2 x^{-2} \alpha_\nu^{-5} - \frac{\alpha_\nu^{-6}}{2} + \frac{3}{8} \alpha_\nu^{-5} \right] \quad , \quad (AD.12)$$

$$\sigma_2 \sim -\sum_{l=1}^{\infty} \sum_{r=0}^{l} a_{lr} \left(r + \frac{l}{2}\right)\left(r + \frac{l}{2} + 1\right)\left(r + \frac{l}{2} + 2\right) \sum_{n=0}^{\infty} n^2 \nu^{2r} \alpha_\nu^{-l-2r-6} \quad . \quad (AD.13)$$

In these formulae, obtained using uniform asymptotic expansions of Bessel functions of non-integer order, n^2 is the degeneracy due to the scalar harmonics appearing in the expansion (AD.3), ν is the order of the Bessel functions defined in (AD.7), and α_ν has been defined in (AD.8). We now re-express ν^2 as $\left(n^2 - \frac{3}{4}\right)$, and $\alpha_\nu(x) \sim \alpha_n(x)\rho_n(x)$ as in (AD.10). Moreover, we use the contour formula (Moss 1989)

$$\sum_{n=0}^{\infty} n^{2k} \alpha_n^{-2k-m} = \frac{\Gamma\left(k + \frac{1}{2}\right)\Gamma\left(\frac{m}{2} - \frac{1}{2}\right)}{2\Gamma\left(k + \frac{m}{2}\right)} x^{1-m} \qquad \forall k = 1, 2, 3, \dots \quad . \quad (AD.14)$$

We then point out that the asymptotic expansion (AD.12) can be cast in the form

$$\sigma_1 \sim \left[-x^{-6} I_\infty^{(1)} + x^{-6} I_\infty^{(2)} + x^{-4} I_\infty^{(3)} + x^{-2} I_\infty^{(4)} - I_\infty^{(5)} + I_\infty^{(6)} \right] \quad , \quad (AD.15)$$

where (see appendix AD.a)

$$I_\infty^{(2)} - I_\infty^{(1)} \equiv -\sum_{n=0}^{\infty} n^3 + \sum_{n=0}^{\infty} n^2 \left[\frac{\nu^2}{\alpha_\nu} - \left(\nu - n\right) \right] \quad , \quad (AD.16)$$

$$\begin{aligned} I_\infty^{(3)} \sim & \frac{1}{2} \sum_{n=0}^{\infty} n^4 \alpha_n^{-3} + \frac{9}{16} \sum_{n=0}^{\infty} n^4 \alpha_n^{-5} + \frac{135}{256} \sum_{n=0}^{\infty} n^4 \alpha_n^{-7} \\ & - \frac{3}{8} \sum_{n=0}^{\infty} n^2 \alpha_n^{-3} - \frac{27}{64} \sum_{n=0}^{\infty} n^2 \alpha_n^{-5} - \frac{405}{1024} \sum_{n=0}^{\infty} n^2 \alpha_n^{-7} \\ & + \frac{1}{2} \sum_{n=0}^{\infty} n^4 \alpha_n^{-3} \mathrm{O}\left(\alpha_n^{-6}\right) - \frac{3}{8} \sum_{n=0}^{\infty} n^2 \alpha_n^{-3} \mathrm{O}\left(\alpha_n^{-6}\right) \quad , \end{aligned} \quad (AD.17)$$

$$
\begin{aligned}
I_{\infty}^{(4)} \sim\ & \frac{3}{8}\sum_{n=0}^{\infty} n^4 \alpha_n^{-5} + \frac{45}{64}\sum_{n=0}^{\infty} n^4 \alpha_n^{-7} + \frac{945}{1024}\sum_{n=0}^{\infty} n^4 \alpha_n^{-9} \\
& - \frac{9}{32}\sum_{n=0}^{\infty} n^2 \alpha_n^{-5} - \frac{135}{256}\sum_{n=0}^{\infty} n^2 \alpha_n^{-7} - \frac{2835}{4096}\sum_{n=0}^{\infty} n^2 \alpha_n^{-9} \\
& + \frac{3}{8}\sum_{n=0}^{\infty} n^4 \alpha_n^{-5} \mathrm{O}\left(\alpha_n^{-6}\right) - \frac{9}{32}\sum_{n=0}^{\infty} n^2 \alpha_n^{-5} \mathrm{O}\left(\alpha_n^{-6}\right) \quad , \qquad (AD.18)
\end{aligned}
$$

$$
\begin{aligned}
I_{\infty}^{(5)} \equiv\ & \frac{1}{2}\sum_{n=0}^{\infty} n^2 \alpha_\nu^{-6} \\
\sim\ & \frac{1}{2}\sum_{n=0}^{\infty} n^2 \alpha_n^{-6} + \frac{9}{8}\sum_{n=0}^{\infty} n^2 \alpha_n^{-8} + \frac{1}{2}\sum_{n=0}^{\infty} n^2 \alpha_n^{-6} \mathrm{O}\left(\alpha_n^{-4}\right) \quad , \qquad (AD.19)
\end{aligned}
$$

$$
\begin{aligned}
I_{\infty}^{(6)} \equiv\ & \frac{3}{8}\sum_{n=0}^{\infty} n^2 \alpha_\nu^{-5} \\
\sim\ & \frac{3}{8}\sum_{n=0}^{\infty} n^2 \alpha_n^{-5} + \frac{45}{64}\sum_{n=0}^{\infty} n^2 \alpha_n^{-7} + \frac{945}{1024}\sum_{n=0}^{\infty} n^2 \alpha_n^{-9} \\
& + \frac{3}{8}\sum_{n=0}^{\infty} n^2 \alpha_n^{-5} \mathrm{O}\left(\alpha_n^{-6}\right) \quad . \qquad (AD.20)
\end{aligned}
$$

It is therefore clear, using (AD.14), that the $\zeta(0)$ value due to σ_1 and σ_2 is given by $\frac{1}{90} = -2\left(-\frac{1}{180}\right)$ (which coincides with the $\zeta(0)$ value corresponding to the Lorentz gauge-averaging functional) plus additional terms due to second sum in (AD.16), third and fifth sum in (AD.17)-(AD.18), denoted by T_1, T_2, T_3, T_4, third sum in (AD.20), denoted by T_5, and finally (AD.13). Note that $I_{\infty}^{(5)}$ defined in (AD.19) does not contribute to the additional terms. The detailed calculation yields

$$
x^{-4} T_1 \equiv \frac{135}{256} x^{-4} \sum_{n=0}^{\infty} n^4 \alpha_n^{-7} = \frac{27}{256} x^{-6} \quad , \qquad (AD.21)
$$

$$
x^{-4} T_2 \equiv -\frac{27}{64} x^{-4} \sum_{n=0}^{\infty} n^2 \alpha_n^{-5} = -\frac{9}{64} x^{-6} \quad , \qquad (AD.22)
$$

$$x^{-2}T_3 \equiv \frac{945}{1024}x^{-2}\sum_{n=0}^{\infty} n^4\alpha_n^{-9} = \frac{27}{512}x^{-6} \quad , \tag{AD.23}$$

$$x^{-2}T_4 \equiv -\frac{135}{256}x^{-2}\sum_{n=0}^{\infty} n^2\alpha_n^{-7} = -\frac{9}{128}x^{-6} \quad , \tag{AD.24}$$

$$T_5 \equiv \frac{945}{1024}\sum_{n=0}^{\infty} n^2\alpha_n^{-9} = \frac{9}{128}x^{-6} \quad . \tag{AD.25}$$

We now focus on (AD.13) and (AD.16), and we first study the asymptotic expansion (AD.13), since (AD.16) gives rise to severe technical difficulties (see below). For this purpose, we remark that, studying for all integer values $l \in [1,\infty[$, $r \in [1,l]$ the function (see appendix AD.a)

$$\begin{aligned} I_{lr}(x) &\sim \sum_{n=0}^{\infty} n^2\nu^{2r}\alpha_\nu^{-l-2r-6} \\ &\sim \sum_{n=0}^{\infty} n^2\left(n^2-\frac{3}{4}\right)^r \alpha_n^{-l-2r-6}\left[1 + A_{lr}\alpha_n^{-2} + B_{lr}\alpha_n^{-4} + \mathrm{O}\left(\alpha_n^{-6}\right)\right] \\ &\sim \left[I_{lr}^{(1)} + I_{lr}^{(2)} + I_{lr}^{(3)} + I_{lr}^{(4)}\right](x) \quad , \end{aligned} \tag{AD.26}$$

one finds

$$I_{lr}^{(1)}(x) \sim \sum_{n=0}^{\infty} n^2\left(n^2-\frac{3}{4}\right)^r \alpha_n^{-l-2r-6} \quad , \tag{AD.27}$$

$$I_{lr}^{(2)}(x) \sim A_{lr}\sum_{n=0}^{\infty} n^2\left(n^2-\frac{3}{4}\right)^r \alpha_n^{-l-2r-8} \quad , \tag{AD.28}$$

$$I_{lr}^{(3)}(x) \sim B_{lr}\sum_{n=0}^{\infty} n^2\left(n^2-\frac{3}{4}\right)^r \alpha_n^{-l-2r-10} \quad , \tag{AD.29}$$

$$I_{lr}^{(4)}(x) \sim \sum_{n=0}^{\infty} n^2\left(n^2-\frac{3}{4}\right)^r \alpha_n^{-l-2r-6}\mathrm{O}\left(\alpha_n^{-6}\right) \quad , \tag{AD.30}$$

where A_{lr} and B_{lr} are coefficients which only depend on l and r. The case $r = 0$ is easier. Using (AD.13)-(AD.14), $r = 0$ leads to a contribution to $\zeta(0)$ related to

$$T_6 \equiv -a_{10} \frac{15}{8} \frac{21}{8} \frac{\Gamma\left(\frac{3}{2}\right)\Gamma(3)}{2\Gamma\left(\frac{9}{2}\right)} x^{-6} = -\frac{3}{8} a_{10} x^{-6} \quad , \tag{AD.31}$$

where $\frac{21}{8}$ is the coefficient of α_n^{-2} in the asymptotic expansion of $\rho_n^{-7}(x)$ (see appendix AD.a). If the integer r is ≥ 1, one has to study (AD.26)-(AD.30), where

$$\left(n^2 - \frac{3}{4}\right)^r = n^{2r} - \frac{3}{4} r n^{2r-2} + \ldots + r n^2 \left(-\frac{3}{4}\right)^{r-1} + \left(-\frac{3}{4}\right)^r \quad . \tag{AD.32}$$

Inserting (AD.32) into (AD.27)-(AD.30), and using (AD.14), a lengthy calculation yields a contribution to $\zeta(0)$ related to (see appendix AD.a)

$$T_7 \equiv \left[\frac{3}{4} a_{11} - \frac{9}{8} a_{11}\right] x^{-6} = -\frac{3}{8} a_{11} x^{-6} \quad . \tag{AD.33}$$

Note that the two terms on the r.h.s. of (AD.33) are due to the asymptotic expansions of $I_{lr}^{(1)}(x)$ and $I_{lr}^{(2)}(x)$ respectively, whereas (AD.29)-(AD.30) do not affect the $\zeta(0)$ value, since they do not involve x^{-6}. It now remains to evaluate the contribution of (AD.16). Indeed, defining

$$J_\infty \equiv \sum_{n=0}^{\infty} n^2 \nu \left(\frac{\nu}{\alpha_\nu} - 1\right) \tag{AD.34}$$

we point out that multiplying and dividing the round bracket by $\left(\nu + \alpha_\nu\right)$, and then adding and subtracting α_ν in the numerator of the corresponding expression, one finds by virtue of (AD.8) the useful identity

$$J_\infty = -x^2 \sum_{n=0}^{\infty} n^2 \left[\frac{1}{\alpha_\nu} - \frac{1}{\left(\nu + \alpha_\nu\right)}\right] = J_\infty^{(1)} + J_\infty^{(2)} \quad . \tag{AD.35}$$

Moreover, (AD.10) and (AD.14) show that the contribution to $\zeta(0)$ due to $J_{\infty}^{(1)}$ is related to

$$T_8 \equiv -\frac{27}{128}x^{-4}\sum_{n=0}^{\infty} n^2 \alpha_n^{-5} = -\frac{9}{128}x^{-6} \quad . \tag{AD.36}$$

A further contribution is due to

$$T_9 \equiv \frac{1}{120}x^{-6} \quad , \tag{AD.37}$$

originating from $\sum_{n=0}^{\infty} n^3$ in (AD.16). However, we do not yet know how to deal properly with the divergent sum

$$J_{\infty}^{(2)} \equiv x^2 \sum_{n=0}^{\infty} \frac{n^2}{\left(\nu + \alpha_\nu\right)} \quad . \tag{AD.38}$$

We should now add up the numerical coefficients appearing in (AD.21)-(AD.25), (AD.31), (AD.33), (AD.36)-(AD.37), divide them by two, and finally multiply by -2 since the ghost is fermionic and complex. This leads to the following *partial* contribution to the $\zeta(0)$ value for the ghost field :

$$\zeta_{gh}^{(I)}(0) = \frac{1}{90} - \frac{9}{512} + \frac{3}{8}\left(a_{10} + a_{11}\right) + \frac{9}{128} - \frac{1}{120} = \frac{1}{360} + \frac{11}{512} \quad , \tag{AD.39}$$

where $\frac{1}{90}$ is added for the reasons described following (AD.20), and we have used the values $a_{10} = \frac{1}{8}$, $a_{11} = -\frac{5}{24}$ appearing in equation (26) of Moss 1989.

By contrast, if the Lorentz gauge-averaging functional is chosen, one finds

$$\Phi(A) \equiv \Phi_2(A) \equiv {}^{(4)}\nabla^\mu A_\mu = \frac{\partial A_0}{\partial \tau} + {}^{(4)}\nabla^i A_i \quad , \tag{AD.40}$$

which implies

$$\delta(\Phi_2(A)) \equiv \Phi_2(A) - \Phi_2({}^{\epsilon}A) = \sum_{n=1}^{\infty} Q^{(n)}(x)\left[-\frac{d^2}{d\tau^2} - \frac{3}{\tau}\frac{d}{d\tau} + \frac{(n^2-1)}{\tau^2}\right]\epsilon_n(\tau) \quad , \tag{AD.41}$$

where we have used the property ${}^{(4)}\nabla_i\epsilon = {}^{(3)}\nabla_i\epsilon = \epsilon_{|i} = \partial_i\epsilon$, $\forall i = 1,2,3$. Thus, as we anticipated, the familiar one-dimensional Laplace operator appears in the ghost action, so that the ghost contributions to the full $\zeta(0)$ value are more easily computed as $-2\left(-\frac{1}{180}\right)$ and $-2\left(\frac{29}{180}\right)$ in the Dirichlet and Neumann cases, respectively. However, if $\Phi_2(A)$ is chosen as gauge-averaging functional, the form of the action quadratic in the gauge modes becomes $\forall n \geq 2$

$$I_E^{(n)}(g,R) = \frac{1}{2}\int_0^1 \frac{\tau g_n}{(n^2-1)}\left[-\frac{d^2 g_n}{d\tau^2} - \frac{1}{\tau}\frac{dg_n}{d\tau} + \frac{(n^2-1)}{\alpha\tau^2}g_n\right] d\tau$$

$$+\frac{1}{2}\int_0^1 \tau^3 R_n \left[\frac{1}{\alpha}\left(-\frac{d^2 R_n}{d\tau^2} + \frac{3}{\tau}\frac{dR_n}{d\tau}\right) + \left(n^2 - 1 + \frac{9}{\alpha}\right)\frac{R_n}{\tau^2}\right] d\tau$$

$$+\left(1-\frac{1}{\alpha}\right)\int_0^1 \tau g_n \dot{R}_n \, d\tau + \left(1-\frac{3}{\alpha}\right)\int_0^1 g_n R_n \, d\tau$$

$$-\Big[\tau g_n R_n\Big]_0^1 + \frac{1}{2\alpha}\Big[\tau^3 \dot{R}_n R_n\Big]_0^1 \quad . \qquad (AD.42)$$

Thus, the second-order differential operator acting on R_n-modes is no longer the one-dimensional Laplace operator for scalars, and the calculation becomes more involved. For example, if we set $\alpha = 1$, the contribution of $R_1(\tau)$ to $\zeta(0)$ involves a Bessel function of order $\sqrt{13}$. Moreover, a non-vanishing boundary term $I_B^{(n)} \equiv \frac{a^3}{2\alpha}\dot{R}_n(a)R_n(a) = -\frac{3a^2}{2\alpha}R_n^2(a)$ survives in the action, if the whole functional $\Phi_2(A)$ is required to vanish on the boundary in the magnetic case (cf. Moss and Poletti 1990a). Thus, one has to add to the action a boundary term equal to $-I_B^{(n)}$, if the whole of $\Phi_2(A)$ is set to zero on S^3.

Of course, since the theory is gauge-invariant, *infinitely* many other choices for $\Phi(A)$ (but not all choices) are still possible. A very relevant class of choices can be cast in the form

$$\Phi^{(b)}(A) \equiv {}^{(4)}\nabla^{\mu}A_{\mu} + bK^i_{\ i}A_0 \quad , \qquad (AD.43)$$

where b is a real number. With our parametrization, $b = -1$ leads to $\Phi_1(A)$, and $b = 0$ leads to $\Phi_2(A)$. Note that, even if we set $\alpha = 1$, it does not seem possible to decouple gauge modes using $\frac{\Phi^2}{2\alpha}$ *and* obtain a well-defined ghost action, since the decoupling of g_n and R_n, $\forall n \geq 2$, is then obtained setting

$$\Phi(A) \equiv \Phi_3(A) \equiv \sum_{n=2}^{\infty} \sqrt{\left[\left(-\frac{g_n}{\tau^2} + \dot{R}_n\right)^2 + \frac{2}{\tau^2}\frac{d}{d\tau}(g_n R_n)\right]}\; Q^{(n)}(x) \quad . \qquad (AD.44)$$

However, the ghost action should be derived by functionally differentiating the infinite sum of square roots on the right-hand side of (AD.44) as in (AD.4) and (AD.41), and this does not lead to a linear, second-order differential operator. This is why we believe that the coupling of gauge modes is an intrinsic property of problems with boundaries, as well as the choice of gauge-averaging functionals of the form (AD.43), which all reduce to the Lorentz choice in the absence of boundaries.

In light of (5.9.61), (7.3.59) and (AD.39), the full $\zeta(0)$ value for vacuum Euclidean Maxwell theory in the case of magnetic boundary conditions on S^3 takes the form

$$\zeta(0)=\zeta_B^{(PDF)}(0)+\zeta_{R_1}(0)+\zeta_{GM}(0)+\zeta_{gh}(0)=-\frac{243}{360}+\frac{11}{512}+\zeta_{GM}(0)+\zeta_{gh}^{(II)}(0) \quad , \quad (AD.45)$$

where $\zeta_{GM}(0)$ and $\zeta_{gh}^{(II)}(0)$ are the as yet unknown contributions to $\zeta(0)$ arising from coupled gauge modes (section 6.5) and from (AD.38) respectively. We have been unable to evaluate $\zeta_{GM}(0)$ since we do not know explicitly the uniform asymptotic expansion as $\lambda_n \rightarrow \infty$ of the power series in (6.5.25), which is not (obviously) related to well-known special functions (see Appendix AD.b). Moreover, the regularized contribution $\zeta_{gh}^{(II)}(0)$ of (AD.38) to $\zeta(0)$ involves $\nu \equiv +\sqrt{n^2-\frac{3}{4}}$, which is a source of complication. However, it should be emphasized that all divergences are only *fictitious*, since the starting point for the derivation of (AD.12) is the identity

$$\left(\frac{1}{2x}\frac{d}{dx}\right)^3 \log\left(\frac{1}{\nu+\alpha_\nu}\right)=\left(\nu+\alpha_\nu\right)^{-3}\left[-\alpha_\nu^{-3}-\frac{9}{8}\nu\alpha_\nu^{-4}-\frac{3}{8}\nu^2\alpha_\nu^{-5}\right] \quad . \quad (AD.46)$$

This proves that by summing over all integer values of n from 0 to ∞ one gets a convergent series.

Indeed, we have not studied the case of Neumann boundary conditions for the ghost field, i.e. the electric case. This complicated calculation may be, by itself, the object of a paper. However, interestingly, in light of (5.9.61) and (7.3.59)-(7.3.60) one finds (see (6.5.42)-(6.5.43))

$$\zeta_B^{(PDF)}(0)+\zeta_{R_1}(0)=\zeta_E^{(PDF)}(0)+\widetilde{\zeta}_{R_1}(0)=-\frac{61}{90} \quad . \quad (AD.47)$$

In other words, if the gauge-averaging functional of (AD.1) is chosen, physical degrees of freedom and decoupled gauge mode give the same partial contribution to the full $\zeta(0)$, i.e. $-\frac{61}{90}$, both in the magnetic and in the electric case.

One-loop quantum cosmology may add further evidence in favour of different approaches to quantizing gauge theories being inequivalent (Gitman and Tyutin 1990, Govaerts and Troost 1991, Govaerts 1991, Guven and Ryan 1992, Kunstatter 1992, Henneaux and Teitelboim 1992, Barvinsky 1993). Studying flat Euclidean backgrounds bounded by a three-sphere, for vacuum Maxwell theory the PDF method yields $\zeta(0) = -\frac{77}{180}$ and $\zeta(0) = \frac{13}{180}$ in the magnetic and electric cases respectively (Louko 1988 and our section 5.9), whereas the *indirect* method (by this we mean that one-loop amplitudes are expressed using the boundary-counterterms technique and evaluating the various coefficients as in Poletti 1990, Moss and Poletti 1990a) was found to yield $\zeta(0) = -\frac{38}{45}$ in both cases in Poletti 1990. For $N = 1$ supergravity, the PDF method yields *partial* cancellations between spin 2 and spin $\frac{3}{2}$ (chapter nine), whereas the *indirect* method yields a one-loop amplitude which is even more divergent than in the pure-gravity case (Poletti 1990). Finally, for pure gravity, the PDF method yields $\zeta(0) = -\frac{278}{45}$ in the Dirichlet case, whereas the *indirect* method yields $\zeta(0) = -\frac{803}{45}$ (Poletti 1990). Moreover, within the PDF method, it is possible to set to zero on S^3 the linearized magnetic curvature. This yields a well-defined one-loop calculation, and the corresponding $\zeta(0)$ value is $\frac{112}{45}$ (section 7.3). By contrast, using the Faddeev-Popov formula, magnetic boundary conditions for pure gravity are ruled out (Poletti 1990). It was therefore necessary to get a better understanding of the manifestly gauge-invariant formulae for one-loop amplitudes used so far in the literature, by performing a mode-by-mode analysis of the eigenvalue equations, rather than relying on general formulae which contain no explicit information about degeneracies and eigenvalue conditions.

This detailed analysis has been attempted here in the simplest case, i.e. vacuum Maxwell theory at one-loop about a flat Euclidean background bounded by a three-sphere. Our results are here summarized for the sake of clarity:

(1) In light of (5.9.61) and (7.3.59)-(7.3.60) the physical degrees of freedom, and the decoupled gauge mode, give a contribution to the full $\zeta(0)$ equal to $-\frac{61}{90}$ both in the

magnetic and in the electric case, if the gauge-averaging functional $\Phi_1(A)$ of (AD.1) is chosen.

(2) Remaining gauge modes g_n and R_n always obey a coupled system of linear, second-order ordinary differential equations, $\forall n \geq 2$. The solution of such a system corresponding to $\Phi_1(A)$ has been given in section 6.5 and appendix AD.b.

(3) If $\Phi_1(A)$ is chosen, the ghost eigenfunctions involve Bessel functions of non-integer order. The corresponding contribution to the full $\zeta(0)$ can be obtained using the method described in this addendum. Such a technical point appears interesting, since to our knowledge no previous mode-by-mode analysis for the ghost is appearing in the literature in the case of Bessel functions of non-integer order.

It now remains to evaluate the contribution to the full $\zeta(0)$ of the divergent sum in (AD.38), and the uniform asymptotic expansion of g_n- and R_n-modes as $\lambda_n \rightarrow \infty$ at the end of section 6.5. Unfortunately, the generalization of the method described in Barvinsky et al. 1992a-b, Kamenshchik and Mishakov 1992, Kamenshchik and Mishakov 1993 is highly nontrivial. By contrast, a simpler form of the ghost eigenfunctions is obtained using the Lorentz gauge-averaging functional $\Phi_2(A)$. However, this leads to a further complication of the calculations involving gauge modes, since the decoupled mode $R_1(\tau)$ involves a Bessel function of order $\sqrt{13}$ (this implies a contribution to $\zeta(0)$ proportional to $\sqrt{13}$, which we find very puzzling), and coupled gauge modes require the addition to the action, in the magnetic case, of a boundary term equal to $\frac{3a^2}{2\alpha} \sum_{n=2}^{\infty} R_n^2(a)$.

Thus, although some evidence exists that different $\zeta(0)$ values for gauge fields in the presence of boundaries are due to inequivalent quantization techniques, the most important check, i.e. the mode-by-mode analysis of eigenvalue equations for gauge modes and ghost fields, remains a very difficult problem. We hope that our addendum, through its detailed (although incomplete) analysis, may contribute to shed new light on this longstanding problem in quantum field theory.

APPENDIX AD.a

The derivation of (AD.17)-(AD.20), (AD.31), (AD.33) and (AD.36) has been obtained using the following asymptotic expansions :

$$\rho_n^{-1}(x) \sim 1 + \frac{3}{8}\alpha_n^{-2} + \frac{27}{128}\alpha_n^{-4} + O\left(\alpha_n^{-6}\right) \quad , \tag{AD.a.1}$$

$$\rho_n^{-2}(x) \sim 1 + \frac{3}{4}\alpha_n^{-2} + \frac{9}{16}\alpha_n^{-4} + O\left(\alpha_n^{-6}\right) \quad , \tag{AD.a.2}$$

$$\rho_n^{-3}(x) \sim 1 + \frac{9}{8}\alpha_n^{-2} + \frac{135}{128}\alpha_n^{-4} + O\left(\alpha_n^{-6}\right) \quad , \tag{AD.a.3}$$

$$\rho_n^{-4}(x) \sim 1 + \frac{3}{2}\alpha_n^{-2} + \frac{27}{16}\alpha_n^{-4} + O\left(\alpha_n^{-6}\right) \quad , \tag{AD.a.4}$$

$$\rho_n^{-5}(x) \sim 1 + \frac{15}{8}\alpha_n^{-2} + \frac{315}{128}\alpha_n^{-4} + O\left(\alpha_n^{-6}\right) \quad , \tag{AD.a.5}$$

$$\rho_n^{-6}(x) \sim 1 + \frac{9}{4}\alpha_n^{-2} + \frac{27}{8}\alpha_n^{-4} + O\left(\alpha_n^{-6}\right) \quad , \tag{AD.a.6}$$

$$\rho_n^{-7}(x) \sim 1 + \frac{21}{8}\alpha_n^{-2} + \frac{567}{128}\alpha_n^{-4} + O\left(\alpha_n^{-6}\right) \quad , \tag{AD.a.7}$$

$$\rho_n^{-8}(x) \sim 1 + 3\alpha_n^{-2} + \frac{45}{8}\alpha_n^{-4} + O\left(\alpha_n^{-6}\right) \quad , \tag{AD.a.8}$$

$$\rho_n^{-9}(x) \sim 1 + \frac{27}{8}\alpha_n^{-2} + \frac{891}{128}\alpha_n^{-4} + O\left(\alpha_n^{-6}\right) \quad . \tag{AD.a.9}$$

Note that these expansions are valid uniformly in the integer n, $\forall n \geq 1$, as $|\, x \,| \to \infty$. They are obtained using repeatedly (AD.10) and the well-known expansion of $(1+Y)^{-1}$ as $Y \to 0$.

APPENDIX AD.b

Following section 6.5, coupled gauge modes can be written as

$$g_n^{(j)}(\tau) = \sum_{k=0}^{\infty} a_{n,k}^{(j)}\left(n, k, \lambda_n^{(j)}\right)\tau^{k+\mu} \quad , \tag{AD.b.1}$$

$$R_n^{(j)}(\tau) = \sum_{k=0}^{\infty} b_{n,k}^{(j)}\left(n, k, \lambda_n^{(j)}\right)\tau^{k+\mu-1} \quad . \tag{AD.b.2}$$

The label j is introduced because, for each integer value of $n \geq 2$, there is a whole family $\left\{\lambda_n^{(j)}\right\}$ of eigenvalues labelled by the integer j, say. They are solutions of the equation $g_n(a) = 0$, and their degeneracy $d_j(n) = n^2$, $\forall j \geq 1$ and $\forall n \geq 2$ (Louko 1988). Now, defining $\forall k \geq 2$ and $\forall n \geq 2$

$$G(k, n, \mu) \equiv 2 - \frac{\left(2(k+\mu)^2 - 4(k+\mu) + 3\right)}{(n^2 - 1)} \quad , \tag{AD.b.3}$$

one finds $\forall j \geq 1$

$$\frac{a_{n,2}^{(j)}}{a_{n,0}^{(j)}} = -\frac{G(2, n, \mu)}{F(2, n, \mu)}\lambda_n^{(j)} \quad , \tag{AD.b.4}$$

$$F(k, n, \mu)a_{n,k}^{(j)} + G(k, n, \mu)\lambda_n^{(j)} a_{n,k-2}^{(j)} - \frac{\left(\lambda_n^{(j)}\right)^2}{(n^2 - 1)} a_{n,k-4}^{(j)} = 0 \quad \forall k \geq 4 \quad , \tag{AD.b.5}$$

$$b_{n,0}^{(j)} = \left(\frac{\mu^2}{(n^2 - 1)} - 1\right) a_{n,0}^{(j)} \quad , \tag{AD.b.6}$$

$$b_{n,k}^{(j)} = \frac{\lambda_n^{(j)}}{(n^2 - 1)} a_{n,k-2}^{(j)} + \left(\frac{(k+\mu)^2}{(n^2 - 1)} - 1\right) a_{n,k}^{(j)} \quad \forall k \geq 2 \quad , \tag{AD.b.7}$$

$$a_{n,k}^{(j)} = b_{n,k}^{(j)} = 0 \qquad \forall k = (2m+1) \qquad m = 0, 1, 2, \dots \quad . \tag{AD.b.8}$$

Moreover, setting to 1 for simplicity the three-sphere radius a, magnetic boundary conditions (i.e. $g_n(1) = \dot{R}_n(1) = 0$) lead to

$$\sum_{k=0}^{\infty} a_{n,k}^{(j)} = 0 \quad , \tag{AD.b.9}$$

$$\lambda_n^{(j)} = \left(n^2 - 1\right) - \mu(3\mu - 2) - \frac{\sum_{k=0}^{\infty} k^3 a_{n,k}^{(j)}}{\sum_{k=0}^{\infty} k a_{n,k}^{(j)}} - (3\mu - 1)\frac{\sum_{k=0}^{\infty} k^2 a_{n,k}^{(j)}}{\sum_{k=0}^{\infty} k a_{n,k}^{(j)}} \quad . \tag{AD.b.10}$$

Since, $\forall n \geq 2$, there are two values of $\mu > 1$, a further label is necessary to characterize completely the coupled gauge modes as follows :

$$g_{1,n}^{(j)}\left(n, \lambda_{1,n}^{(j)}, \tau\right) \text{ and } R_{1,n}^{(j)}\left(n, \lambda_{1,n}^{(j)}, \tau\right) \text{ if } \mu = \mu_+^{(1)} \quad ,$$

$$g_{2,n}^{(j)}\left(n, \lambda_{2,n}^{(j)}, \tau\right) \text{ and } R_{2,n}^{(j)}\left(n, \lambda_{2,n}^{(j)}, \tau\right) \text{ if } \mu = \mu_+^{(2)} \quad ,$$

(see (6.5.31)-(6.5.32)).

Note that it is extremely difficult (if not impossible) to find the eigenvalues $\lambda_n^{(j)}$ by analytic or numerical methods, since (AD.b.4)-(AD.b.5) imply that a function H exists such that

$$\frac{a_{n,k}^{(j)}}{a_{n,0}^{(j)}} = H(k, n, \mu)\left(\lambda_n^{(j)}\right)^{\frac{k}{2}} \quad , \tag{AD.b.11}$$

for all even values of $k \geq 2$, and $\forall n \geq 2$. Thus, when (AD.b.11) is inserted into (AD.b.9)-(AD.b.10), it is not clear how to find an explicit solution for $\lambda_n^{(j)}$ and $a_{n,k}^{(j)}$.

CHAPTER SEVEN

LOCAL BOUNDARY CONDITIONS FOR THE WEYL SPINOR

Abstract. The imposition of local boundary conditions on S^3 for the Weyl spinor is shown to imply that either the electric curvature E_{ij} or the magnetic curvature B_{ij} should vanish on S^3. The physical degrees of freedom of the problem are picked out imposing the transverse-traceless gauge condition. Thus, taking as background a flat Euclidean space, the perturbed three-metric is expanded in terms of modes $q^n(\tau)$ and of transverse-traceless hyperspherical harmonics. Preserving in time the gauge condition one finds that the perturbative part of the lapse, and the shift functions, vanish $\forall \tau$. We work at linear order in the perturbations in the expansion of E_{ij} and B_{ij}. If the linearized E_{ij} is vanishing on S^3, this implies that :

$$Q^{(n)}(\tau) = \frac{d^2 q^n}{d\tau^2} + A_1^n(\tau)\frac{dq^n}{d\tau} + A_2^n(\tau)q^n(\tau) = 0 \quad \text{on} \quad S^3 \quad , \quad \forall n \geq 3 \quad ,$$

where $A_1^n(\tau)$ has a first-order pole at $\tau = 0$, and $A_2^n(\tau)$ has a second-order pole at $\tau = 0$. However, when one requires the vanishing of $Q^{(n)}$ on S^3, one cannot find a surface term to add to the linearized Einstein action such that the linearized Einstein equations follow from requiring the action to be stationary. Thus we conclude that fixing the linearized electric curvature on S^3 does not lead to a well-posed classical boundary-value problem. This implies that the corresponding Hartle-Hawking path integral for quantum amplitudes cannot be defined, and that electric boundary conditions cannot be used for pure gravity in one-loop quantum cosmology.

When the linearized B_{ij} is vanishing on S^3, the corresponding condition on the modes is found to be : $\frac{dq^n}{d\tau} = 0$ on S^3, $\forall n \geq 3$. The resulting eigenvalue condition is of the type which occurs for a real scalar field subject to Robin boundary conditions on S^3. We can thus generalize a technique originally developed by Moss. This is found to imply the following PDF value : $\zeta(0) = \frac{112}{45}$. This result seems to strengthen the evidence in favour of no matching being possible for the generalized PDF zeta-functions at the origin for fields of various spins. Finally, the Moss technique is compared with the Laplace-transform technique for the heat equation, and they are proved to be in agreement.

7.1 Local Boundary Conditions for the Spin-2 Field Strength

In this chapter we complete the PDF study of the Breitenlohner-Freedman-Hawking local boundary conditions in the case of bosonic fields, by studying their application to one-loop properties of pure gravity about flat Euclidean backgrounds.

In (5.7.1) we already wrote the spinorial representation of the Weyl tensor for a complex space-time. That relation also holds on the Euclidean section where the metric is real and positive-definite. On the three-sphere we require the local boundary conditions (cf. (5.7.3)) :

$$4n^{AA'}n^{BB'}n^{CC'}n^{DD'}\phi_{ABCD} = \epsilon\, \widetilde{\phi}^{A'B'C'D'} \quad , \tag{7.1.1}$$

where $\epsilon = \pm 1$, and we use the fact that the product of 4 Euclidean normals is equal to the product of 4 Lorentzian normals. Let us now recall that the electric and magnetic curvatures, defined respectively as follows (Ellis 1971, pp 129-130) :

$$E_{ij} \equiv C_{i\mu j\nu}n^{\mu}n^{\nu} \quad , \tag{7.1.2}$$

$$B_{ij} \equiv \frac{1}{2}\epsilon_{j\mu}{}^{kl}\, C_{kli\nu}n^{\mu}n^{\nu} \quad , \tag{7.1.3}$$

obey the identities (D'Eath 1986) :

$$E_{jk} + iB_{jk} = 2n^{A}{}_{B'}\, e^{BB'}{}_{j}\, n^{C}{}_{D'}\, e^{DD'}{}_{k}\, \phi_{ABCD} \quad , \tag{7.1.4}$$

$$E_{jk} - iB_{jk} = 2n_{B}{}^{A'} e^{BB'}{}_{j}\, n_{D}{}^{C'}\, e^{DD'}{}_{k}\, \widetilde{\phi}_{A'B'C'D'} \quad . \tag{7.1.5}$$

The insertion of (7.1.1) into (7.1.5) and the use of (A.9) yield therefore :

$$E_{jk} - iB_{jk} = \epsilon(E_{jk} + iB_{jk}) \quad \text{on} \quad S^3 \quad , \tag{7.1.6}$$

which implies :

$$E_{ij} = 0 \quad \text{on} \quad S^3 \quad \text{if} \quad \epsilon = -1 \quad , \tag{7.1.7}$$

$$B_{ij} = 0 \quad \text{on} \quad S^3 \quad \text{if} \quad \epsilon = 1 \quad . \tag{7.1.8}$$

7.2 One Cannot Fix the Linearized Electric Curvature on S^3

In our problem, the lapse and shift functions can be expanded according to the relations (Halliwell and Hawking 1985) :

$$N = N_0 \left[1 + \sum_n \sigma_n(\tau) \tilde{Q}^{(n)}(\phi^k) \right] \quad , \tag{7.2.1}$$

$$N_i = \sum_n \left[\kappa_n^1(\tau) P_i^{(n)}(\phi^k) + \kappa_n^2(\tau) S_i^{(n)}(\phi^k) \right] \quad . \tag{7.2.2}$$

In (7.2.1), N_0 is the flat background value of N and $\tilde{Q}^{(n)}$ are the scalar harmonics. In (7.2.2), $P_i^{(n)}$ and $S_i^{(n)}$ are longitudinal and transverse vector harmonics respectively. Here, we only need to recall the following properties :

$$P_i^{(n)} = \frac{1}{(n^2-1)} \tilde{Q}_{|i}^{(n)} \quad , \tag{7.2.3}$$

$$S_i^{(n)\,|i} = 0 \quad , \tag{7.2.4}$$

where a stroke denotes covariant differentiation on S^3.

In writing (7.2.1-2), we are regarding N and N_i as perturbations of the flat-space values : $N_0 = 1$, $N^i = 0$. Moreover, we know that the metric of our flat Euclidean background is :

$$g_B = d\tau \otimes d\tau + \tau^2 c_{ij} d\phi^i \otimes d\phi^j \quad , \tag{7.2.5}$$

where $\tau^2 c_{ij} = s_{ij}$, c_{ij} being the metric on a unit three-sphere. The four-metric $g_{\mu\nu}$ is now written as (cf. section 4.1) :

$$g_{\mu\nu} = s_{\mu\nu} + \gamma_{\mu\nu} \quad , \tag{7.2.6}$$

where $\gamma_{\mu\nu}$ is a perturbation regular at the origin. We make the transverse-traceless gauge choice (4.1.2-3), expanding the spatial part of the perturbation according to (4.1.4). The idea is now to work at linear order in the perturbations in the expansion of E_{ij} (whereas in the action we work at quadratic order). This implies that all terms of the kind :

$\rho\rho, \rho\nabla\rho, \nabla\rho\nabla\rho$ are neglected, where ρ denotes any of the modes necessary in expanding γ_{ij}, N and N_i (we have suppressed indices for simplicity of notation). From the physical point of view, this means a cut-off in the short-wavelength region. Thus, using the relation : $n^\mu = \frac{1}{N}(1, -N^m)$, and denoting by the $*$ symbol the linearized quantities, we find using (7.1.2) and (7.1.7) :

$$E^*_{ij} = \frac{C^*_{i0j0}}{N^2} = \frac{C^*_{0i0j}}{N^2} = 0 \quad \text{on} \quad S^3 \quad \text{if} \quad \epsilon = -1 \quad . \tag{7.2.7}$$

We now recall the definition of the Weyl tensor (Hawking and Ellis 1973, Lightman et al. 1975), which implies :

$$C_{0i0j} = {}^{(4)}R_{0i0j} + \frac{1}{2}\Big[g_{0j}\left({}^{(4)}R_{0i}\right) - g_{00}\left({}^{(4)}R_{ij}\right) + g_{i0}\left({}^{(4)}R_{j0}\right) - g_{ij}\left({}^{(4)}R_{00}\right)\Big]$$
$$+ \frac{1}{6}\left(g_{00}g_{ij} - g_{0i}g_{0j}\right){}^{(4)}R \quad , \tag{7.2.8}$$

which in turn leads to :

$$C^*_{0i0j} = \left[{}^{(4)}R_{0i0j} - \frac{1}{2}\left({}^{(4)}R_{ij}\right) - \frac{1}{2}s_{ij}\left({}^{(4)}R_{00}\right) + \frac{1}{6}s_{ij}\left({}^{(4)}R\right)\right]^* \quad . \tag{7.2.9}$$

In our calculation, in view of the property :

$$s^{ij}s_{jk} = c^{ij}c_{jk} = \delta^i{}_k \quad , \tag{7.2.10}$$

if we require that :

$$g^{ij}g_{jk} = \delta^i{}_k \quad , \tag{7.2.11}$$

we find in view of (7.2.6) (neglecting the $\gamma\gamma$ term) :

$$s^{ij}\gamma_{jk} + \gamma^{ij}s_{jk} = 0 \quad , \tag{7.2.12}$$

which implies (using (4.1.4) and the relation : $s^{ij} = \tau^{-2}c^{ij}$) :

$$\gamma^{ij} = \sum_{n=3}^{\infty} -\tau^{-4}q^n(\tau)G^{(n)ij}(\phi^k) \quad . \tag{7.2.13}$$

The three-sphere metric raises and lowers indices of the harmonics according to :

$$G^{(n)ij}c_{jk} = c^{ij}G^{(n)}_{jk} = G^{(n)i}{}_{k} \quad . \tag{7.2.14}$$

Thus the transverse-traceless gauge choice (4.1.2-3) can be written in terms of the harmonics as :

$$G^{(n)l}{}_{i|l} = 0 \quad , \tag{7.2.15}$$

$$G^{(n)i}{}_{i} = c^{ij}G^{(n)}_{ij} = c_{ij}G^{(n)ij} = 0 \quad . \tag{7.2.16}$$

In computing (7.2.9), it is now very useful to know the Euclidean form of the extrinsic-curvature tensor K_{ij} and of the ADM formulae for the curvature. Indeed, the Lorentzian formula for K_{ij} is (cf. appendix E) :

$$K^{L}_{ij} \equiv -\frac{1}{2}(L_{n}h)_{ij} = \frac{1}{2N}\left(-\frac{\partial h_{ij}}{\partial t} + 2N_{(i|j)}\right) \quad .$$

The Euclidean formula for $-\frac{1}{2N}\frac{\partial h_{ij}}{\partial t}$ is obtained multiplying by i this function computed when $t \rightarrow -i\tau$ (this corresponds to the Wick rotation $\tau \rightarrow it$). Hence we have :

$$-\frac{1}{2N}\frac{\partial h_{ij}}{\partial t} \rightarrow \frac{1}{2N}\frac{\partial h_{ij}}{\partial \tau} \quad ,$$

and defining :

$$L_{ij} \equiv 2N_{(i|j)} \quad , \tag{7.2.17}$$

we find :

$$K^{E}_{ij} = \frac{1}{2N}\left(\frac{\partial h_{ij}}{\partial \tau} - L_{ij}\right) \quad . \tag{7.2.18}$$

In the same way, we find that :

$$\frac{1}{N}\frac{\partial K^{L}_{ij}}{\partial t} \rightarrow i\left[\frac{1}{N}\frac{\partial K^{L}_{ij}}{\partial t}(t \rightarrow -i\tau)\right] = -\frac{1}{N}\frac{\partial K^{E}_{ij}}{\partial \tau} \quad , \tag{7.2.19}$$

and another formula where we replace K_{ij} by K in (7.2.19). In the Lorentzian ADM formulae of appendix E, we now carefully replace the time derivatives in the way just explained, and introduce a sign change since the form of our metric is no longer (1.1.1), but (cf. Hawking 1984a-b) :

$$g = \left(N^2 + N_i N^i\right) d\tau \otimes d\tau + N_i \left(dx^i \otimes d\tau + d\tau \otimes dx^i\right) + h_{ij} dx^i \otimes dx^j \quad . \tag{7.2.20}$$

By comparison with Hawking 1984a-b, the reader can check for example that the Euclidean version of the formulae of Lightman et al. 1975 can be obtained setting $\epsilon = 1$ on pp 310-311 of that book and using (7.2.18-19). In our more complicated case, we find, defining :

$$\nabla^2 N \equiv h^{ij} N_{|ij} \quad , \tag{7.2.21}$$

the following Euclidean formulae :

$$\begin{aligned} {}^{(4)}R_{0i0j} &= -\frac{1}{N}\frac{\partial K_{ij}}{\partial \tau} + K_i{}^l K_{lj} \\ &+ \frac{1}{N}\left(N^m K_{ij|m} + N^m{}_{|i} K_{jm} + N^m{}_{|j} K_{im}\right) + \frac{N_{|ij}}{N} \quad , \end{aligned} \tag{7.2.22}$$

$${}^{(4)}R^0{}_0 = -\frac{1}{N}\frac{\partial K}{\partial \tau} - K_{ij}K^{ij} + \frac{N^l}{N} K_{|l} + \frac{\nabla^2 N}{N} \quad , \tag{7.2.23}$$

$$\begin{aligned} {}^{(4)}R_{ij} &= {}^{(3)}R_{ij} - \left[K_{ij}K + \frac{1}{N}\frac{\partial K_{ij}}{\partial \tau} - 2K_{im}K_j{}^m\right] \\ &+ \frac{1}{N}\left(N^m K_{ij|m} + N^m{}_{|i} K_{jm} + N^m{}_{|j} K_{im}\right) + \frac{N_{|ij}}{N} \quad , \end{aligned} \tag{7.2.24}$$

$${}^{(4)}R = {}^{(3)}R - K_{ij}K^{ij} - K^2 - \frac{2}{N}\frac{\partial K}{\partial \tau} + \frac{2}{N}\left(\nabla^2 N + N^p K_{|p}\right) \quad . \tag{7.2.25}$$

In linearized theory, defining :

$$\gamma^c{}_b \equiv s^{ca}\gamma_{ab} \quad , \quad \tilde{\gamma}^{lm} \equiv s^{la}s^{mb}\gamma_{ab} = -\gamma^{lm} \quad , \quad \overline{\gamma}^c{}_b \equiv \gamma^c{}_b - \frac{1}{2}\delta^c{}_b\,\gamma^l{}_l \quad , \tag{7.2.26}$$

and denoting by ${}^{(3)}\widetilde{R}_{abcd}$ the unperturbed three-dimensional Riemann tensor, one finds that the Ricci tensor changes by the amount $\delta^{(3)}R_{ab}$ given by :

$$\begin{aligned} 2\delta^{(3)}R_{ab} &= -\nabla_f \nabla^f \gamma_{ab} + \nabla_a \nabla_c \overline{\gamma}^c{}_b + \nabla_b \nabla_c \overline{\gamma}^c{}_a \\ &- {}^{(3)}\widetilde{R}_{almb}\,\tilde{\gamma}^{lm} - {}^{(3)}\widetilde{R}_{blma}\,\tilde{\gamma}^{lm} \\ &+ {}^{(3)}\widetilde{R}_{la}\,\gamma^l{}_b + {}^{(3)}\widetilde{R}_{lb}\,\gamma^l{}_a \quad . \end{aligned} \tag{7.2.27}$$

We now use the transverse-traceless gauge (7.2.15-16), plus (7.2.13) and the identities for the perturbative part of the three-metric :

$$\gamma^i{}_j = \sum_{n=3}^{\infty} \tau^{-2} q^n(\tau) G^{(n)i}{}_j(\phi^k) \quad , \quad \nabla_f \nabla^f \gamma_{ab} = \tau^{-2} D^2 \gamma_{ab} \quad , \tag{7.2.28}$$

$$-D^2 G_{ij}^{(n)} = (n^2 - 3) G_{ij}^{(n)} \quad , \tag{7.2.29}$$

where $-D^2$ is the Laplacian on a unit three-sphere. This is why we find :

$${}^{(3)}R_{ij} = 2c_{ij} + \sum_{n=3}^{\infty} \left[\frac{(n^2 - 1)}{2\tau^2} q^n G_{ij}^{(n)} \right] \quad , \tag{7.2.30}$$

$${}^{(3)}R = g^{ij} \left({}^{(3)}R_{ij} \right) = \frac{6}{\tau^2} \quad . \tag{7.2.31}$$

Therefore, using also (7.2.3), (7.2.17-18), (7.2.21-25) and the relation $N_0 = 1$, we find that (7.2.9) is given by :

$$\begin{aligned} C^*_{0i0j} = & \sum_{n=3}^{\infty} \left[\left(-\frac{1}{4} \frac{d^2 q^n}{d\tau^2} + \frac{3}{4\tau} \frac{dq^n}{d\tau} - \frac{(n^2 - 1)}{4\tau^2} q^n \right) G_{ij}^{(n)} \right] \\ & + \frac{1}{2} \sum_m \left[\sigma_m(\tau) \left(\widetilde{Q}^{(m)}_{|ij} + \frac{(m^2 - 1)}{3} \widetilde{Q}^{(m)} c_{ij} \right) \right] \\ & - \frac{c_{ij}}{12} \left[\left(c^{km} \frac{\partial L_{km}}{\partial \tau} \right) - \frac{c^{km} L_{km}}{\tau} \right] - \frac{1}{4} \left(-\frac{\partial L_{ij}}{\partial \tau} + \frac{L_{ij}}{\tau} \right) \quad . \end{aligned} \tag{7.2.32}$$

The relation (7.2.32) is indeed in agreement with the traceless nature of E^*_{ij} (D'Eath 1986), thus confirming the validity of our previous formulae. It should be emphasized that only ${}^{(4)}R_{0i0j}$ and ${}^{(4)}R_{ij}$ contribute to the part of C^*_{0i0j} which involves the modes q^n. The full linearized contribution of ${}^{(4)}R_{0i0j}$ and ${}^{(4)}R_{ij}$ (cf. (7.2.22) and (7.2.24)) is obtained by raising indices of the linearized K_{ij} with the perturbed three-metric $g^{ij} = s^{ij} + \gamma^{ij}$ according to the rule : $g^{ij} K_{jl} = K^i{}_l$. The reader can easily check that in so doing, $g^{im} g^{jl} K_{ml}$ yields indeed the correct formula for the linearized K^{ij}. In (7.2.32), the term involving the $G_{ij}^{(n)}$ harmonics is completely independent of the sum of the other three. Thus C^{*}_{0i0j} can only vanish on S^3 if :

$$-\frac{1}{4}\frac{d^2q^n}{d\tau^2}+\frac{3}{4\tau}\frac{dq^n}{d\tau}-\frac{(n^2-1)}{4\tau^2}q^n=0 \quad \forall n\geq 3 \quad , \quad \text{when} \quad \tau=a \ , \tag{7.2.33}$$

plus a boundary condition here denoted as :

$$U_{ij}\left(\sigma_n,\kappa_n^1,\kappa_n^2,c_{ij},\widetilde{Q}^{(n)},S_i^{(n)},P_i^{(n)}\right)=0 \quad , \quad \text{when} \quad \tau=a \ . \tag{7.2.34}$$

We now write in a simple and explicit way (7.2.34) taking into account that our analysis is still incomplete since we have not yet shown what happens requiring the preservation in time of the transverse-traceless gauge choice. For this purpose we use the technique already applied in section 5.10. The perturbative expansion of the Hamiltonian to quadratic order can be obtained from Halliwell and Hawking 1985, provided we set $a(t)=t$ (our background is flat) and $\phi=0$ (there is no massive scalar field in our model). The reader should also bear in mind that in the paper of Halliwell and Hawking perturbative calculations are performed in the Lorentzian regime, whereas so far we have been interested in Euclidean formulae. However, this difference will not affect our conclusions. Following Halliwell and Hawking 1985, the perturbed three-metric γ_{ij} can be expanded according to (again, all degeneracy labels are only denoted by n for simplicity) :

$$\gamma_{ij}=t^2\epsilon_{ij} \quad , \tag{7.2.35}$$

$$\epsilon_{ij}=\sum_n\left[\frac{\sqrt{6}}{3}a_n(t)c_{ij}\widetilde{Q}^{(n)}+\sqrt{6}b_n(t)P_{ij}^{(n)}+\sqrt{2}c_n(t)S_{ij}^{(n)}+2d_n(t)G_{ij}^{(n)}\right] \ . \tag{7.2.36}$$

Moreover, we replace $\sigma_n(\tau)$ in (7.2.1) by $\frac{g_n(t)}{\sqrt{6}}$, $\kappa_n^1(\tau)$ and $\kappa_n^2(\tau)$ in (7.2.2) by $\frac{k_n(t)}{\sqrt{6}}$ and $\sqrt{2}j_n(t)$ respectively. It is thus clear that the transverse-traceless gauge choice can be formulated as follows :

$$a_n\approx b_n\approx c_n\approx 0 \quad , \quad \forall n \ . \tag{7.2.37}$$

This implies that the effective Hamiltonian $\widetilde{H}$ is in our case (see comments before (5.10.6))

$$\widetilde{H} \equiv H + \sum_n \sum_{k=1}^{6} \lambda_{k,n} \phi_n^k \quad , \tag{7.2.38}$$

where :

$$H \equiv H_0 + \sum_n \left[H_2^n + g_n H_1^n + k_n \left({}^{(S)}H_{-1}^n \right) + j_n \left({}^{(V)}H_{-1}^n \right) \right] \quad , \tag{7.2.39}$$

$$\phi_n^1 \equiv a_n \quad , \quad \phi_n^2 \equiv b_n \quad , \quad \phi_n^3 \equiv c_n \quad , \tag{7.2.40}$$

$$\phi_n^4 \equiv \pi_{g_n} \quad , \quad \phi_n^5 \equiv \pi_{k_n} \quad , \quad \phi_n^6 \equiv \pi_{j_n} \quad . \tag{7.2.41}$$

The primary constraints in (7.2.41) arise since the Hamiltonian does not contain momenta conjugate to g_n, k_n and j_n (cf. (5.10.8)). The general formulae for $H_0, H_2^n, H_1^n, {}^{(S)}H_{-1}^n$ and ${}^{(V)}H_{-1}^n$ can be found in Halliwell and Hawking 1985, and adapted to our case as explained before. Requiring the preservation in time of the gauge constraints in (7.2.40) and of the primary constraints in (7.2.41) (Dirac 1964), we find the secondary constraints (here $\alpha \equiv \log(t)$) :

$$\phi_n^7 \equiv \left\{ \phi_n^1, \widetilde{H}_n \right\} = e^{-3\alpha} \left[\pi_\alpha (a_n - g_n) - \left(\pi_{a_n} + \frac{k_n}{3} \right) \right] \quad , \tag{7.2.42}$$

$$\phi_n^8 \equiv \left\{ \phi_n^2, \widetilde{H}_n \right\} = e^{-3\alpha} \left[\frac{(n^2-1)}{(n^2-4)} \pi_{b_n} + 4\pi_\alpha b_n + \frac{k_n}{3} \right] \quad , \tag{7.2.43}$$

$$\phi_n^9 \equiv \left\{ \phi_n^3, \widetilde{H}_n \right\} = e^{-3\alpha} \left[\frac{\pi_{c_n}}{(n^2-4)} + 4\pi_\alpha c_n \right] + e^{-\alpha} j_n \quad , \tag{7.2.44}$$

$$\phi_n^{10} \equiv H_1^n = \frac{e^{-3\alpha}}{2} \left[-a_n \pi_\alpha^2 - 2\pi_\alpha \pi_{a_n} - \frac{2}{3} e^{4\alpha} \left((n^2-4) b_n + \left(n^2 + \frac{1}{2} \right) a_n \right) \right] \quad , \tag{7.2.45}$$

$$\phi_n^{11} \equiv {}^{(S)}H_{-1}^n = \frac{e^{-3\alpha}}{3} \left[-\pi_{a_n} + \pi_{b_n} + \left(a_n + \frac{4(n^2-4)}{(n^2-1)} b_n \right) \pi_\alpha \right] \quad , \tag{7.2.46}$$

$$\phi_n^{12} \equiv {}^{(V)}H_{-1}^n = e^{-\alpha} \left[\pi_{c_n} + 4(n^2-4) c_n \pi_\alpha \right] \quad . \tag{7.2.47}$$

This is why, when the TT gauge condition (7.2.37) is preserved in time, its insertion into (7.2.42-47) yields :

$$\pi_{a_n} \approx \pi_{b_n} \approx \pi_{c_n} \approx 0 \quad , \quad \forall n \quad , \tag{7.2.48}$$

$$g_n \approx k_n \approx j_n \approx 0 \quad , \quad \forall n \quad . \tag{7.2.49}$$

Thus the perturbative part of the lapse, and the shift functions, vanish $\forall \tau$, and (7.2.34) holds not only on S^3 but everywhere inside S^3.

We are now interested in deriving the consequence of (7.2.33), because the modes $q^n(\tau)$ are the ones which appear in the expansion of the action. Thus, defining :

$$x^{(n)}(\tau) \equiv q^n(\tau) \quad , \tag{7.2.50}$$

$$Q^{(n)}(\tau) \equiv \frac{d^2 x^{(n)}}{d\tau^2} - \frac{3}{\tau}\frac{dx^{(n)}}{d\tau} + (n^2-1)\frac{x^{(n)}}{\tau^2} \quad , \tag{7.2.51}$$

we remark that (7.2.33) can be cast in the form :

$$Q^{(n)}(\tau = a) = 0 \quad \forall n \geq 3 \quad , \quad \text{if} \quad \epsilon = -1 \quad . \tag{7.2.52}$$

The relation (7.2.51) can also be inverted using the method of variation of parameters for linear inhomogeneous second-order differential equations. Defining :

$$b_n \equiv \sqrt{n^2-5} \quad , \quad G^{(n)}(\tau, y) \equiv -\frac{\tau^2}{b_n y} \sin\left[b_n \log\left(\frac{y}{\tau}\right)\right] \quad , \tag{7.2.53}$$

one finds :

$$x^{(n)}(\tau) = A\tau^2 \sin\Big[B + b_n \log(\tau)\Big] + \int_0^{\tau} Q^{(n)}(y) G^{(n)}(\tau, y)\, dy \quad , \tag{7.2.54}$$

where A and B are constants. We should now recall the formula for the action quadratic in the perturbations in the TT gauge (Schleich 1985) :

$$16\pi G I_2^{TT} = \frac{1}{4}\int_V d^4x \, \sqrt{s}\, \gamma^{TT\,ab}(-\nabla_c \nabla^c)\gamma_{ab}^{TT} \quad . \tag{7.2.55}$$

Thus, denoting by κ a constant and using (7.2.13) and (7.2.50) we find (the Euclidean time τ is denoted by t from now on) :

$$I_2^{TT} = \kappa \sum_{n=3}^{\infty} I_2^{(n)} \quad , \tag{7.2.56}$$

$$I_2^{(n)} = \int_0^1 \left[\frac{x^{(n)}}{t} \frac{d^2 x^{(n)}}{dt^2} - \frac{x^{(n)}}{t^2} \frac{dx^{(n)}}{dt} - (n^2 - 1) \frac{(x^{(n)})^2}{t^3} \right] dt \ . \tag{7.2.57}$$

The following question is now of primary importance : if perturbative modes are required to be regular $\forall t$, and to obey the boundary condition (7.2.52), is there an action I' whose stationarity leads to the linearized Einstein equations ? Indeed, suppressing for simplicity the label n, the following well-known result holds :

Theorem 7.2.1 The action $I' = I + I_b = I - \frac{x\dot{x}}{t}$ has a variation given by :

$$\delta I' = -2 \left[\frac{\dot{x}}{t} \delta x \right]_0^1 + 2 \int_0^1 \delta x \left[\frac{\ddot{x}}{t} - \frac{\dot{x}}{t^2} - (n^2 - 1) \frac{x}{t^3} \right] dt \ . \tag{7.2.58}$$

This is proved integrating by parts in (7.2.57) and using the relation : $\delta \dot{x} = \frac{d}{dt} \delta x$. Now, the linearized vacuum Einstein equations are :

$$\frac{d^2 x^{(n)}}{dt^2} - \frac{1}{t} \frac{dx^{(n)}}{dt} - (n^2 - 1) \frac{x^{(n)}}{t^2} = 0 \quad . \tag{7.2.59}$$

If we look for solutions of (7.2.59) in the form : $x^{(n)}(t) = B_n t^{\mu_n}$, we find : $\mu_n = 1 \pm n$. We want $x^{(n)}$ to be regular at the origin $\forall n \geq 3$. Hence we discard $\mu_n^- = 1 - n$ and we only take $\mu_n^+ = 1 + n$, which implies, in view of (7.2.51) :

$$Q^{(n)}(t) = B_n \Big(2n^2 - 2n - 4 \Big) t^{n-1} \quad . \tag{7.2.60}$$

Note that $Q^{(n)}(t)$ can only vanish on the boundary, $\forall n \geq 3$, if $B_n = 0$, $\forall n \geq 3$, which in turn implies we have $x^{(n)}(t) = 0$, $\forall n \geq 3$. Thus, denoting by G_{ij}^* the linearized Einstein tensor, we find that $G_{ij}^* = 0$ and ($E_{ij}^* = 0$ *on* S^3) imply that $\delta I' = 0$ with vanishing perturbative modes.

However, the requirements $\delta I' = 0$ and ($E^*_{ij} = 0$ *on* S^3) do not imply that $G^*_{ij} = 0$ (cf. (7.2.58)), so that I' cannot be stationary and lead to the linearized Einstein equations (and no improvement is obtained by setting $E^*_{ij} \neq 0$ *on* S^3). This is why we conclude that setting equal to zero on S^3 the linearized electric curvature, one would use a noncanonical change of coordinates which leads to an ill-posed classical boundary-value problem. Thus it is impossible to define the corresponding Hartle-Hawking path integral for quantum amplitudes, and (7.1.7) cannot be used in one-loop quantum cosmology.

7.3 Calculation of the PDF $\zeta(0)$ when the Linearized Magnetic Curvature is Vanishing on S^3

There is a very rich literature on the eigenvalues in Riemannian geometry (e.g. Chavel 1984, Gallot et al. 1987). We now study another problem of interest within this field. In view of (7.1.3) and (7.1.8), we know that :

$$B^*_{ij} = \frac{\epsilon_j^{\ kl}}{2N^2} C^*_{kli0} = 0 \quad \text{on} \quad S^3 \quad \text{if} \quad \epsilon = 1 \quad . \tag{7.3.1}$$

Indeed, using the definition of the Weyl tensor, we have :

$$C_{kli0} = {}^{(4)}R_{kli0} + \frac{1}{2}\Big[g_{k0}\left({}^{(4)}R_{li}\right) - g_{ki}\left({}^{(4)}R_{l0}\right) + g_{li}\left({}^{(4)}R_{k0}\right) - g_{l0}\left({}^{(4)}R_{ki}\right)\Big]$$
$$+ \frac{1}{6}\Big(g_{ki}g_{l0} - g_{k0}g_{li}\Big){}^{(4)}R \quad , \tag{7.3.2}$$

which implies :

$$C^*_{kli0} = -\left[{}^{(4)}R_{0ikl} + \frac{1}{2}g_{ik}\left({}^{(4)}R_{0l}\right) - \frac{1}{2}g_{il}\left({}^{(4)}R_{0k}\right)\right]^* \quad . \tag{7.3.3}$$

Now, the Euclidean ADM formulae for the curvature can be completed by the following :

$${}^{(4)}R^0{}_{ijl} = K_{il|j} - K_{ij|l} \quad . \tag{7.3.4}$$

We use again the TT gauge (4.1.2-3). In so doing we find :

$$-C^{*}_{kli0} = \frac{1}{2}\sum_{n=3}^{\infty}\frac{dq^{n}}{d\tau}\left(G^{(n)}_{il|k} - G^{(n)}_{ik|l}\right) + \frac{1}{4}\left(L_{il|k} - L_{ik|l}\right) \quad . \qquad (7.3.5)$$

But the second term on the right-hand side of (7.3.5) vanishes when we preserve in time the TT gauge, as proved in section 7.2. Hence B^{*}_{ij} can only vanish on S^3 if :

$$\sum_{n=3}^{\infty}\frac{dq^{n}}{d\tau}(a)\epsilon_{j}^{\ kl}\left(G^{(n)}_{il|k} - G^{(n)}_{ik|l}\right) = 0 \quad . \qquad (7.3.6)$$

The only condition on the modes which ensures the validity of (7.3.6) is :

$$\frac{dq^{n}}{d\tau}(a) = 0 \quad , \quad \forall n \geq 3 \quad . \qquad (7.3.7)$$

We are now interested in evaluating $\zeta(0)$ using (7.3.7). Thus, setting $\tau = t$, we study the kernel of the heat equation for the operator (cf. (4.1.9)) :

$$P_{n} \equiv -\left(\frac{d^{2}}{dt^{2}} - \frac{1}{t}\frac{d}{dt} - \frac{(n^{2}-1)}{t^{2}}\right) \quad , \quad \forall n \geq 3 \quad . \qquad (7.3.8)$$

Using (5.9.27), and denoting by $E > 0$ the eigenvalues of P_n, we find that its eigenfunctions regular at the origin are (again, omitting the multiplicative constant) :

$$u_{n}(t) = tJ_{n}(\sqrt{E}t) = q^{n}(t) \quad . \qquad (7.3.9)$$

Thus the boundary condition (7.3.7) implies the eigenvalue condition :

$$J_{n}(\sqrt{E}a) + \sqrt{E}a\dot{J}_{n}(\sqrt{E}a) = 0 \quad , \quad \forall n \geq 3 \quad . \qquad (7.3.10)$$

A new technique is now strictly needed because (7.3.10) is not of the kind $J_n(\sqrt{E}a) = 0$ or $\dot{J}_n(\sqrt{E}a) = 0$. However, it is very similar to (5.9.58), and indeed they are both of the general kind studied in Moss 1989. Setting $a = 1$ for simplicity, we define the function :

$$F_{n}(z) \equiv J_{n}(z) + z\dot{J}_{n}(z) \quad , \quad \forall n \geq 3 \quad . \qquad (7.3.11)$$

Of course, the consideration of such $F_n(z)$ is suggested by (7.3.10). It only has real simple zeros apart from $z = 0$ (Watson 1966, page 482). The basic idea is now the following (Moss 1989). Given the zeta-function at large x :

$$\zeta(s,x^2) \equiv \sum_n \left(\lambda_n + x^2\right)^{-s} \quad , \tag{7.3.12}$$

one has in four dimensions (see theorem 2 on page 6 of Wong 1989) :

$$\Gamma(3)\zeta(3,x^2) = \int_0^\infty t^2 e^{-x^2 t} G(t)\, dt \sim \sum_{n=0}^\infty B_n \Gamma\left(1+\frac{n}{2}\right) x^{-n-2} \quad , \tag{7.3.13}$$

where we have used the asymptotic expansion of the heat kernel for $t \rightarrow 0^+$:

$$G(t) \sim \sum_{n=0}^\infty B_n t^{\frac{n}{2}-2} \quad . \tag{7.3.14}$$

Strictly speaking, since we have not proved general results on the existence of the asymptotic expansion of the heat kernel, our formula (7.3.14) could be initially regarded as an assumption. However, existence theorems hold for the problems studied in this chapter and in the rest of our book (Greiner 1971, Kennedy 1978-1979, Moss 1989, Branson and Gilkey 1990, Moss and Poletti 1990a-b, Poletti 1990).

On the other hand, one also has the identity :

$$\Gamma(3)\zeta(3,x^2) = -\sum_{p=0}^\infty N_p \left(-\frac{1}{2x}\frac{d}{dx}\right)^3 \log\Bigl((ix)^{-p} F_p(ix)\Bigr) \quad , \tag{7.3.15}$$

where N_p is the degeneracy of the problem. Thus the comparison of (7.3.13) and (7.3.15) can yield the coefficients B_n and in particular $\zeta(0) = B_4$, provided we carefully perform a uniform Debye expansion of $F_p(ix)$. It should be emphasized that this technique seems to be the most efficient. In fact, by using this algorithm Moss has been able to compute $\zeta(0)$ for a real scalar field subject to Robin boundary conditions, whereas the technique of Kennedy based on charge layers on the plane tangent to S^3 failed to provide such a value

(Kennedy 1978-1979). Indeed, the eigenvalue condition (7.3.10) is of the Robin type (just set $\beta = 1$ in (22) of Moss 1989). Thus, making the analytic continuation $x \rightarrow ix$ and then defining $\alpha_p \equiv \sqrt{p^2 + x^2}$, $C \equiv -\log(\sqrt{2\pi})$, we can write (see now appendix F) :

$$\log\Bigl((ix)^{-p}F_p(ix)\Bigr) \sim C - p\log(p+\alpha_p) + \frac{1}{2}\log(\alpha_p) + \alpha_p + \sum_{n=1}^{\infty}\sum_{r=0}^{n} a_{nr}p^{2r}\alpha_p^{-n-2r} \ . \quad (7.3.16)$$

The coefficients a_{nr} in (7.3.16) can be computed by comparison using the formula :

$$\sum_{n=1}^{\infty}\sum_{r=0}^{n} a_{nr}t^{2r} = \sum_{m=1}^{\infty} a_m(t) \quad , \quad (7.3.17)$$

because the $a_m(t)$ are known polynomials in t arising from uniform asymptotic expansions of Bessel functions and their first derivatives (cf. appendix F). Thus, setting $\beta = 1$ in the formulae (29-31) of Moss 1989 for the $a_m(t)$, we find in our case that :

$$a_{10} = \frac{5}{8} \quad , \quad a_{11} = \frac{7}{24} \quad , \quad (7.3.18)$$

$$a_{20} = -\frac{3}{16} \quad , \quad a_{21} = \frac{1}{8} \quad , \quad a_{22} = -\frac{7}{16} \quad , \quad (7.3.19)$$

$$a_{30} = \frac{17}{384} \quad , \quad a_{31} = \frac{389}{640} \quad , \quad a_{32} = -\frac{203}{128} \quad , \quad a_{33} = \frac{1463}{1152} \quad , \quad (7.3.20)$$

plus infinitely many other coefficients we do not strictly need here. We can now insert (7.3.16-20) into (7.3.15), apply three times the differential operator $-\frac{1}{2x}\frac{d}{dx}$, and finally use the contour formula for positive integer values of k (Moss 1989) :

$$\sum_{p=0}^{\infty} p^{2k}\alpha_p^{-2k-m} = \frac{\Gamma\left(k+\frac{1}{2}\right)\Gamma\left(\frac{m}{2}-\frac{1}{2}\right)}{2\Gamma\left(k+\frac{m}{2}\right)} x^{1-m} \quad , \quad \forall k = 1,2,3,... \quad , \quad (7.3.21)$$

and the known properties of the gamma function (Abramowitz and Stegun 1964). Now, writing the asymptotic expansion of the right-hand side of (7.3.15) in the form :

$$\Gamma(3)\zeta(3,x^2) \sim \sum_{n=0}^{\infty} b_n x^{-n-2} \quad , \quad (7.3.22)$$

the comparison with (7.3.13) shows that :

$$\zeta(0) = B_4 = \frac{b_4}{2} = \zeta^I(0) + \zeta^{II}(0) \quad , \tag{7.3.23}$$

since it is well-known that the asymptotic expansion, if it exists, is unique. The two contributions to $\zeta(0)$ are obtained from the following formulae :

$$\Gamma(3)\zeta(3, x^2) \sim \left[\sigma_1 + \sigma_2\right] \sim \sum_{n=0}^{\infty} b_n x^{-n-2} \quad , \tag{7.3.24}$$

$$\sigma_1 \sim -\sum_{p=0}^{\infty} N_p \left(-\frac{1}{2x}\frac{d}{dx}\right)^3 \left[-p\log(p+\alpha_p) + \frac{1}{2}\log(\alpha_p) + \alpha_p\right] \quad , \tag{7.3.25}$$

$$\sigma_2 \sim -\sum_{p=0}^{\infty} N_p \left(-\frac{1}{2x}\frac{d}{dx}\right)^3 \sum_{n=1}^{\infty}\sum_{r=0}^{n} a_{nr} p^{2r} \alpha_p^{-n-2r} \quad . \tag{7.3.26}$$

Bearing in mind (7.3.15-16), we write (7.3.24-26) because we can apply theorem 3 on page 7 of Wong 1989, concerning the differentiation of asymptotic expansions.

Thus $\zeta^I(0)$ (respectively $\zeta^{II}(0)$) is half the coefficient of x^{-6} in the asymptotic expansion of σ_1 (respectively σ_2). We first study the asymptotic expansion of σ_2, since it is easier to perform this calculation. In our problem the degeneracy N_p (Schleich 1985) is :

$$N_p = 0 \quad \forall p = 0, 1, 2 \ , \quad N_p = 2(p^2 - 4) \quad \forall p \geq 3 \quad . \tag{7.3.27}$$

This is why we find :

$$\sigma_2 \sim -\sum_{n=1}^{\infty}\sum_{r=0}^{n} a_{nr} \left(r + \frac{n}{2}\right)\left(r + \frac{n}{2} + 1\right)\left(r + \frac{n}{2} + 2\right)\left[(G - H)(r, x, n)\right] \quad , \tag{7.3.28}$$

where, setting $A = -8$, $B = 2$ (cf. (7.3.27)) we have, using also (7.3.21) :

$$\begin{aligned} G(r, x, n) &= \sum_{p=0}^{\infty} (A + Bp^2) p^{2r} \alpha_p^{-n-2r-6} = O(x^{-n-6}) \\ &+ \frac{A}{2} \frac{\Gamma\left(r + \frac{1}{2}\right)\Gamma\left(\frac{n}{2} + \frac{5}{2}\right)}{\Gamma\left(r + \frac{n}{2}\right)} \frac{x^{-5-n}}{\left(r + \frac{n}{2}\right)\left(r + \frac{n}{2} + 1\right)\left(r + \frac{n}{2} + 2\right)} \\ &+ \frac{B}{2} \frac{\Gamma\left(r + \frac{3}{2}\right)\Gamma\left(\frac{n}{2} + \frac{3}{2}\right)}{\Gamma\left(r + \frac{n}{2}\right)} \frac{x^{-3-n}}{\left(r + \frac{n}{2}\right)\left(r + \frac{n}{2} + 1\right)\left(r + \frac{n}{2} + 2\right)} \ , \end{aligned} \tag{7.3.29}$$

$$H(r,x,n)=\sum_{p=0}^{2}2(p^2-4)p^{2r}\alpha_p^{-n-2r-6}=-6x^{-n-2r-6}\left(1+\frac{1}{x^2}\right)^{-\frac{n}{2}-r-3}$$
$$-8\delta_{r0}x^{-n-6} \quad . \tag{7.3.30}$$

Thus $H(r,x,n)$ gives rise to terms in (7.3.28) which contain x^{-k} with $k\geq 7$, and does not contribute to $\zeta^{II}(0)$. This is why (7.3.28-29) lead to :

$$\zeta^{II}(0)=\frac{1}{2}\Big[-A(a_{10}+a_{11})-B(a_{30}+a_{31}+a_{32}+a_{33})\Big] \quad . \tag{7.3.31}$$

The insertion of (7.3.18), (7.3.20) and (7.3.27) into (7.3.31) finally yields :

$$\zeta^{II}(0)=\frac{11}{3}-\frac{121}{360}=\frac{1199}{360} \quad . \tag{7.3.32}$$

The calculation of (7.3.25) is more involved. Performing the three derivatives and using the identity : $\frac{1}{2x}\frac{d\alpha_p}{dx}=\frac{1}{2\alpha_p}$, we find :

$$\left(\frac{1}{2x}\frac{d}{dx}\right)^3\log\left(\frac{1}{p+\alpha_p}\right)=(p+\alpha_p)^{-3}\left[-\alpha_p^{-3}-\frac{9}{8}p\alpha_p^{-4}-\frac{3}{8}p^2\alpha_p^{-5}\right] \quad . \tag{7.3.33}$$

This intermediate step is very important because it proves that by summing over all integer values of p from 0 to ∞ we get a convergent series. However, to be able to perform the $\zeta(0)$ calculation, it is convenient to use the identity :

$$(p+\alpha_p)^{-3}=\frac{(\alpha_p-p)^3}{x^6} \quad . \tag{7.3.34}$$

Inserting (7.3.34) into (7.3.33) and re-expressing p^2 as $\alpha_p^2-x^2$, we obtain :

$$\left(\frac{1}{2x}\frac{d}{dx}\right)^3\Big[-p\log(p+\alpha_p)\Big]=-px^{-6}+p^2x^{-6}\alpha_p^{-1}+\frac{p^2}{2}x^{-4}\alpha_p^{-3}+\frac{3}{8}p^2x^{-2}\alpha_p^{-5}$$
$$\equiv M(x,\alpha_p,p) \quad , \tag{7.3.35}$$

which implies :

$$\sigma_1\sim\left[\sum_{p=0}^{\infty}N_pM(x,\alpha_p,p)\right]+\sigma_1''\sim\left[\sigma_1'+\sigma_1''\right] \quad , \tag{7.3.36}$$

where :

$$\sigma_1'' = -\sum_{p=0}^{\infty} N_p \left(-\frac{\alpha_p^{-6}}{2} - \frac{3}{8}\alpha_p^{-5} \right)$$
$$= \sum_{p=0}^{\infty}(A+Bp^2)\left(\frac{\alpha_p^{-6}}{2} + \frac{3}{8}\alpha_p^{-5} \right)$$
$$+ \sum_{p=0}^{2}(A+Bp^2)\left(-\frac{\alpha_p^{-6}}{2} - \frac{3}{8}\alpha_p^{-5} \right) \quad . \tag{7.3.37}$$

The infinite sum on the right-hand side of (7.3.37) contributes to $\zeta(0)$ only through the following part :

$$\sum_{p=0}^{\infty} \frac{A}{2}\alpha_p^{-6} = \frac{A}{2}\left[\frac{x^{-6}}{2} + \frac{\pi}{2}\frac{3!!}{4!!}x^{-5} \right] \quad . \tag{7.3.38}$$

The result (7.3.38) is proved by applying the Euler-Maclaurin formula in appendix C to the calculation of $\sum_{p=0}^{\infty}(p^2+x^2)^{-3}$, and then using the formula (3.249.1) on page 294 of Gradshteyn and Ryzhik 1965. Also the finite sum on the right-hand side of (7.3.37) contributes to $\zeta(0)$. In fact one finds (we have $x \rightarrow \infty$) :

$$\sum_{p=0}^{2}(A+Bp^2)\left(-\frac{\alpha_p^{-6}}{2} - \frac{3}{8}\alpha_p^{-5} \right) = -\left(\frac{A}{2} + \frac{B}{2} \right) x^{-6}\left[1 - \frac{3}{x^2} + \frac{6}{x^4} + \dots \right]$$
$$- \frac{A}{2}x^{-6} - \frac{3}{8}Ax^{-5}$$
$$- \frac{3}{8}(A+B)x^{-5}\left[1 - \frac{5}{2x^2} + \frac{35}{8x^4} + \dots \right] , \tag{7.3.39}$$

which implies that the total contribution of σ_1'' to $\zeta(0)$ is given by :

$$\zeta^{Ib}(0) = \frac{1}{2}\left(-A - \frac{B}{2} \right) + \frac{A}{8} = \frac{7}{2} - 1 = \frac{5}{2} \quad . \tag{7.3.40}$$

Thus we have so far :

$$\zeta(0) = \zeta^{I}(0) + \zeta^{II}(0) = \zeta^{Ia}(0) + \zeta^{Ib}(0) + \zeta^{II}(0) \quad , \tag{7.3.41}$$

where :

$$\zeta^{Ib}(0)+\zeta^{II}(0)=\frac{5}{2}+\frac{1199}{360} \quad . \tag{7.3.42}$$

It now remains to compute $\zeta^{Ia}(0)$, i.e. the contribution to $\zeta(0)$ due to σ_1' in (7.3.36). Indeed, one has :

$$\sigma_1' \sim \left[A\sum_{p=0}^{\infty}M(x,\alpha_p,p)+B\sum_{p=0}^{\infty}p^2M(x,\alpha_p,p)-\sum_{p=0}^{2}(A+Bp^2)M(x,\alpha_p,p)\right] \quad . \tag{7.3.43}$$

Let us now denote by $\Sigma^{(a)}$, $\Sigma^{(b)}$ and $\Sigma^{(c)}$ the three sums on the right-hand side of (7.3.43). Both $\Sigma^{(a)}$ and $\Sigma^{(b)}$ contain divergent parts in view of (7.3.35). These *fictitious* divergences may be regularized dividing by α_p^{2s} and then taking the limit as s tends to zero, as explained in Moss 1989. It might not appear *a priori* obvious that this technique leads to unambiguous results, since the limit as $s \rightarrow 0$ is a delicate mathematical point. However, a fundamental consistency check is presented in section 7.4 for all PDF one-loop calculations involving bosonic fields, showing that the method is correct. In performing the calculation we must use the contour formula (7.3.21) and also the following asymptotic expansion (Moss 1989) :

$$\sum_{p=0}^{\infty}p\alpha_p^{-1-n} \sim \frac{x^{1-n}}{\sqrt{\pi}}\sum_{r=0}^{\infty}\frac{2^r}{r!}\widetilde{B}_r x^{-r}\frac{\Gamma\left(\frac{r}{2}+\frac{1}{2}\right)\Gamma\left(\frac{n}{2}-\frac{1}{2}+\frac{r}{2}\right)}{2\Gamma\left(\frac{1}{2}+\frac{n}{2}\right)}\cos\left(\frac{r\pi}{2}\right) \quad , \tag{7.3.44}$$

where $\widetilde{B}_0=1$, $\widetilde{B}_1=-\frac{1}{2}$, $\widetilde{B}_2=\frac{1}{6}$, $\widetilde{B}_4=-\frac{1}{30}$ etc. are Bernoulli numbers. Thus, using the label R for the regularized quantities, we define :

$$\Sigma_R^{(a)} \equiv A\Bigg[-x^{-6}\left(\lim_{s\rightarrow 0}\sum_{p=0}^{\infty}p\alpha_p^{-1-(2s-1)}\right)+x^{-6}\left(\lim_{s\rightarrow 0}\sum_{p=0}^{\infty}p^2\alpha_p^{-2-(2s-1)}\right)$$
$$+\frac{x^{-4}}{2}\left(\lim_{s\rightarrow 0}\sum_{p=0}^{\infty}p^2\alpha_p^{-2-(2s+1)}\right)+\frac{3}{8}x^{-2}\left(\lim_{s\rightarrow 0}\sum_{p=0}^{\infty}p^2\alpha_p^{-2-(2s+3)}\right)\Bigg] . \tag{7.3.45}$$

In view of (7.3.44), the first limit in (7.3.45) gives the following contribution to $\zeta(0)$:

$$\delta_1=-\frac{A}{2}\left(-\frac{\widetilde{B}_2}{\sqrt{\pi}}\Gamma\left(\frac{3}{2}\right)\right)=\frac{A}{24}=-\frac{1}{3} \quad , \tag{7.3.46}$$

whereas the other limits in (7.3.45) do not contribute to $\zeta(0)$ in view of (7.3.21), because one only gets terms proportional to x^{-4}.

Moreover, bearing in mind the identity :

$$\sum_{p=0}^{\infty} p^3 \alpha_p^{-2s} = \sum_{p=0}^{\infty} p \alpha_p^{-1-(2s-3)} - x^2 \sum_{p=0}^{\infty} p \alpha_p^{-1-(2s-1)} \quad , \tag{7.3.47}$$

we also define :

$$\Sigma_R^{(b)} \equiv B \Biggl[- x^{-6} \left(\lim_{s \to 0} \sum_{p=0}^{\infty} p^3 \alpha_p^{-2s} \right) + x^{-6} \left(\lim_{s \to 0} \sum_{p=0}^{\infty} p^4 \alpha_p^{-4-(2s-3)} \right)$$
$$+ \frac{x^{-4}}{2} \left(\lim_{s \to 0} \sum_{p=0}^{\infty} p^4 \alpha_p^{-4-(2s-1)} \right) + \frac{3}{8} x^{-2} \left(\lim_{s \to 0} \sum_{p=0}^{\infty} p^4 \alpha_p^{-4-(2s+1)} \right) \Biggr] . \tag{7.3.48}$$

In view of (7.3.44) and (7.3.47), the first limit in (7.3.48) gives the following contribution to $\zeta(0)$:

$$\delta_2 = -\frac{B}{2} \left(-\frac{\widetilde{B}_4}{4} \right) = -\frac{B}{240} = -\frac{1}{120} \quad . \tag{7.3.49}$$

Note that the second sum in (7.3.47) does not contribute to δ_2 because its only constant term contains $\frac{\Gamma(s+1)}{\Gamma(s)}$, which tends to 0 as $s \to 0$. The other limits in (7.3.48) do not contribute to $\zeta(0)$ in view of (7.3.21), because they only yield terms proportional to x^{-2}.

Finally, the sum $\Sigma^{(c)}$ in (7.3.43) has the following asymptotic behaviour as $x \to \infty$:

$$\Sigma^{(c)} \sim \left[(3A + 9B) x^{-6} + \sum_{k=0}^{\infty} (AC_k + BD_k) x^{-7-k} \right] \quad , \tag{7.3.50}$$

which yields the following contribution to $\zeta(0)$:

$$\delta_3 = \frac{(3A + 9B)}{2} = -3 \quad . \tag{7.3.51}$$

To sum up, we find :

$$\zeta^{Ia}(0) = \delta_1 + \delta_2 + \delta_3 = -\frac{1}{3} - \frac{1}{120} - 3 \quad , \tag{7.3.52}$$

so that the full PDF $\zeta(0)$ is given by (cf. (7.3.41-42)) :

$$\zeta(0)=\zeta^{Ia}(0)+\frac{5}{2}+\frac{1199}{360}=\frac{112}{45} \quad . \tag{7.3.53}$$

We are now in a position to understand the result presented in (5.9.59). In fact, $\zeta(0)$ for a real scalar field subject to Neumann boundary conditions on S^3 has the following general structure :

$$\zeta^N(0)=-\frac{\widetilde{B}}{2}\left(\sum_{r=0}^{3}\widetilde{a}_{3r}\right)-\frac{\widetilde{B}}{240} \quad , \tag{7.3.54}$$

where $\widetilde{B}=1$ since the degeneracy $N_p=p^2$, and the coefficients $\widetilde{a}_{3r}$ take into account the different sign of $J_n(\sqrt{E}a)$ appearing in (5.9.58) with respect to (7.3.10). This implies that one has to set $\beta=-1$ in the general formulae (29-31) of Moss 1989, so that :

$$\zeta^N(0)=-\frac{1}{720}-\frac{1}{240}-\frac{(-1)}{6}=\frac{29}{180} \quad . \tag{7.3.55}$$

Also the contributions $-\frac{1}{4}$ and $-\frac{3}{4}$ to (6.5.42-43) can now be explained. In fact, when the eigenvalue condition (6.5.41) holds, the asymptotic expansions (7.3.25-26) are replaced by :

$$\sigma_1 \sim \left(\frac{1}{2x}\frac{d}{dx}\right)^3\left[-\log(1+\widetilde{\alpha})+\frac{1}{2}\log(\widetilde{\alpha})+\widetilde{\alpha}\right] \quad , \tag{7.3.56}$$

$$\sigma_2 \sim \left(\frac{1}{2x}\frac{d}{dx}\right)^3\sum_{n=1}^{\infty}\sum_{r=0}^{n} z_{nr}\widetilde{\alpha}^{-n-2r} \quad , \tag{7.3.57}$$

where $\widetilde{\alpha}\equiv\sqrt{1+x^2}$. A regularization is still needed, because there are infinitely many solutions of (6.5.41). However, the corresponding zeta-function is not a double sum (cf. (B.1)), but a single infinite sum, since the order of the Bessel function is $p=n=1$. Hence no sum over all values of p appears in (7.3.56-57). It is easy to see that the asymptotic expansion of σ_2 in (7.3.57) does not contribute to $\zeta(0)$, whereas one finds :

$$\sigma_1 \sim \left[\frac{3}{8}x^{-5}-x^{-6}+\frac{x^{-6}}{2}+\sum_{k=7}^{\infty}\omega_k x^{-k}\right] \quad , \tag{7.3.58}$$

which implies :

$$\zeta_{R_1}(0) = \frac{1}{2}\left(-1 + \frac{1}{2}\right) = -\frac{1}{4} \quad . \tag{7.3.59}$$

When the eigenvalue condition $J_1(\sqrt{\lambda_1}) = 0$ holds for the R_1-mode, the square bracket on the right-hand side of (7.3.56) contains $-\frac{1}{2}\log(\widetilde{\alpha})$ rather than $\frac{1}{2}\log(\widetilde{\alpha})$. This leads to :

$$\widetilde{\zeta}_{R_1}(0) = \frac{1}{2}\left(-1 - \frac{1}{2}\right) = -\frac{3}{4} \quad , \tag{7.3.60}$$

since, again, the asymptotic expansion of the corresponding σ_2 does not contribute to $\zeta(0)$.

7.4 Comparison of Different Techniques

An important check is now in order. In appendix F we prove that the techniques for the PDF calculation of $\zeta(0)$ based on the Watson transform and on the Euler-Maclaurin formula are equivalent. Indeed, both these algorithms are used when the Laplace transform of the heat equation is applied. However, in section 7.3 we have used the Moss technique, which does not involve the modified Bessel functions I_n and K_n, but just the function which appears in the eigenvalue condition. It is known (Moss 1989) and it can be easily checked that, for the case of a real scalar field subject to Dirichlet boundary conditions on S^3, these two techniques yield the same value of $\zeta(0)$. However, before comparing the PDF $\zeta(0)$ values obtained in chapter five with (7.3.53), we should understand whether the Moss technique agrees with the PDF higher-spin results derived in Schleich 1985, Louko 1988b, and with our result (5.9.53). Indeed, from appendix F we know that when the eigenvalues are given by $J_n(\sqrt{E}) = 0$, a formula of the kind (7.3.16) holds, but with $-\frac{1}{2}$ (rather than $\frac{1}{2}$) multiplied by $\log(\alpha)$. The coefficients a_{nr} are replaced by other coefficients, here denoted by b_{nr} when $B_i = 0$ on S^3, by e_{nr} when $E_i = 0$ on S^3, and g_{nr} for pure gravity in Schleich's problem (hereafter referred to as PG). After our careful application of the Moss technique in the most difficult bosonic case, the reader can easily check the following general formulae :

$$\zeta_B(0) = -\frac{A'}{2}\sum_{r=0}^{1} b_{1r} - \frac{B'}{2}\sum_{r=0}^{3} b_{3r} + \frac{A'}{24} - \frac{B'}{240} - \frac{A'}{8} + \frac{A'}{4} \quad , \tag{7.4.1}$$

$$\zeta_E(0) = -\frac{A'}{2}\sum_{r=0}^{1} e_{1r} - \frac{B'}{2}\sum_{r=0}^{3} e_{3r} + \frac{A'}{24} - \frac{B'}{240} + \frac{A'}{8} - \frac{A'}{4} \quad , \tag{7.4.2}$$

$$\zeta_{PG}(0) = -\frac{A}{2}\sum_{r=0}^{1} g_{1r} - \frac{B}{2}\sum_{r=0}^{3} g_{3r} + \frac{A}{24} - \frac{B}{240} + \frac{(3A+9B)}{2} + \frac{1}{2}\left(A + \frac{B}{2}\right) - \frac{A}{8}. \tag{7.4.3}$$

In (7.4.1-2), $A' = -B' = -2$ because the degeneracy $d(n) = 2(n^2 - 1)$, $\forall n \geq 2$. The contribution $+\frac{A'}{4}$ (respectively $-\frac{A'}{4}$) is due to a sum of the kind (7.3.37), where we subtract the contribution of $p = 0, 1$. It is the only one which survives, because $A' = -B'$. The coefficients b_{1r} and b_{3r} can be derived from (26-28) in Moss 1989, and one has : $b_{1r} = g_{1r}$, $\forall r = 0, 1$; $b_{3r} = g_{3r}$, $\forall r = 0, 1, 2, 3$. The coefficients e_{1r} and e_{3r} can be derived from (29-31) in Moss 1989 by setting $\beta = 0$. This yields :

$$\zeta_B(0) = -\frac{1}{12} - \frac{1}{360} - \frac{1}{12} - \frac{1}{120} + \frac{1}{4} - \frac{1}{2} = -\frac{1}{6} - \frac{1}{90} - \frac{1}{4} = -\frac{77}{180} \quad , \tag{7.4.4}$$

$$\zeta_E(0) = -\frac{1}{12} - \frac{1}{360} - \frac{1}{12} - \frac{1}{120} - \frac{1}{4} + \frac{1}{2} = -\frac{1}{6} - \frac{1}{90} + \frac{1}{4} = \frac{13}{180} \quad , \tag{7.4.5}$$

$$\zeta_{PG}(0) = -\frac{1}{3} - \frac{1}{360} - \frac{1}{3} - \frac{1}{120} - 3 - \frac{7}{2} + 1 = -\frac{278}{45} \quad . \tag{7.4.6}$$

Hence the two techniques are in agreement.

To sum up, in section 7.3 we have found that, when for pure gravity the linearized magnetic curvature is vanishing on S^3, the PDF value of $\zeta(0)$ has the same sign as in the PDF spin-1 calculation when $E_i = 0$ on S^3, but these values are not equal. In all models we have studied there is no matching at the origin of the generalized PDF zeta-functions, and PDF one-loop finiteness is not obtained in the presence of a S^3 boundary. More precisely, we only proved so far that for bosonic fields the PDF zeta-functions do not match at the origin. However, denoting by $\zeta_s(0)$ the values of the zeta-functions for a spin-s field for any $s = 0, \frac{1}{2}, 1, \frac{3}{2}, 2$, we can remark that : $\zeta_0(0) \neq \zeta_1(0)$ implies that either $\zeta_{\frac{1}{2}}(0) \neq \zeta_0(0)$ or $\zeta_{\frac{1}{2}}(0) \neq \zeta_1(0)$. Similarly, we find that $\zeta_1(0) \neq \zeta_2(0)$ implies that either $\zeta_{\frac{3}{2}}(0) \neq \zeta_1(0)$ or $\zeta_{\frac{3}{2}}(0) \neq \zeta_2(0)$. Thus, the failure of the matching at the origin is partially proved also in comparing bosonic and fermionic fields subject to local boundary conditions on S^3. Our PDF proof will achieve completion in chapter nine. However, a more important question is whether $\zeta(0)$ values in the presence of boundaries can be expressed in terms of the Euler number χ. We here anticipate that both our work and recent work by other authors (Poletti 1990) shows that one-loop results are no longer related to χ when boundaries are present. This is an intriguing property, and a satisfactory interpretation is still lacking at present.

It is worth mentioning here recent work on the spin-2 gravitational trace anomaly (Pascual et al. 1989). These authors do not study pure gravity with local boundary conditions on S^3 for the Weyl spinor as we do. They look at the most general Weyl-invariant quadratic Lagrangian in presence of classical gravity. They find that the part proportional to the Euler-Poincaré characteristic of the contribution of spin-2 fields to the gravitational trace anomaly is of the same sign as all lower-spin contributions, thus making anomaly cancellation impossible. They also explain why their result does not contradict known properties of supergravity. However, our result might have finally implications for the one-loop finiteness of supersymmetric models in the presence of a S^3 boundary, if the relation between PDF calculations and *direct* BRST-invariant calculations could be properly understood (cf. sections 6.5 and 9.3).

Interestingly, the boundary conditions we have studied seem to suggest that the arguments of the wave function of the universe might become the curvatures, so that for example $\psi = \psi(..., E_i, ..., B_{ij})$, or $\psi = \psi(..., B_i, ..., B_{ij})$. Moreover, the electric and magnetic parts of the Weyl tensor are related to Ashtekar's new variables for canonical gravity as follows (Ashtekar 1988, page 118) :

$$Tr\left(\widetilde{\sigma}^j F_{ab}\right)\epsilon^{abk} = \sqrt{2}\, h\left(E^{jk} - iB^{jk}\right) \quad . \tag{7.4.7}$$

This shows an intriguing link between perturbative and nonperturbative quantum gravity. It should be also emphasized that our analysis confirms that for bosonic fields, only Dirichlet, Neumann or Robin boundary conditions lead to self-adjoint, second-order differential operators with positive spectrum and to well-posed classical boundary-value problems (see, however, Schröder 1989). The role of these boundary conditions in the quantum theory of bosonic and fermionic fields has been analyzed in great detail in Luckock 1991.

Last but not least, the following comment is in order. In Poletti 1990, it is argued that the perturbative three-metric can only obey Dirichlet boundary conditions on S^3. In fact the author studies invariance properties of the gravitational action under gauge transformations of the background field : $g_{ab} \rightarrow g_{ab} + \nabla_{(a}\epsilon_{b)}$, where ϵ_b is the gauge parameter subject to the condition $n \cdot \epsilon = 0$. Demanding that the ghosts respect this condition, he concludes that alternative boundary conditions for the perturbed three-metric are ruled out. More precisely, Poletti relies on the work in Luckock 1991, where it is argued that alternative boundary conditions are not physically relevant. However, what we have done in this chapter is radically different, since we have restricted *a priori* pure gravity to its physical degrees of freedom (the transverse-traceless modes for the perturbed three-metric), so that no ghost fields appear in the one-loop calculation. In other words, we should emphasize or anticipate that :

(a) Within the PDF approach, magnetic boundary conditions for pure gravity lead to a well-defined one-loop calculation, because the corresponding eigenvalues are all real and positive;

(b) The PDF one-loop calculation of section 7.3 is especially important because the technique can be generalized to fermionic fields subject to local boundary conditions, as shown in chapters eight and nine;

(c) So far, it has not been possible to apply magnetic boundary conditions for pure gravity to supergravity or extended supergravity models, after taking into account supersymmetry transformation rules.

CHAPTER EIGHT

ONE-LOOP RESULTS FOR THE SPIN-$\frac{1}{2}$ FIELD WITH LOCAL BOUNDARY CONDITIONS

Abstract. The PDF one-loop calculation for pure gravity subject to magnetic boundary conditions on S^3 enables one to derive in a similar way the one-loop properties of fermionic fields subject to local boundary conditions on the three-sphere. For this purpose, we first study the function F which occurs in the nonlinear eigenvalue condition for the massless Majorana spin-$\frac{1}{2}$ field subject to these local boundary conditions on S^3. Using the theory of canonical products, we prove that, in terms of squared eigenvalues, F still obeys a relation of the kind used for $\zeta(0)$ calculations in section 7.3. Using the parameter $x \rightarrow \infty$, after the analytic continuation $x \rightarrow ix$, and defining as usual $\alpha_m \equiv \sqrt{m^2+x^2}$, one can again expand asymptotically a formula of the kind $\log(\Sigma)$ as $\sum_{n=1}^{\infty} \frac{A_n(m,\alpha_m)}{(\alpha_m)^n}$. However, the form of the coefficients A_n is more involved. In fact they are polynomials with both even and odd powers of $t \equiv \frac{m}{\alpha_m}$. Five infinite sums can contribute to $\zeta(0)$. Using the contour formulae of section 7.3 and the uniform asymptotic expansions of the regular Bessel functions J_m and their first derivatives J'_m, we find for a massless Majorana field : $\zeta(0) = \frac{11}{360}$. We also prove that no higher-order terms, i.e. $A_n(m,\alpha_m)$ when $n > 3$, can contribute to our value of $\zeta(0)$.

Using two-component spinor techniques, we finally prove that, if the Dirac operator preserves local boundary conditions involving the normal to the boundary and the spin-$\frac{1}{2}$ field, this implies another boundary condition involving the normal derivatives of the Majorana field, and the trace of the extrinsic-curvature tensor of the boundary. This is done so as to understand the relation between our results and previous work in the literature, and its foundations lie in the general theory of the Dirac operator.

8.1 General Structure of the $\zeta(0)$ Calculation for the Spin-$\frac{1}{2}$ Field Subject to Local Boundary Conditions on S^3

In section 5.8 we derived the eigenvalue condition (5.8.48) which, setting $a = 1$ for simplicity, can be written in the nonlinear form (5.8.47) :

$$F(E) \equiv \Big[J_{n+1}(E)\Big]^2 - \Big[J_{n+2}(E)\Big]^2 = 0 \quad , \quad \forall n \geq 0 \quad . \tag{8.1.1}$$

The function F is the product of the entire functions (functions analytic in the whole complex plane) $F_1 \equiv J_{n+1} - J_{n+2}$ and $F_2 \equiv J_{n+1} + J_{n+2}$, which can be written in the form :

$$F_1(z) \equiv J_{n+1}(z) - J_{n+2}(z) = \gamma_1 z^{(n+1)} e^{g_1(z)} \prod_{i=1}^{\infty} \left(1 - \frac{z}{\mu_i}\right) e^{\frac{z}{\mu_i}} \quad , \tag{8.1.2}$$

$$F_2(z) \equiv J_{n+1}(z) + J_{n+2}(z) = \gamma_2 z^{(n+1)} e^{g_2(z)} \prod_{i=1}^{\infty} \left(1 - \frac{z}{\nu_i}\right) e^{\frac{z}{\nu_i}} \quad . \tag{8.1.3}$$

In (8.1.2-3), γ_1 and γ_2 are constants, g_1 and g_2 are entire functions, the μ_i are the (real) zeros of F_1 and the ν_i are the (real) zeros of F_2. In fact, using the terminology in Ahlfors 1966 (see pp 194-195 therein), F_1 and F_2 are entire functions whose canonical product has genus 1. Namely, in light of the asymptotic behaviour of the eigenvalues (cf. (5.8.50-51)), we know that $\sum_{i=1}^{\infty} \frac{1}{|\mu_i|} = \infty$ and $\sum_{i=1}^{\infty} \frac{1}{|\nu_i|} = \infty$, whereas $\sum_{i=1}^{\infty} \frac{1}{|\mu_i|^2}$ and $\sum_{i=1}^{\infty} \frac{1}{|\nu_i|^2}$ are convergent. This is why $e^{\frac{z}{\mu_i}}$ and $e^{\frac{z}{\nu_i}}$ must appear in (8.1.2-3), which are called the canonical-product representations of F_1 and F_2. The genus of the canonical product for F_1 is the minimum integer h such that $\sum_{i=1}^{\infty} \frac{1}{|\mu_i|^{h+1}}$ converges, and similarly for F_2, replacing μ_i with ν_i. If the genus is equal to 1, this ensures that no higher powers of $\frac{z}{\mu_i}$ and $\frac{z}{\nu_i}$ are needed in the argument of the exponential. However, there is a very simple relation between μ_i and ν_i. In fact, if J_{n+1} is an even function (i.e. if n is odd), then J_{n+2} is an odd function, and viceversa. Thus :

$$J_{n+1}(-z) - J_{n+2}(-z) = J_{n+1}(z) + J_{n+2}(z) \ \text{if } n \text{ is odd} \ , \tag{8.1.4}$$

$$J_{n+1}(-z) - J_{n+2}(-z) = -J_{n+1}(z) - J_{n+2}(z) \text{ if } n \text{ is even} \ . \tag{8.1.5}$$

The relations (8.1.4-5) imply that the zeros of F_1 are minus the zeros of F_2 : $\mu_i = -\nu_i, \forall i$. This is of course just the symmetry property of the eigenvalues pointed out in section 5.8, following (5.8.94). This implies in turn that (cf. (8.1.1)) :

$$F(z) = \widetilde{\gamma} z^{2(n+1)} e^{(g_1+g_2)(z)} \prod_{i=1}^{\infty} \left(1 - \frac{z^2}{\mu_i^2}\right) \quad , \tag{8.1.6a}$$

where $\widetilde{\gamma} = \gamma_1 \gamma_2$, and μ_i^2 are the positive zeros of $F(z)$. It turns out that the function $(g_1 + g_2)$ is actually a constant, so that we can write :

$$F(z) = F(-z) = \gamma z^{2(n+1)} \prod_{i=1}^{\infty} \left(1 - \frac{z^2}{\mu_i^2}\right) \quad , \tag{8.1.6b}$$

In fact, the following theorem holds (see Titchmarsh 1939, pp 250-251, and in particular Ivić 1985, pp 12-17).

Theorem 8.1.1 Let f be an entire function. If $\forall \epsilon > 0 \ \exists \ A_\epsilon$ such that :

$$\log \max\left\{1, \mid f(z) \mid\right\} \leq A_\epsilon \mid z \mid^{1+\epsilon} \quad , \tag{8.1.7}$$

then f can be expressed in terms of its zeros as :

$$f(z) = e^{A+Bz} \prod_{i=1}^{\infty} \left(1 - \frac{z}{\widetilde{\nu}_i}\right) e^{\frac{z}{\widetilde{\nu}_i}} \quad . \tag{8.1.8}$$

If we now apply theorem 8.1.1 to the functions $F_1(z)z^{-(n+1)}$ and $F_2(z)z^{-(n+1)}$ (cf. (8.1.2-3)), we discover that the well-known formula (Abramowitz and Stegun 1964, relation 9.1.21 on page 360) :

$$J_n(z) = \frac{i^{-n}}{\pi} \int_0^\pi e^{iz\cos\theta} \cos(n\theta) \, d\theta \quad , \tag{8.1.9}$$

leads to the fulfillment of (8.1.7) for $F_1(z)z^{-(n+1)}$ and $F_2(z)z^{-(n+1)}$. Hence these functions satisfy (8.1.8) with constants A_1 and B_1 for $F_1(z)z^{-(n+1)}$, and constants A_2 and $B_2 = -B_1$

for $F_2(z)z^{-(n+1)}$. The fact that $B_2 = -B_1$ is well-understood if we look again at (8.1.4-5). We are most indebted to Dr. R. Pinch for providing this argument.

A relation of the kind (8.1.6b) is also obeyed by the function defined in (7.3.11) for the $\zeta(0)$ calculation of chapter seven, but in that case one has p rather than the even number $2(n+1)$. The application of the method of section 7.3 makes it necessary to re-express the function F in (8.1.1) in terms of Bessel functions and their first derivatives of the same order. Thus, defining $m \equiv n+2$, and using the identity :

$$J_l'(x) = J_{l-1}(x) - \frac{l}{x}J_l(x) \quad , \tag{8.1.10}$$

we find :

$$\begin{aligned} J_{m-1}^2(x) - J_m^2(x) &= \left(J_m' + \frac{m}{x}J_m - J_m\right)\left(J_m' + \frac{m}{x}J_m + J_m\right) \\ &= J_m'^{\,2} + \left(\frac{m^2}{x^2} - 1\right)J_m^2 + 2\frac{m}{x}J_mJ_m' \quad . \end{aligned} \tag{8.1.11}$$

This is why, making the analytic continuation $x \rightarrow ix$, and then defining $\alpha_m \equiv \sqrt{m^2+x^2}$ and using the uniform asymptotic expansions (F.13-14) we obtain :

$$J_{m-1}^2(ix) - J_m^2(ix) \sim \frac{(ix)^{2(m-1)}}{2\pi}\alpha_m e^{2\alpha_m} e^{-2m\log(m+\alpha_m)}\left[\Sigma_1^2 + \Sigma_2^2 + 2\frac{m}{\alpha_m}\Sigma_1\Sigma_2\right] . \tag{8.1.12}$$

Thus we only need to multiply the left-hand side of (8.1.12) by $(ix)^{-2(m-1)}$ so as to get a function of the kind : $\frac{\alpha_m}{2\pi}e^{2\alpha_m}e^{-2m\log(m+\alpha_m)}\widetilde{\Sigma}$, similarly to section 7.3. We now define :

$$t \equiv \frac{m}{\alpha_m} \quad , \quad \widetilde{\Sigma} \equiv \Sigma_1^2 + \Sigma_2^2 + 2t\Sigma_1\Sigma_2 \quad , \tag{8.1.13}$$

and study the asymptotic expansion of $\log(\widetilde{\Sigma})$ in the relation :

$$\log\left[(ix)^{-2(m-1)}\left(J_{m-1}^2 - J_m^2\right)(ix)\right] \sim -\log(2\pi) + \log(\alpha_m) + 2\alpha_m - 2m\log(m+\alpha_m) + \log(\widetilde{\Sigma}). \tag{8.1.14}$$

This is obtained bearing in mind that the functions Σ_1 and Σ_2 in (F.13-14) and (8.1.12-13) have asymptotic series given by :

$$\Sigma_1 \sim \sum_{k=0}^{\infty} \frac{u_k(\frac{m}{\alpha_m})}{m^k} \sim \left[1 + \frac{a_1(t)}{\alpha_m} + \frac{a_2(t)}{\alpha_m^2} + \frac{a_3(t)}{\alpha_m^3} + \ldots\right] \quad , \tag{8.1.15}$$

$$\Sigma_2 \sim \sum_{k=0}^{\infty} \frac{v_k(\frac{m}{\alpha_m})}{m^k} \sim \left[1 + \frac{b_1(t)}{\alpha_m} + \frac{b_2(t)}{\alpha_m^2} + \frac{b_3(t)}{\alpha_m^3} + \ldots\right] \quad , \tag{8.1.16}$$

so that :

$$a_i(t) = \frac{u_i(\frac{m}{\alpha_m})}{(\frac{m}{\alpha_m})^i} = \frac{u_i(t)}{t^i} \quad , \quad b_i(t) = \frac{v_i(\frac{m}{\alpha_m})}{(\frac{m}{\alpha_m})^i} = \frac{v_i(t)}{t^i} \quad , \quad \forall i \geq 0 \quad , \tag{8.1.17}$$

where $u_i(t)$ and $v_i(t)$ are the polynomials already defined in (4.4.17-22). The relations (8.1.13-16) lead to the asymptotic expansion :

$$\widetilde{\Sigma} \sim \left[c_0 + \frac{c_1}{\alpha_m} + \frac{c_2}{\alpha_m^2} + \frac{c_3}{\alpha_m^3} + \ldots\right] \quad , \tag{8.1.18}$$

where :

$$c_0 = 2(1+t) \quad , \quad c_1 = 2(1+t)(a_1 + b_1) \quad , \tag{8.1.19}$$

$$c_2 = a_1^2 + b_1^2 + 2(1+t)(a_2 + b_2) + 2ta_1 b_1 \quad , \tag{8.1.20}$$

$$c_3 = 2(1+t)(a_3 + b_3) + 2(a_1 a_2 + b_1 b_2) + 2t(a_1 b_2 + a_2 b_1) \quad . \tag{8.1.21}$$

Higher-order terms have not been computed in (8.1.18) because they do not affect the result for $\zeta(0)$, as we will prove in detail in section 8.4. Since t is variable, it is necessary to define :

$$\Sigma \equiv \frac{\widetilde{\Sigma}}{c_0} \sim \left[1 + \frac{(\frac{c_1}{c_0})}{\alpha_m} + \frac{(\frac{c_2}{c_0})}{\alpha_m^2} + \frac{(\frac{c_3}{c_0})}{\alpha_m^3} + \ldots\right] \quad . \tag{8.1.22}$$

We can now expand $\log(\Sigma)$ according to the usual algorithm valid as $\omega \rightarrow 0$:

$$\log(1+\omega) = \omega - \frac{\omega^2}{2} + \frac{\omega^3}{3} - \frac{\omega^4}{4} + \frac{\omega^5}{5} + \ldots \quad , \tag{8.1.23}$$

as we did in section 7.3 when we wrote (7.3.16) following Moss 1989. Hence we find :

$$\log(\widetilde{\Sigma}) = \Big[\log(c_0) + \log(\Sigma)\Big] \sim \Big[\log(c_0) + \frac{A_1}{\alpha_m} + \frac{A_2}{\alpha_m^2} + \frac{A_3}{\alpha_m^3} + ...\Big] \quad , \tag{8.1.24}$$

where :

$$A_1 = \left(\frac{c_1}{c_0}\right) \quad , \quad A_2 = \left(\frac{c_2}{c_0}\right) - \frac{\left(\frac{c_1}{c_0}\right)^2}{2} \quad , \tag{8.1.25}$$

$$A_3 = \left(\frac{c_3}{c_0}\right) - \left(\frac{c_1}{c_0}\right)\left(\frac{c_2}{c_0}\right) + \frac{\left(\frac{c_1}{c_0}\right)^3}{3} \quad . \tag{8.1.26}$$

Using (4.4.17-22) and (8.1.17-26), we find after a lengthy calculation the following fundamental result :

$$A_1 = \sum_{r=0}^{2} k_{1_r} t^r \quad , \quad A_2 = \sum_{r=0}^{4} k_{2_r} t^r \quad , \quad A_3 = \sum_{r=0}^{6} k_{3_r} t^r \quad , \tag{8.1.27}$$

where :

$$k_{10} = -\frac{1}{4} \quad , \quad k_{11} = 0 \quad , \quad k_{12} = \frac{1}{12} \quad , \tag{8.1.28}$$

$$k_{20} = 0 \quad , \quad k_{21} = -\frac{1}{8} \quad , \quad k_{22} = k_{23} = \frac{1}{8} \quad , \quad k_{24} = -\frac{1}{8} \quad , \tag{8.1.29}$$

$$k_{30} = \frac{5}{192} \quad , \quad k_{31} = -\frac{1}{8} \quad , \quad k_{32} = \frac{9}{320} \quad , \quad k_{33} = \frac{1}{2} \quad , \tag{8.1.30}$$

$$k_{34} = -\frac{23}{64} \quad , \quad k_{35} = -\frac{3}{8} \quad , \quad k_{36} = \frac{179}{576} \quad . \tag{8.1.31}$$

The relations (8.1.13-14), (8.1.19) and (8.1.27-31) finally lead to the formula :

$$\log\Big[(ix)^{-2(m-1)}\left(J_{m-1}^2 - J_m^2\right)(ix)\Big] \sim \sum_{i=1}^{5} S_i(m, \alpha_m(x)) + \text{higher}-\text{order terms} \; , \tag{8.1.32}$$

where :

$$S_1(m, \alpha_m(x)) \equiv -\log(\pi) + 2\alpha_m \quad , \tag{8.1.33}$$

$$S_2(m,\alpha_m(x)) \equiv -(2m-1)\log(m+\alpha_m) \quad , \tag{8.1.34}$$

$$S_3(m,\alpha_m(x)) \equiv \sum_{r=0}^{2} k_{1_r} m^r \alpha_m^{-r-1} \quad , \tag{8.1.35}$$

$$S_4(m,\alpha_m(x)) \equiv \sum_{r=0}^{4} k_{2_r} m^r \alpha_m^{-r-2} \quad , \tag{8.1.36}$$

$$S_5(m,\alpha_m(x)) \equiv \sum_{r=0}^{6} k_{3_r} m^r \alpha_m^{-r-3} \quad . \tag{8.1.37}$$

This is why, defining :

$$W_\infty \equiv \sum_{m=0}^{\infty} (m^2-m)\left(\frac{1}{2x}\frac{d}{dx}\right)^3 \left[\sum_{i=1}^{5} S_i(m,\alpha_m(x))\right] = \sum_{i=1}^{5} W_\infty^{(i)} \quad , \tag{8.1.38}$$

we find (see comments following (7.3.24-26)) :

$$\Gamma(3)\zeta(3,x^2) = \sum_{m=2}^{\infty} (m^2-m)\left(\frac{1}{2x}\frac{d}{dx}\right)^3 \log\left[(ix)^{-2(m-1)}\left(J_{m-1}^2 - J_m^2\right)(ix)\right]$$

$$\sim W_\infty + \sum_{n=5}^{\infty} \hat{q}_n x^{-2-n} \quad . \tag{8.1.39}$$

In deriving the fundamental formulae (8.1.29-31), the relevant intermediate steps are the following (see again (8.1.17-26)) :

$$\frac{c_2}{c_0} = a_2 + b_2 + a_1 b_1 + \frac{(a_1-b_1)^2}{2(1+t)} \quad , \tag{8.1.40}$$

$$(a_1-b_1)^2 = \frac{1}{4}(1-t)^2(1+t)^2 \quad , \tag{8.1.41}$$

$$\frac{c_2}{c_0} = \frac{1}{32} - \frac{t}{8} + \frac{5}{48}t^2 + \frac{t^3}{8} - \frac{35}{288}t^4 \quad , \tag{8.1.42}$$

$$-\left(\frac{c_1}{c_0}\right)\left(\frac{c_2}{c_0}\right) + \frac{\left(\frac{c_1}{c_0}\right)^3}{3} = \frac{1}{384} - \frac{t}{32} + \frac{11}{384}t^2 + \frac{t^3}{24} - \frac{47}{1152}t^4 - \frac{t^5}{96} + \frac{107}{10368}t^6 \, , \tag{8.1.43}$$

$$\frac{c_3}{c_0} = a_3 + b_3 + a_1 b_2 + a_2 b_1 + \frac{\Big[(a_2 - b_2)(a_1 - b_1)\Big]}{(1+t)} \quad , \tag{8.1.44}$$

$$a_3 + b_3 + a_1 b_2 + a_2 b_1 = -\frac{9}{128} + \frac{293}{640}t^2 - \frac{787}{1152}t^4 + \frac{3115}{10368}t^6 \quad , \tag{8.1.45}$$

$$\frac{\Big[(a_2 - b_2)(a_1 - b_1)\Big]}{(1+t)} = \frac{3}{32} - \frac{3}{32}t - \frac{11}{24}t^2 + \frac{11}{24}t^3 + \frac{35}{96}t^4 - \frac{35}{96}t^5 \quad . \tag{8.1.46}$$

8.2 Contribution of $W_\infty^{(1)}$ and $W_\infty^{(2)}$

The term $W_\infty^{(1)}$ does not contribute to $\zeta(0)$. In fact, using (8.1.33), (8.1.38), (7.3.21) and (7.3.44) we find :

$$W_\infty^{(1)} = \frac{3}{4}\sum_{m=0}^{\infty}\left(m^2 - m\right)\alpha_m^{-5}$$

$$\sim \frac{x^{-2}}{4} - \frac{3}{4}\frac{x^{-3}}{\Gamma(\frac{1}{2})}\sum_{r=0}^{\infty}\frac{2^r}{r!}\widetilde{B}_r x^{-r}\frac{\Gamma\left(\frac{r}{2}+\frac{1}{2}\right)\Gamma\left(\frac{r}{2}+\frac{3}{2}\right)}{2\Gamma\left(\frac{5}{2}\right)}\cos\left(\frac{r\pi}{2}\right) \quad , \tag{8.2.1}$$

which implies that x^{-6} does not appear.

Now the series for $W_\infty^{(2)}$ is convergent, as may be checked using (8.1.34), (8.1.38) and (7.3.33). However, when the sum over m is rewritten using the splitting (7.3.34), the individual pieces become divergent. As in section 7.3, these *fictitious* divergences may be regularized dividing by α_m^{2s}, summing using (7.3.21) and (7.3.44), and then taking the limit $s \rightarrow 0$. With this understanding, and using the sums ρ_i defined in (F.15-18) we find :

$$W_\infty^{(2)} = -2x^{-6}\rho_1 + 2x^{-6}\rho_2 + x^{-4}\rho_3 + \frac{3}{4}x^{-2}\rho_4 + 3x^{-6}\rho_5 - 3x^{-6}\rho_6$$

$$-\frac{3}{2}x^{-4}\rho_7 - \frac{9}{8}x^{-2}\rho_8 - x^{-6}\rho_9 + x^{-6}\rho_{10} + \frac{x^{-4}}{2}\rho_{11} + \frac{3}{8}x^{-2}\rho_{12} \quad . \tag{8.2.2}$$

When odd powers of m greater than 1 occur, we can still apply (7.3.44) after re-expressing m^2 as $\alpha_m^2 - x^2$. Thus, applying again the contour formulae (7.3.21) and (7.3.44), only ρ_1 and ρ_9 are found to contribute to $\zeta^{(2)}(0)$, leading to :

$$\zeta^{(2)}(0) = -\frac{1}{120} + \frac{1}{24} = \frac{1}{30} \quad . \tag{8.2.3}$$

8.3 Effect of $W_\infty^{(3)}, W_\infty^{(4)}$ and $W_\infty^{(5)}$

The term $W_\infty^{(3)}$ does not contribute to $\zeta(0)$. In fact, using the relations (8.1.35) and (8.1.38) we find :

$$W_\infty^{(3)} = -\frac{1}{8}\sum_{r=0}^{2} k_{1r}(r+1)(r+3)(r+5)\left[\sum_{m=0}^{\infty}\left(m^2-m\right)m^r\alpha_m^{-r-7}\right] \quad . \tag{8.3.1}$$

In light of (7.3.21) and (7.3.44), x^{-6} does not appear in the asymptotic expansion of (8.3.1) at large x.

For the term $W_\infty^{(4)}$ a remarkable cancellation occurs. In fact, using (8.1.36) and (8.1.38) we find :

$$W_\infty^{(4)} = -\frac{1}{8}\sum_{r=0}^{4} k_{2r}(r+2)(r+4)(r+6)\left[\sum_{m=0}^{\infty}\left(m^2-m\right)m^r\alpha_m^{-r-8}\right] \quad . \tag{8.3.2}$$

The application of (7.3.21), (7.3.44) and (8.1.29) leads to :

$$\zeta^{(4)}(0) = \frac{1}{2}\sum_{r=0}^{4} k_{2r} = 0 \quad . \tag{8.3.3}$$

Finally, using (8.1.37-38) we find :

$$W_\infty^{(5)} = -\frac{1}{8}\sum_{r=0}^{6} k_{3r}(r+3)(r+5)(r+7)\left[\sum_{m=0}^{\infty}\left(m^2-m\right)m^r\alpha_m^{-r-9}\right] \quad . \tag{8.3.4}$$

Again, the contour formulae (7.3.21) and (7.3.44) lead to :

$$\zeta^{(5)}(0) = -\frac{1}{2}\sum_{r=0}^{6} k_{3r} = -\frac{1}{360} \quad , \tag{8.3.5}$$

in light of (8.1.30-31).

8.4 Vanishing Effect of Higher-Order Terms

We now prove the statement made after (8.1.21), i.e. that we do not need to compute the explicit form of c_k in (8.1.18), $\forall k > 3$. In fact, the formulae (8.1.25-26) can be completed by :

$$A_4 = \left(\frac{c_4}{c_0}\right) - \frac{\left(\frac{c_2}{c_0}\right)^2}{2} - \left(\frac{c_1}{c_0}\right)\left(\frac{c_3}{c_0}\right) + \left(\frac{c_1}{c_0}\right)^2\left(\frac{c_2}{c_0}\right) - \frac{\left(\frac{c_1}{c_0}\right)^4}{4} \quad , \tag{8.4.1}$$

plus infinitely many others, and the general term has the structure :

$$A_n = \sum_{p=1}^{l} h_{np}(1+t)^{-p} + \sum_{r=0}^{2n} k_{nr}t^r \quad , \quad \forall n \geq 1 \quad , \tag{8.4.2}$$

where $l < n$, the h_{np} are constants, and r assumes both odd and even values. The integer n appearing in (8.4.2) should not be confused with the one occurring in (8.1.1) and in the definition of m. We have indeed proved that $h_{11} = h_{21} = h_{31} = 0$, but the calculation of h_{np} for all values of n is not obviously feasible. However, we will show that the exact value of h_{np} does not affect the $\zeta(0)$ value. Thus, $\forall n > 3$, we must study :

$$H_{\infty}^{(n)} \equiv \sum_{m=0}^{\infty} (m^2 - m)\left(\frac{1}{2x}\frac{d}{dx}\right)^3 \left[\frac{A_n}{(\alpha_m)^n}\right] = H_{\infty}^{n,A} + H_{\infty}^{n,B} \quad , \tag{8.4.3}$$

where, defining :

$$a_{np} \equiv (p-n)(p-n-2)(p-n-4) \ , \ b_{np} \equiv 3\Big(-p^3 + (3+2n)p^2 - (n^2+3n+1)p\Big) \ , \tag{8.4.4a}$$

$$c_{np} \equiv 3\Bigr(p^3 - (p^2+p)\,n - p\Bigr) \quad , \quad d_{np} \equiv -p(p+1)(p+2) \quad , \tag{8.4.4b}$$

one has :

$$H_{\infty}^{n,A} \equiv \sum_{p=1}^{l} h_{np} \sum_{m=0}^{\infty} (m^2 - m) \left(\frac{1}{2x}\frac{d}{dx}\right)^3 \Bigr[\alpha_m^{p-n}(m+\alpha_m)^{-p}\Bigr]$$

$$= \sum_{p=1}^{l} \frac{h_{np}}{8} \sum_{m=0}^{\infty} (m^2 - m) \Bigr[a_{np}\alpha_m^{p-n-6}(m+\alpha_m)^{-p} + b_{np}\alpha_m^{p-n-5}(m+\alpha_m)^{-p-1}$$

$$+ c_{np}\alpha_m^{p-n-4}(m+\alpha_m)^{-p-2} + d_{np}\alpha_m^{p-n-3}(m+\alpha_m)^{-p-3}\Bigr] \quad , \tag{8.4.5}$$

$$H_{\infty}^{n,B} \equiv -\frac{1}{8}\sum_{r=0}^{2n} k_{nr}(r+n)(r+n+2)(r+n+4)\left[\sum_{m=0}^{\infty}(m^2-m)\, m^r \alpha_m^{-r-n-6}\right] \ . \tag{8.4.6}$$

Because we are only interested in understanding the behaviour of (8.4.5) as a function of x, the application of the Euler-Maclaurin formula (C.2) is more useful than the splitting (7.3.34). In so doing, we find that the part of (C.2) involving the integral on the left-hand side, when $n = \infty$, contains the least negative power of x. Thus, if we prove that the conversion of (8.4.5) into an integral only contains x^{-l} with $l > 6$, $\forall n > 3$, we have proved that $H_{\infty}^{n,A}$ does not contribute to $\zeta(0)$, $\forall n > 3$. This is indeed the case, because in so doing we deal with the integrals defined in (F.19-26), where $I_1^{(np)}, I_3^{(np)}, I_5^{(np)}$ and $I_7^{(np)}$ are proportional to x^{-3-n}, and $I_2^{(np)}, I_4^{(np)}, I_6^{(np)}$ and $I_8^{(np)}$ are proportional to x^{-4-n}, where $n > 3$.

Finally, in (8.4.6) we must study the case when r is even and the case when r is odd. In so doing, defining :

$$\Sigma_{(I)} \equiv \sum_{m=0}^{\infty} m^{2+r}\alpha_m^{-r-n-6} \quad , \quad \Sigma_{(II)} \equiv \sum_{m=0}^{\infty} m^{1+r}\alpha_m^{-r-n-6} \quad , \tag{8.4.7}$$

we find for $r = 2k > 0$ $(k = 1, 2, ...)$:

$$\Sigma_{(I)} = \frac{x^{-3-n}}{2}\frac{\Gamma\left(k+\frac{3}{2}\right)\Gamma\left(\frac{n}{2}+\frac{3}{2}\right)}{\Gamma\left(3+k+\frac{n}{2}\right)} \quad , \tag{8.4.8}$$

and for $r = 2k+1$ $(k = 0,1,2,...)$:

$$\Sigma_{(I)} \sim \frac{x^{-3-n}}{\Gamma\left(\frac{1}{2}\right)} \sum_{l=0}^{\infty} \left\{ \frac{2^l}{l!} \frac{\widetilde{B}_l}{2} x^{-l} \Gamma\left(\frac{l}{2}+\frac{1}{2}\right) \cos\left(\frac{l\pi}{2}\right) \right.$$

$$\left. \left[\frac{\Gamma\left(\frac{n}{2}+\frac{3}{2}+\frac{l}{2}\right)}{\Gamma\left(\frac{n}{2}+\frac{5}{2}\right)} + ... + (-1)^{(1+k)} \frac{\Gamma\left(\frac{n}{2}+\frac{5}{2}+k+\frac{l}{2}\right)}{\Gamma\left(\frac{n}{2}+\frac{7}{2}+k\right)} \right] \right\} . \tag{8.4.9}$$

Moreover, we find for $r = 2k > 0$ $(k = 1,2,...)$:

$$\Sigma_{(II)} \sim \frac{x^{-4-n}}{\Gamma\left(\frac{1}{2}\right)} \sum_{l=0}^{\infty} \left\{ \frac{2^l}{l!} \frac{\widetilde{B}_l}{2} x^{-l} \Gamma\left(\frac{l}{2}+\frac{1}{2}\right) \cos\left(\frac{l\pi}{2}\right) \right.$$

$$\left. \left[\frac{\Gamma\left(\frac{n}{2}+2+\frac{l}{2}\right)}{\Gamma\left(\frac{n}{2}+3\right)} + ... + (-1)^{k} \frac{\Gamma\left(\frac{n}{2}+2+k+\frac{l}{2}\right)}{\Gamma\left(\frac{n}{2}+3+k\right)} \right] \right\} , \tag{8.4.10}$$

and for $r = 2k+1$ $(k = 0,1,2,...)$:

$$\Sigma_{(II)} = \frac{x^{-4-n}}{2} \frac{\Gamma\left(k+\frac{3}{2}\right)\Gamma\left(\frac{n}{2}+2\right)}{\Gamma\left(\frac{7}{2}+k+\frac{n}{2}\right)} . \tag{8.4.11}$$

Once more, in deriving (8.4.8) and (8.4.11) we used (7.3.21), and in deriving (8.4.9-10) we used (7.3.44). Thus also $H_{\infty}^{n,B}$ does not contribute to $\zeta(0)$, $\forall n > 3$, and our proof is completed.

8.5 $\zeta(0)$ Value

In light of (8.2.3), (8.3.3), (8.3.5), and using the result proved in section 8.4, we conclude that for the complete Majorana field $\left(\psi^A, \widetilde{\psi}^{A'}\right)$:

$$\zeta(0) = \frac{1}{30} - \frac{1}{360} = \frac{11}{360} . \tag{8.5.1}$$

Remarkably, this coincides with the result (4.5.3) obtained in the case of global boundary conditions. Note that, if we study the classical boundary-value problem for the Majorana field obeying the Weyl equation and subject to homogeneous local boundary conditions :

$$\sqrt{2}\,{}_e n_A^{\ A'}\psi^A - \epsilon\,\widetilde{\psi}^{A'} = \Phi^{A'} = 0 \quad \text{on} \quad S^3 \quad ,$$

regularity of the solution implies the same conditions on the modes m_{np}, r_{np} as in the spectral case (cf. section 4.6), even though the conditions on the modes are quite different if $\Phi^{A'}$ does not vanish, or for one-loop calculations. We also know from section 6.5 that Moss and Poletti find the same spin-1 $\zeta(0)$ values for electric and magnetic boundary conditions, and we pointed out this seems to respect a fundamental $(E \rightarrow B)$ symmetry of the classical Maxwell theory. Interestingly, these results, combined with the $\zeta(0)$ values for fermionic fields derived in chapter four and in this chapter, seem to add evidence in favour of quantum amplitudes for various spins respecting the fundamental properties of classical theory, with the corresponding classical boundary-value problems.

Note also that for our problem the degeneracy is half the one occurring in the case of global boundary conditions, since we need twice as many modes to get the same number of eigenvalue conditions ($J_{n+1}(E) = 0$, $\forall n \geq 0$, in the global case, or (8.1.1) in the local case). If there are N massless Majorana fields, the full $\zeta(0)$ in (8.5.1) should be multiplied by N. After section 7.3, our result (8.5.1) could add further evidence in favour of no cancellation being possible of one-loop divergences in the presence of boundaries for fermionic and bosonic fields. However, it is worth pointing out that our $\zeta(0)$ for spin $\frac{1}{2}$ disagrees with the result in Poletti 1990, where the following value is found : $\zeta(0) = \frac{17}{180}$. In deriving his result, Poletti relies on Moss 1989 and Moss and Poletti 1990a; on the other hand, our *direct* $\zeta(0)$ calculation is indeed necessary, and our derivation enables the reader to check all intermediate steps. Thus we hope that our *direct* $\zeta(0)$ calculation can give an useful contribution to the study of one-loop divergences of physical theories in the presence of boundaries (see the end of section 11.3).

8. One-Loop Results for the Spin-$\frac{1}{2}$ Field with Local Boundary Conditions

The approach to the boundary conditions for fermions in Poletti 1990 can be described as follows. Denoting by A the Dirac operator, Poletti squares it up and studies the second-order differential operator $A^{\dagger}A$. Thus, setting $P_{\pm} = \frac{1}{2}\left(1 \pm i\gamma_5\gamma_\mu n^\mu\right)$, he has to impose the local mixed boundary conditions :

$$P_{-}\phi = 0 \quad , \tag{8.5.2}$$

$$\left(n \cdot \nabla + \frac{(TrK)}{2}\right) P_{+}\phi = 0 \quad . \tag{8.5.3}$$

One can easily check that (8.5.2) coincides with our local boundary conditions studied in sections 4.6 and 5.8 : $\sqrt{2}\ {}_e n_A^{\ A'} \psi^A = \epsilon\ \widetilde{\psi}^{A'}$. However, one now deals with the extra condition (8.5.3) involving the normal derivatives of the Majorana field. Condition (8.5.3) is obtained by requiring that the image of the first-order operator C defined in (5.8.64) also obeys the boundary conditions (5.8.3), so that on S^3 :

$$\sqrt{2}\ {}_e n_A^{\ A'} \left(\nabla^A_{\ B'} \widetilde{\psi}^{B'}\right) = \epsilon\left(\nabla_B^{\ A'} \psi^B\right) \quad . \tag{8.5.4}$$

We then use (5.8.72), (A.5-6), (A.8-9), (A.11), and we also apply the well-known relations

$$^{(4)}\nabla_i \psi^A = {}^{(3)}\nabla_i \psi^A - {}_e n^A_{\ B'}\ e^{BB'l} K_{il}\ \psi_B \quad , \tag{8.5.5}$$

$$e_{AB'}^{\ \ \ i}\ e^{BB'l} = \left[-\frac{1}{2} h^{il} \epsilon_A^{\ B} + const.\ \epsilon^{ilk} \sqrt{h}\ {}_e n_{AB'} e^{BB'}_{\ \ \ k}\right] \quad , \tag{8.5.6}$$

the identities : ${}_e n_B^{\ B'} e^{BC'l} = {}_e n_B^{\ C'} e^{BB'l}$, ${}_e n^A_{\ B'} e^{BB'l} = {}_e n^B_{\ B'} e^{AB'l}$, and of course (5.8.3). One then finds that the contributions of $^{(3)}\nabla_i$ add up to zero, using also the property that $^{(3)}\nabla_i$ annihilates ${}_e n_{AB'}$. Moreover, the second term on the right-hand sides of (A.11) and (8.5.6) plays no role because for our torsion-free model the extrinsic-curvature tensor of the boundary is symmetric. Thus (8.5.4) leads to the following boundary conditions :

$$\left({}_e n^{BB'} \nabla_{BB'} + \frac{(TrK)}{2}\right)\left(\sqrt{2}\ {}_e n_A^{\ A'} \psi^A + \epsilon\ \widetilde{\psi}^{A'}\right) = 0 \ \text{ on } S^3 \quad , \tag{8.5.7}$$

which are clearly of the type (8.5.3). We hope the reader will find useful the translation into two-component spinor language of the approach used in Poletti 1990. We should emphasize that (8.5.7) (or equivalently (8.5.3)) does not change the spectrum determined by (5.8.3) (or equivalently (8.5.2)). In fact, as we said before, (8.5.7) only shows that the operator C defined in (5.8.64) preserves the boundary condition (5.8.3). This is automatically guaranteed, in the approach used in section 5.8, by the eigenvalue equations (5.8.56-57), which confirm that A maps $D(A)$ (cf. (5.8.87)) to itself.

At a deeper mathematical level, the situation can be described using the formalism in Luckock 1991. Given the vector bundles U and V over a compact Riemannian manifold (M, g_R) with boundary ∂M, we denote by P_+ and P_- complementary projection operators acting on U at ∂M, and by Q_+ and Q_- complementary projection operators acting on V at ∂M. They are assumed to map each fibre to itself, and can be extended to the interior of M by parallel transport along geodesics normal to ∂M. Dirichlet boundary conditions for sections of U and V are : $P_+ u = Q_+ v = 0$ on ∂M, and similarly for dual bundles. Dirichlet conditions ensure that the dual of a differential operator is itself a differential operator. Given the operator $\widetilde{A} : C^\infty(U) \rightarrow C^\infty(V)$, if we insist that the image of $\widetilde{A}$ should still be contained in $C^\infty(V)$, we must reduce the original domain of $\widetilde{A}$ to only include those $u \in C^\infty(U)$ which also obey the additional condition : $Q_+(\widetilde{A}u) = 0$ on ∂M. In Luckock 1991 it is shown under which conditions this boundary condition assumes the Robin form (called Neumann by the author). In the particular case of the Dirac operator A (when $dim\ U = dim\ V$), one finds that for u to be eigenfunction of $A^\dagger A$, Au must satisfy the boundary conditions appropriate to V. Thus the eigenfunctions of $A^\dagger A$ obey an extra differential boundary condition, which assumes the Robin form if a criterion discussed in section 5 of Luckock 1991 is satisfied. In the same way one is led to Robin boundary conditions for the eigenfunctions of $AA^\dagger$.

As remarked also in Poletti 1990, the reason for the disagreement between our result (8.5.1) and his result remains unclear, despite several attempts to understand what gives rise to the discrepancy. Unfortunately a direct comparison is not possible, because Poletti relies on an *indirect* technique where degeneracies and eigenvalue condition play no role (e.g. (6.5.5-6)). Moreover, it remains to be seen whether there is exact cancellation of one-loop divergences for extended supergravity theories. For example, in $N = 8$ supergravity the full $\zeta(0)$ is given by the formula :

$$\zeta(0) = \zeta_2(0) - 8\zeta_{\frac{3}{2}}(0) + 28\zeta_1(0) - 56\zeta_{\frac{1}{2}}(0) + 70\zeta_0(0) \quad , \tag{8.5.8}$$

subject to corrections due to antisymmetric tensor fields. Thus, the value of $\zeta_{\frac{3}{2}}(0)$ is needed before reaching a (final) conclusion.

CHAPTER NINE

LOCAL SUPERSYMMETRY IN PERTURBATIVE QUANTUM COSMOLOGY

Abstract. In Euclidean supergravity, the spin-$\frac{3}{2}$ potential has the pair of spatial components $\left(\psi_i^A,\ \widetilde{\psi}_i^{A'}\right)$. We perform a one-loop calculation for gravitinos subject to the following local boundary conditions on S^3 : $\sqrt{2}\ {}_e n_A{}^{A'} \psi_i^A = \pm \widetilde{\psi}_i^{A'}$. As in chapter five, the background is flat Euclidean space, and the physical degrees of freedom (PDF) are picked out imposing the supersymmetry constraints and choosing the gauge condition : $e_{AA'}{}^i\ \psi_i^A = 0$, $e_{AA'}{}^i\ \widetilde{\psi}_i^{A'} = 0$. The boundary conditions are then found to imply the following eigenvalue condition : $\Big[J_{n+2}(E)\Big]^2 - \Big[J_{n+3}(E)\Big]^2 = 0$, $\forall n \geq 0$, with degeneracy $(n+4)(n+1)$. Hence we can apply again the technique of chapter eight. The $\zeta(0)$ value is given by the one for the massless Majorana spin-$\frac{1}{2}$ field plus two other terms, leading to the PDF result : $\zeta(0) = -\frac{289}{360}$. Thus, for the gravitino field the PDF method and local boundary conditions lead to a result for $\zeta(0)$ which is equal to the PDF value one obtains setting equal to zero on S^3 all untwiddled coefficients of ψ_i^A and $\widetilde{\psi}_i^{A'}$.

Finally, we show that also extended supergravity theories do not lead to a vanishing result for the total PDF $\zeta(0)$. The boundary conditions used in this calculation are Dirichlet for scalar fields, magnetic for spin 1, Dirichlet for the perturbed three-metric for pure gravity, plus the other local boundary conditions for fermionic fields used in chapter eight and in this chapter. However, the question of the one-loop finiteness of extended supergravity, in the presence of boundaries, has not yet been settled. In fact the *direct* calculation of the *full* $\zeta(0)$, including gauge-averaging and ghost terms for gauge fields, remains a formidable problem.

9.1 Local Boundary Conditions for the Spin-$\frac{3}{2}$ Potential

In Euclidean supergravity (cf. section 1.3), the mathematical description of the gravitino leads to the introduction of the independent spinor-valued one-forms $\left(\psi_\mu^A, \widetilde{\psi}_\mu^{A'}\right)$ with spatial components $\left(\psi_i^A, \widetilde{\psi}_i^{A'}\right)$. We are here interested in a generalization to simple supergravity of the calculations of chapter eight for the spin-$\frac{1}{2}$ field. Thus we consider a flat Euclidean background, requiring on the bounding S^3 that :

$$\sqrt{2}\ {}_e n_A^{\ A'} \psi_i^A = \epsilon\, \widetilde{\psi}_i^{A'} \quad , \tag{9.1.1}$$

where $\epsilon = \pm 1$. Note that in so doing we are not using the local boundary conditions (5.7.4) involving field strengths. In fact it is not yet clear whether, and eventually how, (5.7.4) can be used to perform one-loop calculations.

As already explained in section 1.4, the consideration of (9.1.1) is suggested by the work in Luckock and Moss 1989, where it is shown that : (a) the spatial tetrad $e^{AA'}_{\ \ \ i}$ and the projection $\left(\pm \widetilde{\psi}_i^{A'} - \sqrt{2}\ {}_e n_A^{\ A'} \psi_i^A\right)$ transform into each other under half of the local supersymmetry transformations at the boundary; (b) after adding a suitable boundary term, the supergravity action is invariant under these local supersymmetry transformations. Now, from chapter five we already know that, imposing the supersymmetry constraints and choosing the gauge condition (5.3.1) (see comment thereafter), ψ_i^A finally assumes the form (5.3.6). In so doing, we are using again the PDF method, which may be subject to the criticisms we presented in sections 6.4-5. However, our calculations of chapters five and seven for gauge fields are based on the PDF method, and it appears more appropriate to compare them with another PDF result for $\zeta(0)$, so as to complete this line of research. Moreover, in section 6.5 we have seen that gauge-invariant *direct* $\zeta(0)$ calculations are already extremely difficult for Abelian gauge theories, so that we could not be more successful in the case of supergravity.

In the PDF form of ψ_i^A given by (5.3.6), $\beta^{nqABB'}$ and $\overline{\mu}^{nqABB'}$ involve the completely symmetric harmonics ρ^{nqABC} and $\overline{\sigma}^{nqABC}$. As we learned in the simpler cases of sections

5.8 and 5.9, it is useful to derive at first identities relating barred to unbarred harmonics, generalizing the technique in D'Eath and Halliwell 1987. This is achieved by using the relations (cf. (A.22)) :

$$\int d\mu \, \rho^{np}_{ABC} \, n^{AA'} n^{BB'} n^{CC'} \overline{\rho}^{mq}_{A'B'C'} = \delta^{nm} H^{pq}_{n} \quad , \tag{9.1.2}$$

$$\int d\mu \, \rho^{np}_{ABC} \, \epsilon^{AD} \epsilon^{BE} \epsilon^{CF} \rho^{mq}_{DEF} = \delta^{nm} A^{pq}_{n} \quad , \tag{9.1.3}$$

and the expansion of the totally symmetric field strength :

$$\phi_{ABC}(x) = \sum_{np} \Bigl(a_{np} \rho^{np}_{ABC}(x) + b_{np} \overline{\sigma}^{np}_{ABC}(x) \Bigr) \quad . \tag{9.1.4}$$

Thus we can express the a_{np} coefficients in two equivalent ways using (9.1.4), and (9.1.2) or (9.1.3). The equality of the two resulting formulae leads to :

$$n^{AA'} n^{BB'} n^{CC'} \sum_{q} \overline{\rho}^{nq}_{A'B'C'} (H^{-1}_{n})^{qp} = \epsilon^{AD} \epsilon^{BE} \epsilon^{CF} \sum_{q} \rho^{nq}_{DEF} (A^{-1}_{n})^{qp} \quad , \tag{9.1.5}$$

which is finally cast in the form :

$$\overline{\rho}^{np}_{D'E'F'} = -8 n^{D}_{\ D'} \, n^{E}_{\ E'} \, n^{F}_{\ F'} \sum_{q} \rho^{nq}_{DEF} (A^{-1}_{n} H_{n})^{qp} \quad , \tag{9.1.6}$$

in light of (A.9). In a similar way, we obtain :

$$\overline{\sigma}^{np}_{DEF} = -8 n_{D}^{\ D'} n_{E}^{\ E'} n_{F}^{\ F'} \sum_{q} \sigma^{nq}_{D'E'F'} (A^{-1}_{n} H_{n})^{qp} \quad . \tag{9.1.7}$$

The form of the matrices A^{pq}_{n} and H^{pq}_{n} is obtained taking the complex conjugate of (9.1.6), and then inserting the form of ρ^{np}_{DEF} so obtained into the right-hand side of (9.1.6). In so doing we find the consistency condition :

$$A^{-1}_{n} H_{n} A^{-1}_{n} H_{n} = -\frac{1}{8} 1_{n} \quad , \tag{9.1.8}$$

which is solved by : $A_n^{-1} = \frac{1}{2\sqrt{2}}\begin{pmatrix} 0 & -1 \\ 1 & 0 \end{pmatrix}$, $H_n = \begin{pmatrix} 1 & 0 \\ 0 & 1 \end{pmatrix}$, so that $A_n = 2\sqrt{2}\begin{pmatrix} 0 & 1 \\ -1 & 0 \end{pmatrix}$.

We can now remark that (9.1.1), together with (5.1.9), (5.1.12) and the Euclidean form of (5.3.6), where $\overline{r}_{np}^{(\mu)}$ is replaced by $\widetilde{r}_{np}^{(\mu)}$ (and similarly for $\widetilde{\psi}_i^{A'}$ containing $\widetilde{m}_{np}^{(\beta)}$), implies the following relations :

$$-i\sqrt{2}\sum_{pq} \alpha_n^{pq} m_{np}^{(\beta)}(a)\rho^{nqABD} n_A{}^{A'} n_D{}^{B'} = \epsilon \sum_{pq} \alpha_n^{pq} \widetilde{m}_{np}^{(\beta)}(a)\overline{\rho}^{nqA'B'D'} n^B{}_{D'} \quad , \tag{9.1.9}$$

$$-i\sqrt{2}\sum_{pq} \alpha_n^{pq} \widetilde{r}_{np}^{(\mu)}(a)\overline{\sigma}^{nqABD} n_A{}^{A'} n_D{}^{B'} = \epsilon \sum_{pq} \alpha_n^{pq} r_{np}^{(\mu)}(a)\sigma^{nqA'B'D'} n^B{}_{D'} \quad . \tag{9.1.10}$$

This is why the relations (9.1.6-7), (9.1.9-10) and the formulae for $A_n^{-1}H_n$ lead to boundary conditions which can be obtained from (5.8.6) and (5.8.12) replacing ϵ by $-\epsilon$, ρ^{nqA} by ρ^{nqABC} (and similarly for the σ harmonics), so that no substantial change occurs. In other words we find for example (cf. (5.8.10-11) and (5.8.14-15)) :

$$i m_{n1}^{(\beta)}(a) = \epsilon \, \widetilde{m}_{n2}^{(\beta)}(a) \quad , \tag{9.1.11}$$

$$-i m_{n2}^{(\beta)}(a) = \epsilon \, \widetilde{m}_{n1}^{(\beta)}(a) \quad , \tag{9.1.12}$$

$$i \widetilde{r}_{n1}^{(\mu)}(a) = \epsilon \, r_{n2}^{(\mu)}(a) \quad , \tag{9.1.13}$$

$$-i \widetilde{r}_{n2}^{(\mu)}(a) = \epsilon \, r_{n1}^{(\mu)}(a) \quad . \tag{9.1.14}$$

The coupled eigenvalue equations can be obtained from a system of the kind (5.8.24-27), where $\kappa_n = n + \frac{5}{2}$ rather than $n + \frac{3}{2}$, in light of (5.5.1). Thus we can repeat the procedure leading to (5.8.24-48) replacing n with $n+1$, so that (9.1.11-14) lead to the eigenvalue condition :

$$\Big[J_{n+2}(E)\Big]^2 - \Big[J_{n+3}(E)\Big]^2 = 0 \quad , \quad \forall n \geq 0 \quad , \tag{9.1.15}$$

where we set $a = 1$ for simplicity.

9.2 Physical-Degrees-of-Freedom Contribution to $\zeta(0)$

The eigenvalue condition (9.1.15) is very similar to the formula (8.1.1) we used for the calculations of the previous chapter. Thus the same technique can be now applied to derive the PDF contribution to $\zeta(0)$ in the case of gravitinos. As we know from section 5.5, the completely symmetric harmonics have degeneracy $d(n) = (n+4)(n+1)$, $\forall n \geq 0$. This is the full degeneracy in the case of local boundary conditions (9.1.1), for the same reason described after (8.5.1). Thus, defining $m \equiv n+3$, we find (see (8.1.39) and comments following (7.3.24-26) and (8.1.38)) :

$$\Gamma(3)\zeta(3,x^2) = \sum_{m=3}^{\infty}(m+1)(m-2)\left(\frac{1}{2x}\frac{d}{dx}\right)^3 \log\left[(ix)^{-2(m-1)}\left(J_{m-1}^2 - J_m^2\right)(ix)\right]$$

$$\sim \sum_{m=0}^{\infty}\left(m^2 - m\right)\left(\frac{1}{2x}\frac{d}{dx}\right)^3 \left[\sum_{i=1}^{5} S_i(m,\alpha_m(x))\right]$$

$$+ Z_1 + Z_2 + \sum_{n=5}^{\infty} q_n x^{-2-n} \quad , \tag{9.2.1}$$

where the $S_i(m,\alpha(x))$ are the ones defined in (8.1.33-37), and we define :

$$Z_1 \equiv -2\sum_{m=0}^{\infty}\left(\frac{1}{2x}\frac{d}{dx}\right)^3 \left[\sum_{i=1}^{5} S_i(m,\alpha_m(x))\right] = \sum_{i=1}^{5} X_{\infty}^{(i)} \quad , \tag{9.2.2}$$

$$Z_2 \equiv 2\sum_{m=0}^{1}\left(\frac{1}{2x}\frac{d}{dx}\right)^3 \left[\sum_{i=1}^{5} S_i(m,\alpha_m(x))\right] = \sum_{i=1}^{5} Y_{\infty}^{(i)} \quad . \tag{9.2.3}$$

In light of (8.1.38-39), the formula (9.2.1) shows that $\zeta(0)$ is given by the result (8.5.1) plus the contributions due to Z_1 and Z_2. These contributions should be finally divided by 2 (cf. (7.3.23)).

Indeed, from (9.2.2) we find :

$$X_{\infty}^{(1)} = -\frac{3}{2}\sum_{m=0}^{\infty} \alpha_m^{-5} \quad , \tag{9.2.4}$$

which does not contain x^{-6} and thus does not contribute to $\zeta(0)$. However, (7.3.35) and (8.1.34) imply :

$$X_{\infty}^{(2)} = 4x^{-6}\beta_1 - 4x^{-6}\beta_2 - 2x^{-4}\beta_3 - \frac{3}{2}x^{-2}\beta_4 - 2x^{-6}\beta_5 + 2x^{-6}\beta_6 + x^{-4}\beta_7 + \frac{3}{4}x^{-2}\beta_8 . \quad (9.2.5)$$

In (9.2.5), one has $\beta_1 = \rho_9, \beta_2 = \rho_{10}, \beta_3 = \rho_{11}, \beta_4 = \rho_{12}$ (cf. (F.17-18)), whereas the other sums are :

$$\beta_5 \equiv \lim_{s \to 0} \sum_{m=0}^{\infty} \alpha_m^{-2s} \quad , \quad \beta_6 \equiv \lim_{s \to 0} \sum_{m=0}^{\infty} m \alpha_m^{-1-2s} \quad , \quad (9.2.6)$$

$$\beta_7 \equiv \sum_{m=0}^{\infty} m \alpha_m^{-3} \quad , \quad \beta_8 \equiv \sum_{m=0}^{\infty} m \alpha_m^{-5} \quad . \quad (9.2.7)$$

As we know (see comment before (8.2.2)), the regularization is needed so as to get rid of the *fictitious* divergences introduced by the use of (7.3.34). In light of (7.3.21), (7.3.44) and the Euler-Maclaurin formula, only β_1 and β_5 contribute to $\zeta(0)$. The formula (7.3.44) implies that β_1 gives the contribution :

$$\delta^{(a)} = 2\cos(\pi) \frac{\Gamma\left(\frac{3}{2}\right)}{\Gamma\left(\frac{1}{2}\right)} \widetilde{B}_2 = -\frac{1}{6} \quad , \quad (9.2.8)$$

and the Euler-Maclaurin formula shows that β_5 contributes :

$$\delta^{(b)} = -\frac{1}{2} \quad . \quad (9.2.9)$$

From (8.1.28), (8.1.35) and (9.2.2), we also find that :

$$X_{\infty}^{(3)} = \frac{15}{4} k_{10} \sum_{m=0}^{\infty} \alpha_m^{-7} + \frac{105}{4} k_{12} \sum_{m=0}^{\infty} m^2 \alpha_m^{-9} \quad . \quad (9.2.10)$$

Thus, using (7.3.21) and (8.1.28), we derive the following contribution to $\zeta(0)$:

$$\delta^{(c)} = (k_{10} + k_{12}) = -\frac{1}{6} \quad . \quad (9.2.11)$$

Finally, using (8.1.36-37) and (9.2.2) we obtain :

$$X_{\infty}^{(4)} = \frac{1}{4}\sum_{r=0}^{4} k_{2r}(r+2)(r+4)(r+6)\left[\sum_{m=0}^{\infty} m^r \alpha_m^{-r-8}\right] \quad , \tag{9.2.12}$$

$$X_{\infty}^{(5)} = \frac{1}{4}\sum_{r=0}^{6} k_{3r}(r+3)(r+5)(r+7)\left[\sum_{m=0}^{\infty} m^r \alpha_m^{-r-9}\right] \quad , \tag{9.2.13}$$

and in light of (7.3.21) and (7.3.44) we derive that the asymptotic behaviour of $X_{\infty}^{(4)}$ is $O(x^{-7})$, and the asymptotic form of $X_{\infty}^{(5)}$ is $O(x^{-8})$. Hence they do not affect the $\zeta(0)$ value.

Moreover, the whole of Z_2 (cf. (9.2.3)) does not affect $\zeta(0)$. In fact one finds :

$$Y_{\infty}^{(1)} = \frac{3}{2}x^{-5}\left[1 + \left(1 + x^{-2}\right)^{-\frac{5}{2}}\right] \quad , \tag{9.2.14}$$

$$Y_{\infty}^{(2)} = 2x^{-7}\left(1+x^{-2}\right)^{-\frac{1}{2}} + x^{-7}\left(1+x^{-2}\right)^{-\frac{3}{2}} + \frac{3}{4}x^{-7}\left(1+x^{-2}\right)^{-\frac{5}{2}} \quad , \tag{9.2.15}$$

$$Y_{\infty}^{(3)} = -\frac{15}{4}k_{10}x^{-7}\left[1 + \left(1+x^{-2}\right)^{-\frac{7}{2}}\right] - \frac{105}{4}k_{12}x^{-9}\left(1+x^{-2}\right)^{-\frac{9}{2}} \quad , \tag{9.2.16}$$

$$Y_{\infty}^{(4)} = -\frac{1}{4}\sum_{r=1}^{4} x^{-r-8}k_{2r}(r+2)(r+4)(r+6)\left(1+x^{-2}\right)^{-\frac{r}{2}-4} \quad , \tag{9.2.17}$$

$$Y_{\infty}^{(5)} = -\frac{105}{4}k_{30}x^{-9} - \frac{1}{4}\sum_{r=0}^{6} x^{-r-9}k_{3r}(r+3)(r+5)(r+7)\left(1+x^{-2}\right)^{-\frac{r}{2}-\frac{9}{2}} \quad , \tag{9.2.18}$$

and the reader can now easily see that the formulae (9.2.14-18) do not contain terms proportional to x^{-6}.

At the end, we must consider the effect of higher-order terms in the expansion of $\log(\widetilde{\Sigma})$ as we did in chapter 8. In light of section 8.4 and of (9.2.1-3) we study, $\forall n > 3$:

$$\widetilde{H}_{\infty}^{n,A} \equiv -\frac{1}{4}\sum_{p=1}^{l} h_{np}\sum_{m=0}^{\infty}\Big[a_{np}\alpha_m^{p-n-6}(m+\alpha_m)^{-p} + b_{np}\alpha_m^{p-n-5}(m+\alpha_m)^{-p-1}$$

$$+ c_{np}\alpha_m^{p-n-4}(m+\alpha_m)^{-p-2} + d_{np}\alpha_m^{p-n-3}(m+\alpha_m)^{-p-3}\Big] \quad , \tag{9.2.19}$$

$$\widetilde{H}_{\infty}^{n,B} \equiv \frac{1}{4}\sum_{r=0}^{2n} k_{nr}(r+n)(r+n+2)(r+n+4)\sum_{m=0}^{\infty} m^r \alpha_m^{-r-n-6} \quad , \tag{9.2.20}$$

$$\widetilde{H}_{\infty}^{n,C} \equiv \frac{1}{4}\sum_{p=1}^{l} h_{np}\sum_{m=0}^{1}\Big[a_{np}\alpha_m^{p-n-6}(m+\alpha_m)^{-p} + b_{np}\alpha_m^{p-n-5}(m+\alpha_m)^{-p-1}$$
$$+ c_{np}\alpha_m^{p-n-4}(m+\alpha_m)^{-p-2} + d_{np}\alpha_m^{p-n-3}(m+\alpha_m)^{-p-3}\Big] \quad , \tag{9.2.21}$$

$$\widetilde{H}_{\infty}^{n,D} \equiv -\frac{1}{4}\sum_{r=0}^{2n} k_{nr}(r+n)(r+n+2)(r+n+4)\sum_{m=0}^{1} m^r \alpha_m^{-r-n-6} \quad . \tag{9.2.22}$$

In (9.2.19-22), n should not be confused with the integer appearing in (9.1.15) and in the definition of m. Again, the Euler-Maclaurin formula (C.2) is very useful in studying $\widetilde{H}_{\infty}^{n,A}$. The equivalent of $f(0)$ in (C.2) gives a contribution proportional to x^{-n-6}. Bernoulli numbers and derivatives of odd order give a contribution proportional to x^{-n-7} plus higher-order terms. The conversion of (9.2.19) into an integral yields a term proportional to x^{-n-5}, as it is evident looking at the integrals defined in (F.27-30).

The effect of $\widetilde{H}_{\infty}^{n,B}$ is derived by using (7.3.21) and (7.3.44). When $r = 0$ we must consider $\sum_{m=0}^{\infty} \alpha_m^{-n-6}$, which does not contain x^{-6}. When $r = 2k > 0$, (7.3.21) leads to a contribution proportional to x^{-n-5}, and when $r = 2k+1$, (7.3.44) leads to a contribution proportional to x^{-n-5} plus higher-order terms. One also finds that :

$$\widetilde{H}_{\infty}^{n,C} = \frac{x^{-n-6}}{4}\sum_{p=1}^{l} h_{np}\Big[(a_{np}+b_{np}+c_{np}+d_{np})$$
$$+ a_{np}\left(1+x^{-2}\right)^{\frac{p}{2}-\frac{n}{2}-3}\left(x^{-1}+\sqrt{1+x^{-2}}\right)^{-p}$$
$$+ b_{np}\left(1+x^{-2}\right)^{\frac{p}{2}-\frac{n}{2}-\frac{5}{2}}\left(x^{-1}+\sqrt{1+x^{-2}}\right)^{-p-1}$$
$$+ c_{np}\left(1+x^{-2}\right)^{\frac{p}{2}-\frac{n}{2}-2}\left(x^{-1}+\sqrt{1+x^{-2}}\right)^{-p-2}$$
$$+ d_{np}\left(1+x^{-2}\right)^{\frac{p}{2}-\frac{n}{2}-\frac{3}{2}}\left(x^{-1}+\sqrt{1+x^{-2}}\right)^{-p-3}\Big] \quad , \tag{9.2.23}$$

$$\widetilde{H}_{\infty}^{n,D} = -\frac{1}{4}k_{n0}\, n(n+2)(n+4)x^{-n-6}\left[1+\left(1+x^{-2}\right)^{-\frac{n}{2}-3}\right]$$

$$-\frac{1}{4}\sum_{r=1}^{2n} k_{nr}(r+n)(r+n+2)(r+n+4)x^{-r-n-6}\left(1+x^{-2}\right)^{-\frac{r}{2}-\frac{n}{2}-3} . \quad (9.2.24)$$

This is why $\widetilde{H}_{\infty}^{n,A}$, $\widetilde{H}_{\infty}^{n,B}$, $\widetilde{H}_{\infty}^{n,C}$ and $\widetilde{H}_{\infty}^{n,D}$ do not contain terms proportional to x^{-6}, and thus do not contribute to $\zeta(0)$.

To sum up, in light of (8.5.1), (9.2.1), (9.2.8-9), (9.2.11) and (9.2.19-24) we find :

$$\zeta(0) = \frac{11}{360} - \frac{5}{6} = -\frac{289}{360} \quad , \quad (9.2.25)$$

which is equal to the PDF value we found in (5.5.23) when we set equal to zero on S^3 all untwiddled coefficients of ψ_i^A and $\widetilde{\psi}_i^{A'}$.

9.3 Problems for Extended Supergravity Theories

Extended supergravity theories (van Nieuwenhuizen 1981) contain a number N of gravitinos between 2 and 8, and have been considered in the literature as candidates for a unified theory of fundamental interactions, despite severe problems due to a cosmological constant 10^{120} times larger than the experimental upper bound (Breitenlohner and Freedman 1982). No theory exists with $N > 8$, because one would then have particles with spins > 2. But it is known that the free-field spin-$\frac{5}{2}$ action is unique and cannot be coupled in a consistent way to either gravity or to simple matter systems (van Nieuwenhuizen 1981). There is indeed a rich literature on the problem of the one-loop vacuum energy in gauged extended supergravity, especially in the case of anti-de Sitter space (Allen and Davis 1983, Davis 1985, Burges et al. 1986). This is a background solution obtained under the assumption that spinor fields and vector potentials vanish (Breitenlohner and Freedman 1982). However, for our work we are interested in problems where an Euclidean background is bounded by a three-sphere. Note that, as emphasized in Gibbons 1983, in the absence of boundaries the study of flat backgrounds would be *inconsistent.* In fact, a peculiar property of gauged supergravity theories is the existence of a cosmological constant $\Lambda \neq 0$, whereas a flat background cannot be an *exact* solution of the Euclidean Einstein field equations with a nonvanishing Λ.

However, in the Hartle-Hawking program we restrict the path integral to compact Riemannian four-geometries. Moreover, the choice of S^3-boundaries is especially relevant. As explained in Schleich 1985, if the boundary three-geometry is a three-sphere of radius a such that $a\sqrt{\frac{\Lambda}{3}} < 1$, the compact extrema of the action inducing the three-sphere metric are sections of the Riemannian version (also called Euclidean) of de Sitter space. One then finds that the flat-space one-loop result is a good *approximation* up to corrections of order $\frac{a^2\Lambda}{3}$, provided the radius a of the boundary three-sphere is sufficiently small (Schleich 1985). Thus our flat-space supergravity results should be actually interpreted as *approximations* of a more complicated calculation involving curved backgrounds. Interestingly, more recent work in the literature seems to add evidence in favour of the flat-space limit being physically relevant. In fact, in Louko 1991a quantum cosmological models consisting of the closed Friedmann model with any *finite* number of inhomogeneous perturbations were studied. In that paper, the author claimed that, using zeta-function regularization and a scale-invariant measure, the semiclassical prefactor has different small-three-geometry behaviour in the cases $\Lambda > 0$ and $\Lambda = 0$. However, it was later realized in Louko 1991b this conclusion is wrong, since it was relying on the work in Louko 1988a, where the last line of equation (A.17) should contain an additional term. Accordingly, other amendments are in order in sections 4b, 5 and 6 of Louko 1988a (we are grateful to Dr. J. Louko for sending us the Erratum). The problem now remains to repeat Louko's analysis in the case of *infinitely many* perturbative modes (Louko 1991a), and for *negative values* of the cosmological constant, which play a key role in higher-N supergravity models. Note that in the $k = 1$ FRW minisuperspace model there are classical solutions with the Hartle-Hawking boundary data also for negative values of Λ. The corresponding Euclidean actions are real, so the semiclassical wave function is not rapidly oscillating, and is thus not relevant to Lorentzian cosmological models (cf. end of section 1.2). The semiclassical approximation with $\Lambda < 0$ appears, however, well-defined within the Hartle-Hawking approach. We are grateful to Dr. J. Louko for discussing this point at length.

On the other hand, the reader should bear in mind that one-loop supergravity calculations in the presence of boundaries with *arbitrary* choices of curved backgrounds can be *inconsistent* as well. In fact, the corresponding classical boundary-value problems might fail to have an unique smooth solution for topological reasons (cf. section 4.2). Thus the whole problem which motivated our research is subject to very severe consistency restrictions, which also apply to the general one-loop formulae for curved backgrounds appearing in Poletti 1990.

Denoting by s the spin of the various fields and by N the number of gravitinos, we have (but see below) the following formula for the total zeta-function at the origin :

$$\zeta_T^{(N)}(0) = \sum_{s=0}^{2} (-1)^{p(s)} M_s(N) \zeta_s(0) \quad . \tag{9.3.1}$$

In (9.3.1), the case $N = 1$ corresponds of course to simple supergravity, $p(s) = 2$ if $s = 0, 1, 2$, $p(s) = 1$ if $s = \frac{1}{2}, \frac{3}{2}$, and $\zeta_s(0)$ is the $\zeta(0)$ value for a field of spin s. The coefficients $M_s(N)$ depend on the number N of gravitinos present, but $M_2(N) = 1, \forall N$. In $O(N)$-gauged supergravity theories the nonvanishing $M_s(N)$ are (we are grateful to Dr. G. Gibbons for clarifying this point) :

$$M_{\frac{3}{2}}(1) = 1 \quad , \quad M_{\frac{3}{2}}(2) = 2 \quad , \quad M_1(2) = 1 \quad , \tag{9.3.2}$$

$$M_{\frac{3}{2}}(3) = M_1(3) = 3 \quad , \quad M_{\frac{1}{2}}(3) = 1 \quad , \tag{9.3.3}$$

$$M_{\frac{3}{2}}(4) = 4 \quad , \quad M_1(4) = 6 \quad , \quad M_{\frac{1}{2}}(4) = 4 \quad , \quad M_0(4) = 2 \quad , \tag{9.3.4}$$

$$M_{\frac{3}{2}}(5) = 5 \quad , \quad M_1(5) = 10 \quad , \quad M_{\frac{1}{2}}(5) = 11 \quad , \quad M_0(5) = 10 \quad , \tag{9.3.5}$$

$$M_{\frac{3}{2}}(6) = 6 \quad , \quad M_1(6) = 16 \quad , \quad M_{\frac{1}{2}}(6) = 26 \quad , \quad M_0(6) = 30 \quad , \tag{9.3.6}$$

$$M_{\frac{3}{2}}(7) = 8 \quad , \quad M_1(7) = 28 \quad , \quad M_{\frac{1}{2}}(7) = 56 \quad , \quad M_0(7) = 70 \quad , \tag{9.3.7}$$

$$M_s(8) = M_s(7) \quad , \quad \forall s \quad . \tag{9.3.8}$$

Indeed, following Gibbons 1983, we warn the reader that the existence of the gauged version of these theories with $N > 4$ remains an open problem, even though the numbers appearing in (9.3.5-8) are correct (but see below).

Boundary conditions compatible with supersymmetry transformation rules are local Dirichlet for bosonic fields, and local of the kind (5.8.3) and (9.1.1) for fermionic fields. This choice follows from the work in Luckock and Moss 1989 (cf. sections 1.4 and 9.1), and has also been made in Poletti 1990. In so doing, the results in Stewartson and Waechter 1971, Schleich 1985, Louko 1988b, jointly with (8.5.1), (9.2.25), (9.3.1) and the use of scale-invariant measures yield (D'Eath and Esposito 1991c) :

$$\zeta_T^{(1)}(0) = -\frac{43}{8} \ , \ \zeta_T^{(2)}(0) = -5 \ , \ \zeta_T^{(3)}(0) = -\frac{61}{12} \ , \ \zeta_T^{(4)}(0) = -\frac{17}{3} \ , \tag{9.3.9}$$

$$\zeta_T^{(5)}(0) = -\frac{41}{6} \ , \ \zeta_T^{(6)}(0) = -\frac{55}{6} \ , \ \zeta_T^{(7)}(0) = \zeta_T^{(8)}(0) = -\frac{83}{6} \ . \tag{9.3.10}$$

Here the fermions are Majorana, and all spin-0 particles are assumed to be described by real scalar fields. In so doing, our PDF analysis is *incomplete*, because (cf. Poletti 1990) in $N = 8$ supergravity the effects of two-forms and three-forms should be taken into account, thus reducing the numbers of scalars appearing in (9.3.4-8). Unfortunately we did not know the corresponding modified numbers of scalar fields, plus numbers of two-forms and three-forms to be used for models with less than 7 gravitinos. In other words, (9.3.9-10) are the PDF results *before* the effects of two-forms and three-forms are considered. However, as explained on page 240 of Duff 1982, after making a topologically nontrivial duality transformation, one does indeed obtain a theory where the number of scalar fields in (9.3.4-8) remains unchanged.

The alternative possibility is the one where spin-0 particles consist of an equal number of scalars and pseudoscalars (the latter would obey Neumann boundary conditions). However, this is not necessarily true for models obtained by dimensional reduction, as explained in Poletti 1990.

At present it remains unclear whether the preliminary results (9.3.9-10) can add evidence in favour of no exact cancellation of one-loop divergences being possible for physical theories in the presence of boundaries. We should also bear in mind that (9.3.9-10) are obtained using the PDF values for spin 1, $\frac{3}{2}$ and 2. Thus, as we remarked in sections 6.4-5, other quantization techniques might lead to different values. This is indeed what happens in Poletti 1990, where, studying the effect of ghost fields and gauge degrees of freedom, the author finds : $\zeta_{\frac{3}{2}}(0) = \frac{197}{180}$. In this case the difference with respect to the PDF value (9.2.25) is substantial, at least because the signs are opposite. However, one should bear in mind that the discrepancy found in section 8.5 for the spin-$\frac{1}{2}$ result also affects the spin-$\frac{3}{2}$ calculation. Moreover, it is also worth remarking that in Poletti 1990 the gravitino contribution to $\zeta(0)$ in simple supergravity makes the one-loop amplitude even more divergent, when perturbative modes for the three-metric are set equal to zero on S^3. By contrast, within the PDF approach, the gravitino contribution to $\zeta(0)$ in $N = 1$ supergravity partially cancels the contribution of the gravitational field in such a case. Thus it seems we can make the following concluding remarks :

(1) If a *negative* result holds (i.e. that $\zeta_T^{(N)}(0)$ does not vanish), a preliminary indication of this property might be given from the PDF analysis, the only one where *partial* cancellations occur;

(2) The PDF analysis was needed so as to complete earlier work appearing in the literature for bosonic fields in Schleich 1985 and Louko 1988b. In other words, at our present level of understanding of quantum cosmology, it can still be useful, as a first step, to compare the results of different techniques (cf. Henneaux 1985, pp 37-38) in the calculation of one-loop divergences in the presence of boundaries;

(3) Of course, since the $\zeta(0)$ value for a PDF calculation might depend on the choice of gauge conditions, and since gauge invariance is the underlying physical principle, the ultimate aim is to obtain a gauge-invariant result for $\zeta(0)$ using *direct* techniques. Unfortunately, this is not possible at present, in light of severe technical difficulties (e.g. the end of section 6.5).

PART III:

CLASSICAL GRAVITY

CHAPTER TEN

LORENTZIAN GEOMETRY, U_4 THEORIES AND SINGULARITIES IN COSMOLOGY

Abstract. At first, the space-time manifold is introduced and some aspects of Riemannian and Lorentzian geometry such as the distance function and the relations between topology and curvature are compared. We then define spinor structures in general relativity, and the conditions for their existence are discussed. The causality conditions are studied through an analysis of strong causality, stable causality and global hyperbolicity. In looking at the asymptotic structure of space-time, we focus on conformal infinity, the asymptotic symmetry group of Bondi, Metzner and Sachs, and the b-boundary construction of Schmidt, with a mention of recent work which tries to improve Schmidt's definition. The Hamiltonian structure of space-time is also analyzed, with emphasis on Ashtekar's spinorial variables.

Finally, the problem of a rigorous theory of singularities in space-times with torsion is studied, describing in detail recent work by the author. We define geodesics as curves whose tangent vector moves by parallel transport. This is different from what other authors do, because their definition of geodesics only involves Christoffel symbols, though studying theories with torsion. We then prove how to extend Hawking's singularity theorem without causality assumptions to the space-time of the ECSK theory. This is achieved by studying the generalized Raychaudhuri equation in the ECSK theory, the conditions for the existence of conjugate points and properties of maximal timelike geodesics. Our result can also be interpreted as a no-singularity theorem if the torsion tensor does not obey some additional conditions. Namely, it seems that the occurrence of singularities in closed cosmological models based on the ECSK theory can be less generic than in general relativity. Our work should be compared with important previous papers. There are some relevant differences, because we rely on a different definition of geodesics, we keep the field equations of the ECSK theory in their original form rather than casting them in a form similar to general relativity with a modified energy-momentum tensor, and we emphasize the role played by the full extrinsic-curvature tensor and by the variation formulae.

10. Lorentzian Geometry, U_4 Theories and Singularities in Cosmology

10.1 Introduction

The space-time manifold plays still a vital role in modern relativity theory, and we here examine in detail its mathematical structures. Our first aim is to present an unified description of some aspects of Lorentzian and Riemannian geometry, of the theory of spinors, and of causal, asymptotic and Hamiltonian structure. Thus in section 10.2, after defining the space-time manifold, following Beem and Ehrlich 1981 we discuss the distance function and the relations between topology and curvature in Lorentzian and Riemannian geometry. In section 10.3 we use two-component spinor language which is more familiar to relativists, and we present the results of Geroch on spinor structures. This section can be seen in part as complementary to important recent work in Budinich and Trautman 1987. In section 10.4, after some basic definitions, we study three fundamental causality conditions such as strong causality, stable causality and global hyperbolicity. In section 10.5, the asymptotic structure of space-time is studied focusing on conformal infinity, the asymptotic symmetry group of Bondi-Metzner-Sachs (hereafter referred to as BMS) and the boundary of space-time. This choice of arguments is motivated by the second part of our chapter, where singularity theory in cosmology for space-times with torsion is studied. In fact the Poincaré group can be seen as the subgroup of the BMS group which maps good cuts into good cuts, and it is also known that the gauge theory of the Poincaré group leads to theories with torsion (Hehl et al. 1976, Trautman 1980, Hehl et al. 1980, Nester 1983, Awada et al. 1986). Thus it appears important to clarify these properties. The boundary of space-time is studied in section 10.5.3 defining the b-boundary of Schmidt (Schmidt 1971), discussing its construction and the attempts to improve it appearing in Geroch et al. 1972, Szabados 1988-1989. In section 10.6 we present Ashtekar's spinorial variables for canonical gravity (Ashtekar 1988). In agreement with the aims of our chapter, we only emphasize the classical aspects of Ashtekar's theory. This section presents a striking application of the concepts defined in section 10.3, and it illustrates the modern approach to the Hamiltonian formulation of general relativity. Finally, section 10.7 is devoted to the clarification of our recent work (Esposito 1990) on the singularity problem for space-times with torsion, using also concepts defined in sections 10.2, 10.4 and 10.5.

So far, the singularity problem for theories with torsion had been studied defining geodesics as extremal curves. However, a rigorous theory of geodesics in general relativity can be based on the concept of autoparallel curves (Hawking and Ellis 1973, Beem and Ehrlich 1981). Thus, it appears rather important to develop the mathematical theory of singularities when geodesics are defined as curves whose tangent vector moves by parallel transport. This definition involves the full connection with torsion, whereas extremal curves just involve the Christoffel symbols. In so doing one appreciates the role of the full extrinsic-curvature tensor and of the variation formulae, two important concepts which were not considered in Hehl et al. 1974. One can also see that one can keep the field equations of the Einstein-Cartan-Sciama-Kibble (hereafter referred to as ECSK) theory in their original form, rather than casting them (as done in Hehl et al. 1974) in a form similar to general relativity but with a modified energy-momentum tensor. We then follow and clarify Esposito 1990 in proving how to extend Hawking's singularity theorem without causality assumptions to the space-time of the ECSK theory. Concluding remarks are presented in section 10.8.

10.2 Lorentzian and Riemannian Geometry

10.2.1 The Space-Time Manifold

A space-time (M, g) is the following collection of mathematical objects (Hawking and Ellis 1973, Beem and Ehrlich 1981) :

(1) A connected, four-dimensional, Hausdorff, C^∞ manifold M;

(2) A Lorentz metric g on M, i.e. the assignment of a symmetric, nondegenerate bilinear form $g_{|p} : T_pM \times T_pM \rightarrow R$, with diagonal form $(-,+,+,+)$, to each tangent space. Thus g has signature $+2$ and is not positive-definite;

(3) A time orientation, given by a globally defined, timelike vector field $X : M \rightarrow TM$. A timelike or null tangent vector $v \in T_pM$ is said to be future-directed if $g(X(p), v) < 0$, or past-directed if $g(X(p), v) > 0$.

Some remarks are now in order :

(a) Condition (1) can be formulated for any number of space-time dimensions ≥ 2;

(b) Also the convention $(+, -, -, -)$ for the diagonal form of the metric can be chosen (Penrose 1983). This convention seems to be preferred in the study of spinors, and can be adopted also in using tensors as Penrose does so as to avoid a change of conventions. The definitions of timelike and spacelike become then opposite to our definitions : X is timelike if $g(X(p), X(p)) > 0$, $\forall p \in M$, and X is spacelike if $g(X(p), X(p)) < 0$, $\forall p \in M$;

(c) The pair (M, g) is only defined up to equivalence. Two pairs (M, g) and (M', g') are equivalent if there is a diffeomorphism $\alpha : M \rightarrow M'$ such that : $\alpha_* g = g'$. Thus we are really dealing with an equivalence class of pairs (Hawking and Ellis 1973).

The fact that the metric is not positive-definite is the source of several mathematical problems. This is why mathematicians generally focused their attention on Riemannian geometry. We now summarize some basic results of Riemannian geometry, and formulate their counterpart (when possible) in Lorentzian geometry. This comparison is also very useful for gravitational physics. In fact Riemannian geometry is related to the Euclidean path-integral approach to quantum gravity (Hawking 1979a-b), whereas Lorentzian geometry is the framework of general relativity.

10.2.2 Comparison between Riemannian and Lorentzian Geometry

A Riemannian metric g_0 on a manifold M is a smooth and positive-definite section of the bundle of symmetric bilinear two-forms on M. A fundamental result in Riemannian geometry is the Hopf-Rinow theorem. It can be formulated as follows (Beem and Ehrlich 1981) :

Theorem 10.2.1 For any Riemannian manifold (M, g_0) the following properties are equivalent :

(1) Metric completeness : M together with the Riemannian distance function (cf. section 10.2.2.1) is a complete metric space;

(2) Geodesic completeness : $\forall v \in TM$, the geodesic $c(t)$ in M such that $c'(0) = v$ is defined $\forall t \in R$;

(3) For some $p \in M$, the exponential map exp_p is defined on the entire tangent space T_pM;

(4) Finite compactness : any subset K of M such that $sup\{d_0(p,q) : p,q \in K\} < \infty$ has compact closure.

Moreover, if any of these properties holds, we also know that :

(5) $\forall p, q \in M$, there exists a smooth geodesic segment c from p to q with $L_0(c) = d_0(p,q)$ (i.e. any two points can be joined by a minimal geodesic).

In Lorentzian geometry there is no sufficiently strong analogue to the Hopf-Rinow theorem. However, one can learn a lot by comparing the definitions of distance function and the relations between topology and curvature in the two cases.

10.2.2.1 Distance Function in Riemannian Geometry

Let Ω_{pq} be the set of piecewise smooth curves in M from p to q. Given $c : [0,1] \to M$ and belonging to Ω_{pq}, there is a finite partition of $[0,1]$ such that c restricted to the sub-interval $[t_i, t_{i+1}]$ is smooth $\forall i$. The Riemannian arc length of c with respect to g_0 is defined by :

$$L_0(c) \equiv \sum_{i=1}^{k-1} \int_{t_i}^{t_{i+1}} \sqrt{g_0(c'(t), c'(t))}\, dt \quad . \tag{10.2.1}$$

The Riemannian distance function $d_0 : M \times M \to [0, \infty)$ is then defined by (Beem and Ehrlich 1981) :

$$d_0(p,q) \equiv \inf \{L_0(c) : c \in \Omega_{pq}\} \quad . \tag{10.2.2}$$

Thus d_0 has the following properties :

(1) $d_0(p,q) = d_0(q,p) \quad \forall p, q \in M$;

(2) $d_0(p,q) \leq d_0(p,r) + d_0(r,q) \quad \forall p,q,r \in M$;

(3) $d_0(p,q) = 0$ if and only if $p = q$;

(4) d_0 is continuous and, $\forall p \in M$ and $\epsilon > 0$, the family of metric balls $B(p,\epsilon) \equiv \{q \in M : d_0(p,q) < \epsilon\}$ is a basis for the manifold topology.

10.2.2.2 Distance Function in Lorentzian Geometry

Let $\widetilde{\Omega}_{pq}$ be the space of all future-directed nonspacelike curves $\gamma : [0,1] \rightarrow M$ with $\gamma(0) = p$ and $\gamma(1) = q$. Given $\gamma \in \widetilde{\Omega}_{pq}$ we choose a partition of $[0,1]$ such that γ restricted to $[t_i, t_{i+1}]$ is smooth $\forall i = 0, 1, ..., n-1$. The Lorentzian arc length is then defined as (Beem and Ehrlich 1981) :

$$L(\gamma) \equiv \sum_{i=0}^{n-1} \int_{t_i}^{t_{i+1}} \sqrt{-g(\gamma'(t), \gamma'(t))}\, dt \quad . \tag{10.2.3}$$

The Lorentzian distance function $d : M \times M \rightarrow R \cup \{\infty\}$ is thus defined as follows. Given $p \in M$, if q does not belong to the causal future of p (cf. section 10.4) : $q \notin J^+(p)$, we set $d(p,q) = 0$. Otherwise, if $q \in J^+(p)$, we set (Beem and Ehrlich 1981) :

$$d(p,q) \equiv \sup \left\{ L_g(\gamma) : \gamma \in \widetilde{\Omega}_{pq} \right\} \quad . \tag{10.2.4}$$

Thus such $d(p,q)$ may not be finite, if timelike curves from p to q attain arbitrarily large arc lengths. It also fails to be symmetric in general, and one has : $d(p,q) \geq d(p,r) + d(r,q)$ if there are future-directed nonspacelike curves from p to r and from r to q. Finally, we should recall the definition of timelike diameter diam(M,g) of a space-time (M,g) (Beem and Ehrlich 1981) :

$$\mathrm{diam}(M,g) \equiv \sup \{d(p,q) : p,q \in M\} \quad . \tag{10.2.5}$$

10.2.2.3 Topology and Curvature in Riemannian Geometry

A classical result is the Myers-Bonnet theorem which shows how the properties of the Ricci curvature may influence the topological properties of the manifold. In fact one has (Gallot et al. 1987) :

Theorem 10.2.2 Let (M, g) be a complete n-dimensional Riemannian manifold with Ricci curvature $Ric(v,v)$ such that : $Ric(v,v) \geq \frac{(n-1)}{r}$. Then $\text{diam}(M,g) \leq \text{diam}(S^n(r))$, $\text{diam}(M,g) \leq \pi\sqrt{r}$, and M is compact. Moreover, M has finite fundamental homotopy group.

10.2.2.4 Topology and Curvature in Lorentzian Geometry

The Lorentzian analogue of the Myers-Bonnet theorem can be formulated in the following way (Beem and Ehrlich 1981) :

Let (M, g) be a n-dimensional globally hyperbolic space-time (cf. section 10.4.3) such that either :

(1) All timelike sectional curvatures are $\leq -l < 0$, or :

(2) $Ric(v,v) \geq (n-1)l > 0$, $\forall$ unit timelike vectors $v \in TM$.

Then $\text{diam}(M,g) \leq \frac{\pi}{\sqrt{l}}$.

The proof of this theorem, together with the discussion of the Lorentzian analogue of the index and Rauch I,II comparison theorems can be found in chapter ten of Beem and Ehrlich 1981. For another recent treatise on Riemannian geometry, see Francaviglia 1988. An enlightening comparison of Riemannian and Lorentzian geometry can also be found in O'Neill 1983.

10.3 Spinor Structure

The basic ideas of two-component spinor calculus have been summarized in appendix A. We now focus our attention on some more general aspects, following Geroch 1968a.

In defining spinors at a point of space-time, we may start by addressing the question of how an array of complex numbers $\mu_{CD'}^{AB'}$ is transformed in going from a tetrad v at p to a tetrad w at p. The mapping $L : v \rightarrow w$ between v and w is realized by an element L of the restricted Lorentz group L_0 (which preserves temporal direction and spatial parity). Now, to each L there correspond two elements $\pm U^A{}_B$ of $SL(2,C)$. Thus the transformation law contains a sign ambiguity :

$$\mu^{AB'}_{CD'}(w) = \pm U^A{}_E \, \overline{U}^{B'}{}_{F'} \left(U^{-1}\right)^G{}_C \, \overline{U^{-1}}^{H'}{}_{D'} \, \mu^{EF'}_{GH'}(v) \quad .$$

In order to remove this sign ambiguity, let us consider the six-dimensional space : $\psi \equiv \Big\{$ set of all tetrads at p $\Big\}$. We then move to the universal covering manifold $\widetilde{\psi}$ of ψ :

$$\widetilde{\psi} \equiv \{(v,\alpha) : v \in \psi, \alpha = \text{path in } \psi \text{ from } v \text{ to } w\} \quad .$$

Definition 10.3.1 (v,α) is equivalent to (u,β) if $u = v$ and if we can continuously deform α into β keeping fixed the terminal points.

An important property (usually described by the Dirac-scissors argument) is that the tetrad at p changes after a 2π rotation, but remains unchanged after a 4π rotation. The advantage of considering $\widetilde{\psi}$ is that in so doing, $\forall v, w \in \widetilde{\psi}$, there is an unique element $U^A{}_B$ of $SL(2,C)$ which transforms v into w. Thus we give (Geroch 1968a) :

Definition 10.3.2 A spinor at p is a rule which assigns to each $v \in \widetilde{\psi}$ an array $\mu^{AB'}_{CD'}$ of complex numbers such that, given $v, w \in \widetilde{\psi}$ related by $U^A{}_B \in SL(2,C)$, then :

$$\mu^{AB'}_{CD'}(w) = U^A{}_E \, \overline{U}^{B'}{}_{F'} \left(U^{-1}\right)^G{}_C \, \overline{U^{-1}}^{H'}{}_{D'} \, \mu^{EF'}_{GH'}(v) \quad . \tag{10.3.1}$$

In defining spinor structures on M, we begin by considering the principal fibre bundle B of oriented orthonormal tetrads on M. The structure group of B is the restricted Lorentz group, and the fibre at $p \in M$ is the collection ψ of tetrads at p with given temporal and spatial orientation. The sign ambiguity is corrected by taking a fibre bundle whose fibre is the universal covering space $\widetilde{\psi}$ (Geroch 1968a).

Definition 10.3.3 A spinor structure on M is a principal fibre bundle $\widetilde{B}$ on M with group $SL(2,C)$, together with a $2-1$ application $\phi : \widetilde{B} \rightarrow B$ such that :

(1) ϕ realizes the mapping of each fibre of $\widetilde{B}$ into a single fibre of B;

(2) ϕ commutes with the group operations. Thus, $\forall U \in SL(2,C)$ we have : $\phi U = E(U)\phi$, where $E : SL(2,C) \rightarrow L_0$ is the covering group of L_0.

Definition 10.3.4 A spinor field on M is a mapping μ of $\widetilde{B}$ into arrays of complex numbers such that (10.3.1) holds.

The basic theorems about spinor structures are the following (Geroch 1968a) :

Theorem 10.3.1 If a space-time (M,g) has a spinor structure, for this structure to be unique M must be simply connected.

Theorem 10.3.2 A space-time (M,g) oriented in space and time has spinor structure if and only if the second Stiefel-Whitney class vanishes.

Remark : Stiefel-Whitney classes w_i can be defined for each vector bundle ξ by means of a sequence of cohomology classes $w_i(\xi) \in H^i(B(\xi); Z_2)$, where we denote by $H^i(B(\xi); Z_2)$ the i-th singular cohomology group of $B(\xi)$ with coefficients in Z_2, the group of integers modulo 2 (Milnor and Stasheff 1974). If $w_2 \neq 0$, one cannot define parallel transport of spinors on M. The orientability of space-time assumed in theorem 10.3.2 and in section 10.2.1 implies that also the first Stiefel-Whitney class must vanish.

Theorem 10.3.3 (M,g) has a spinor structure if and only if the fundamental homotopy groups of B and M are related by :

$$\pi_1(B) \approx \pi_1(M) \oplus \pi_1(\psi) = \pi_1(M) \oplus Z_2 \quad . \tag{10.3.2}$$

Theorem 10.3.4 A space-time (M,g) space- and time-oriented has spinor structure if and only if each of its covering manifolds has spinor structure.

Theorem 10.3.5 Let M be noncompact. Then (M,g) has spinor structure if and only if a global system of orthonormal tetrads exists on M (Geroch 1968a).

When we unwrap ψ, we annihilate $\pi_1(\psi)$. The existence of a spinor structure implies we can unwrap all fibres on B. Spinor structures are related to the second homotopy group of M, whereas covering spaces are related to the first homotopy group. However, it is wrong to think that a spinor structure can be created simply by taking a covering manifold. In a space-time (M, g) which does not have spinor structure, there must be some closed curve γ which lies in the fibre over $p \in M$ such that (Geroch 1968a) :

(a) γ is not homotopically zero in the fibre;

(b) γ can be contracted to a point in the whole bundle of frames.

A very important application of the spinorial formalism in general relativity is studied in section 10.6, where we define Ashtekar's spinorial variables for canonical gravity.

10.4 Causal Structure

Let (M, g) be a space-time, and let $p \in M$. The chronological future of p is defined as (Hawking and Ellis 1973, Beem and Ehrlich 1981) :

$$I^+(p) \equiv \{q \in M : p << q\} \quad , \tag{10.4.1}$$

i.e. $I^+(p)$ is the set of all points q of M such that there is a future-directed timelike curve from p to q. Similarly, we define the chronological past of p :

$$I^-(p) \equiv \{q \in M : q << p\} \quad . \tag{10.4.2}$$

The causal future of p is then defined by :

$$J^+(p) \equiv \{q \in M : p \leq q\} \quad , \tag{10.4.3}$$

and similarly for the causal past :

$$J^-(p) \equiv \{q \in M : q \leq p\} \quad , \tag{10.4.4}$$

where $a \leq b$ means there is a future-directed nonspacelike curve from a to b. The causal structure of (M, g) is the collection of past and future sets at all points of M together with their properties. Following Penrose 1968 and Geroch 1970, we here recall the following definitions, which will then be useful in sections 10.4.3 and 10.7.

Definition 10.4.1 A set Σ is achronal if no two points of Σ can be joined by a timelike curve.

Definition 10.4.2 A point p is an endpoint of the curve λ if λ enters and remains in any neighbourhood of p.

Definition 10.4.3 Let Σ be a spacelike or null achronal three-surface in M. The future Cauchy development (or future domain of dependence) $D^+(\Sigma)$ of Σ is the set of points $p \in M$ such that every past-directed timelike curve from p without past endpoint intersects Σ.

Definition 10.4.4 The past Cauchy development $D^-(\Sigma)$ of Σ is defined interchanging future and past in definition 10.4.3. The total Cauchy development of Σ is then given by $D(\Sigma) = D^+(\Sigma) \cup D^-(\Sigma)$.

Definition 10.4.5 The future Cauchy horizon $H^+(\Sigma)$ of Σ is given by :

$$H^+(\Sigma) \equiv \{X : X \in D^+(\Sigma), I^+(X) \cap D^+(\Sigma) = \phi\} \quad . \tag{10.4.5}$$

Similarly, the past Cauchy horizon $H^-(\Sigma)$ is defined as :

$$H^-(\Sigma) \equiv \{X : X \in D^-(\Sigma), I^-(X) \cap D^-(\Sigma) = \phi\} \quad . \tag{10.4.6}$$

Definition 10.4.6 The edge of an achronal set Σ is given by all points $p \in \overline{\Sigma}$ such that any neighbourhood U of p contains a timelike curve from $I^-(p, U)$ to $I^+(p, U)$ that does not meet Σ (O'Neill 1983).

Our definitions of Cauchy developments differ indeed from the ones in Hawking and Ellis 1973, in that Hawking and Ellis look at past-inextendible curves which are timelike or null, whereas we agree with Penrose and Geroch in not including null curves in the definition. We now discuss three fundamental causality conditions : strong causality, stable causality and global hyperbolicity.

10.4.1 Strong Causality

The underlying idea for the definition of strong causality is that there should be no point p such that every small neighbourhood of p intersects some timelike curve more than once (Hawking 1971). In other words, the space-time (M, g) does not *almost contain* closed timelike curves. In rigorous terms, strong causality is defined as follows (Penrose 1968) :

Definition 10.4.7 Strong causality holds at $p \in M$ if arbitrarily small neighbourhoods of p exist which each intersect no timelike curve in a disconnected set.

A very important characterization of strong causality can be given by defining at first the Alexandrov topology (Hawking and Ellis 1973).

Definition 10.4.8 In the Alexandrov topology, a set is open if and only if it is the union of one or more sets of the form : $I^+(p) \cap I^-(q)$, $p, q \in M$.

Thus any open set in the Alexandrov topology is open in the manifold topology. Now, the following fundamental result holds (Penrose 1983) :

Theorem 10.4.1 The following three requirements are equivalent :

(1) the space-time (M, g) is strongly causal;

(2) the Alexandrov topology agrees with the manifold topology;

(3) the Alexandrov topology is Hausdorff.

10.4.2 Stable Causality

Strong causality is not enough to ensure that space-time is not just about to violate causality (Hawking 1971). The situation can be considerably improved if stable causality holds. For us to be able to properly define this concept, we discuss the problem of putting a topology on the space of all Lorentz metrics on a four-manifold M. Essentially three possible topologies seem to be of major interest (Hawking 1971) : compact-open topology, open topology, fine topology.

10.4.2.1 Compact-Open Topology

$\forall i = 0, 1, ..., r$, let ϵ_i be a set of continuous positive functions on M, U be a compact set $\subset M$ and g the Lorentz metric we are studying. We then define : $G(U, \epsilon_i, g) \equiv$ set of all Lorentz metrics $\tilde{g}$ such that :

$$\left| \frac{\partial^i \tilde{g}}{\partial x^i} - \frac{\partial^i g}{\partial x^i} \right| < \epsilon_i \quad \text{on } U \quad , \quad \forall i = 0, 1, ..., r \quad .$$

In the compact-open topology, open sets are obtained from the $G(U, \epsilon_i, g)$ through the operations of arbitrary union and finite intersection.

10.4.2.2 Open Topology

We no longer require U to be compact, and we take $U = M$ in section 10.4.2.1.

10.4.2.3 Fine Topology

We define : $H(U, \epsilon_i, g) \equiv$ set of all Lorentz metrics $\tilde{g}$ such that :

$$\left| \frac{\partial^i \tilde{g}}{\partial x^i} - \frac{\partial^i g}{\partial x^i} \right| < \epsilon_i \quad ,$$

and $\tilde{g} = g$ out of the compact set U. Moreover, we consider : $G'(\epsilon_i, g) \equiv \cup H(U, \epsilon_i, g)$. A sub-basis for the fine topology is then given by the neighbourhoods $G'(\epsilon_i, g)$ (Hawking 1971).

Now, the underlying idea for stable causality is that space-time must not contain closed timelike curves, and we still fail to find closed timelike curves if we open out the null cones. In view of the former definitions, this idea can be formulated as follows :

Definition 10.4.9 A metric g satisfies the stable causality condition if, in the C^0 open topology (cf. section 10.4.2.2), an open neighbourhood of g exists no metric of which has closed timelike curves.

The Minkowski, FRW, Schwarzschild and Reissner-Nordström space-times are all stably causal. If stable causality holds, the differentiable and conformal structure can be determined from the causal structure, and space-time cannot be compact (because in a compact space-time there are closed timelike curves). A very important characterization of stable causality is given by the following theorem (Hawking and Ellis 1973) :

Theorem 10.4.2 A space-time (M, g) is stably causal if and only if a cosmic time function exists on M, i.e. a function whose gradient is everywhere timelike.

10.4.3 Global Hyperbolicity

Global hyperbolicity plays a key role in developing a rigorous theory of geodesics in Lorentzian geometry and in proving singularity theorems. Its ultimate meaning can be seen as requiring the existence of Cauchy surfaces, i.e. spacelike hypersurfaces which each nonspacelike curve intersects exactly once. In fact some authors (Wald 1984) take this property as the starting point in discussing global hyperbolicity. Indeed, Leray's original idea was that the set of nonspacelike curves from p to q must be compact in a suitable topology (Leray 1952). We here follow Geroch 1970, Hawking and Ellis 1973 and Wald 1984 defining and proving in part what follows.

Definition 10.4.10 A space-time (M, g) is globally hyperbolic if :

(a) strong causality holds;

(b) $J^+(p) \cap J^-(q)$ is compact $\forall p, q \in M$.

Theorem 10.4.3 In a globally hyperbolic space-time, the following properties hold :

(1) $J^+(p)$ and $J^-(p)$ are closed $\forall p$;

(2) $\forall p, q$, the space $C(p, q)$ of all nonspacelike curves from p to q is compact in a suitable topology;

(3) there are Cauchy surfaces.

Proof of (1). It is well-known that, if (X, F) is a Hausdorff space and $A \subset X$ is compact, then A is closed. In our case, this implies that $J^+(p) \cap J^-(q)$ is closed. Moreover, it is not difficult to see that $J^+(p)$ itself must be closed. In fact, otherwise we could find a point $r \in \overline{J^+(p)}$ such that $r \notin J^+(p)$. Let us now choose $q \in I^+(r)$. We would then have : $r \in \overline{J^+(p) \cap J^-(q)}$ but $r \notin J^+(p) \cap J^-(q)$, which implies that $J^+(p) \cap J^-(q)$ is not closed, not in agreement with what we found before. Similarly we also prove that $J^-(p)$ is closed.

Remark : a stronger result can also be proved. Namely, if (M, g) is globally hyperbolic and $K \subset M$ is compact, then $J^+(K)$ is closed (Wald 1984).

Proof of (3). The proof relies on the following ideas :

Step 1. We define a function f^+, and we prove that global hyperbolicity implies continuity of f^+ on M (Hawking and Ellis 1973).

Step 2. We consider the function :

$$f : p \in M \rightarrow f(p) \equiv \frac{f^-(p)}{f^+(p)} \quad , \tag{10.4.7}$$

and we prove that the $f = constant$ surfaces are Cauchy surfaces (Geroch 1970).

Step 1

The function f^+ we are looking for is given by $f^+ : p \in M \rightarrow$ volume of $J^+(p, M)$. This can only be done with a suitable choice of measure. The measure is chosen in such a way that the total volume of M is equal to 1. For f^+ to be continuous on M, it is sufficient to show that f^+ is continuous on any nonspacelike curve γ. In fact, let $r \in \gamma$, and let $\{x_n\}$ be a sequence of points on γ in the past of r. We now define :

$$T \equiv \cap J^+(x_n, M) \quad . \tag{10.4.8}$$

If f^+ were not upper semi-continuous on γ in r, there would be a point $q \in T - J^+(r, M)$, with $r \notin J^-(q, M)$. On the other hand, since $x_n \in J^-(q, M)$ one should find that $r \in \overline{J^-(q, M)}$, which is impossible in view of global hyperbolicity. The absurd proves that f^+ is upper semi-continuous. In the same way (exchanging the role of past and future) we can prove lower semi-continuity, and thus continuity. It becomes then trivial to prove the continuity of the function $f^+ : p \in M \rightarrow$ volume of $I^+(p, M)$. From now on, we denote by f^+ the volume function of $I^+(p, M)$.

Step 2

Let Σ be the set of points where $f = 1$, and let $p \in M$ be such that $f(p) > 1$. The idea is to prove that every past-directed timelike curve from p intersects Σ, so that $p \in D^+(\Sigma)$. In a similar way, if $f(p) < 1$, one can then prove that $p \in D^-(\Sigma)$ (which finally implies that Σ is indeed a Cauchy surface). The former result can be proved as follows (Geroch 1970).

Step 2a

We consider any past-directed timelike curve μ without past endpoint from p. In view of the continuity of f proved in step 1, such a curve μ must intersect Σ, provided one can show that there is $\epsilon \rightarrow 0^+ : f_{on\ \mu} = \epsilon$, where ϵ is arbitrary.

Step 2b

Given $q \in M$, we denote by U a subset of M such that $U \subset I^+(q)$. The subsets U of this form cover M. Moreover, any U cannot be in $I^-(r)\ \forall r \in \mu$. This is forbidden by global hyperbolicity. In fact, suppose for absurd that $q \in \cap_{r \in \mu} I^-(r)$. We then choose a sequence $\{t_i\}$ of points on μ such that :

$$t_{i+1} \in I^-(t_i) \quad ; \quad \exists i : z \in I^-(t_i) \quad , \quad \forall z \in \mu \quad .$$

$\forall i$, we also consider a timelike curve μ' such that :

(a) μ' begins at p;

(b) $\mu' = \mu$ to t_i;

(c) μ' continues to q.

Global hyperbolicity plays a role in ensuring that the sequence $\{t_i\}$ has a limit curve Ω, which by construction contains μ. On the other hand, we know this is impossible. In fact, if μ were contained in a causal curve from p to q, it should have a past endpoint, which is not in agreement with the hypothesis. Thus, having proved that $\exists r \in \mu : U \not\subset I^-(r)$, we find that $f^-(r) \rightarrow 0$ when r continues into the past on μ, which in turn implies that μ intersects Σ as we said in step 2a (Geroch 1970).

The proof of (2) is not given here, and can be found in Hawking and Ellis 1973. Global hyperbolicity plays a key role in proving singularity theorems because, if p and q lie in a globally hyperbolic set and $q \in J^+(p)$, there is a nonspacelike geodesic from p to q whose length is greater than or equal to that of any other nonspacelike curve from p to q. The proof that arbitrary, sufficiently small variations in the metric do not destroy global hyperbolicity can be found for example in Geroch 1970. Globally hyperbolic space-times are also peculiar since for them the Lorentzian distance function defined in section 10.2.2.2 is finite and continuous as the Riemannian distance function (see Beem and Ehrlich 1981, page 81). The relation between strong causality, finite distance function and global hyperbolicity is proved on page 107 of Beem and Ehrlich 1981. More recent work on causal structure of Lorentzian manifolds can be found in Levichev 1989 and references therein.

10.5 Conformal Structure and Asymptotic Properties

Under this name one can discuss black-holes theory, gravitational radiation, positive-mass theorems (for the ADM and Bondi's mass), the singularity problem. Here we choose to focus on three topics : conformal infinity, the asymptotic symmetry group of space-time and the definition of boundary of space-time.

10.5.1 Conformal Infinity

In the course of studying massless free-field equations and asymptotic properties of space-time, one faces the problem of giving a precise mathematical meaning to otherwise ill-defined limits as $r \rightarrow \infty$. This is achieved bringing the infinity down to a finite distance, after a careful definition of boundary of space-time. We here describe how this can be obtained in the simplest case, i.e. four-dimensional Minkowski space-time.

Minkowski space-time (M_4, η) is an affine space isomorphic to R^4. In Cartesian coordinates, its flat Lorentzian metric is :

$$g_1 = -dt \otimes dt + dx \otimes dx + dy \otimes dy + dz \otimes dz \quad , \tag{10.5.1}$$

whereas in polar coordinates one has :

$$g_2 = -dt \otimes dt + dr \otimes dr + r^2 \Omega \quad , \tag{10.5.2}$$

where Ω is the unit-two-sphere metric : $\Omega = d\theta \otimes d\theta + (\sin\theta)^2 d\phi \otimes d\phi$, and r, θ, ϕ lie in the intervals : $r \in]0, \infty[$, $\theta \in]0, \pi[$, $\phi \in]0, 2\pi[$. Thus the space-time (M_4, g_1) has topology R^4, whereas the space-time (M_4, g_2) has topology $R \times \left(R^3 - \{0\}\right)$, since we are using a space-time foliation which is singular at $r = 0$. Note also that we have 2 trivial polar-coordinates singularities at $\theta = 0, \pi$. One thus needs *two* coordinate neighbourhoods of the kind described following (10.5.2) to cover the whole of Minkowski space-time.

We now introduce retarded and advanced null coordinates in $c = 1$ units :

$$w \equiv t - r \quad , \quad v \equiv t + r \quad , \tag{10.5.3}$$

so that a new metric is obtained :

$$g_3 = -\frac{1}{2}\left(dv \otimes dw + dw \otimes dv\right) + r^2 \Omega \quad . \tag{10.5.4}$$

The cross-term appearing in (10.5.4) has to be eliminated by yet another change of coordinates, implicitly obtained by the relations :

$$\tan p \equiv v \quad , \quad \tan q \equiv w \quad , \quad p - q \geq 0 \quad . \tag{10.5.5}$$

In fact, the coordinates (v, w) defined in (10.5.3) still lie in the infinite interval $]-\infty, \infty[$, whereas p and q in (10.5.5) lie in the interval $]-\frac{\pi}{2}, \frac{\pi}{2}[$. One thus obtains the metric :

$$g_4 = (\sec p)^2 (\sec q)^2 \left[-\frac{1}{2}\left(dp \otimes dq + dq \otimes dp\right) + \frac{1}{4}(\sin(p-q))^2 \Omega\right] = \omega^{-2} \widetilde{g} \quad , \tag{10.5.6}$$

where :

$$\omega \equiv (\cos p)(\cos q) \quad , \tag{10.5.7}$$

and $\widetilde{g}$ is the metric in square brackets in (10.5.6). This $\widetilde{g}$ is more conveniently re-written after defining :

$$t' \equiv \frac{(p+q)}{2} \quad , \quad r' \equiv \frac{(p-q)}{2} \quad , \tag{10.5.8}$$

which implies :

$$\widetilde{g} = -dt' \otimes dt' + \left[dr' \otimes dr' + \frac{1}{4}(\sin(2r'))^2 \Omega \right] \quad . \tag{10.5.9}$$

10.5.1.1 Einstein's Static Universe

With the terminology of section 10.5.1.4, we can say that, in light of (10.5.6-7), a *conformal rescaling* exists relating the metrics g_4 and $\widetilde{g}$ in the *same* coordinates. However, what have we gained from this ?

The answer can be found after a more careful study of (10.5.9) and of the procedure we briefly outlined. Note first that (10.5.9) is *locally* identical to the metric of the so-called Einstein static universe. For us to be able to go beyond a local analysis, an analytic extension is needed. In other words, we *analytically extend* (10.5.9) to the *whole* of the Einstein static universe, so that the coordinates are extended to cover the manifold with $R^1 \times S^3$ topology. This means that t' is finally taken to lie in the interval $]-\infty, \infty[$, whereas r', θ, ϕ are *regarded* as coordinates on the three-sphere S^3. The coordinate singularities at $r' = 0, \pi$ and at $\theta = 0, \pi$ can be removed by transforming to other local coordinates in a neighbourhood of points where (10.5.9) is singular (Hawking and Ellis 1973). One thus finds that the whole of Minkowski space-time is conformal (cf. section 10.5.1.4) to the region :

$$(t' + r') \in]-\pi, \pi[\quad , \quad (t' - r') \in]-\pi, \pi[\quad , \quad r' \geq 0 \quad , \tag{10.5.10}$$

of the Einstein static universe. The boundary of this region is then thought of (or defined) as representing the *conformal structure of infinity* of Minkowski space-time. It consists of

(1) The null surface $SCRI^- \equiv \left\{t' - r' = q = -\frac{\pi}{2}\right\}$, i.e. the future light cone of the point $r' = 0, t' = -\frac{\pi}{2}$;

(2) The null surface $SCRI^+ \equiv \left\{t' + r' = p = \frac{\pi}{2}\right\}$, i.e. the past light cone of the point $r' = 0, t' = \frac{\pi}{2}$;

(3) Past timelike infinity, i.e. the point

$$\iota^- \equiv \left\{r' = 0, t' = -\frac{\pi}{2}\right\} \Rightarrow p = q = -\frac{\pi}{2} \quad ;$$

(4) Future timelike infinity, i.e. the point

$$\iota^+ \equiv \left\{r' = 0, t' = \frac{\pi}{2}\right\} \Rightarrow p = q = \frac{\pi}{2} \quad ;$$

(5) Spacelike infinity (also called *SPI*), i.e. the point

$$\iota^0 \equiv \left\{r' = \frac{\pi}{2}, t' = 0\right\} \Rightarrow p = -q = \frac{\pi}{2} \quad .$$

It can be easily seen that ι^-, ι^+ and ι^0 are points rather than two-spheres, since $(\sin(2r'))^2$ vanishes at $\iota^0, \iota^-, \iota^+$ (cf. (10.5.9)).

Second, it is worth emphasizing what follows (Penrose 1974). It is sometimes convenient to accompany a conformal rescaling with a coordinate change. This is what we have done at the beginning of this section, and is merely in order that the newly adjoined points be assigned finite coordinates. No transformation of space-time points is involved. If desired, the new coordinates (which are a matter of convenience) may be introduced first, before any conformal rescaling takes place. This serves to overemphasize that the *conformal rescaling has nothing to do with a change of coordinates*. We thus reassure the reader that conformal-infinity calculations are of the coordinate-free type.

This remark enables one to present the conformal-infinity construction as usually preferred in the original literature, while avoiding some confusion which might arise. More details will be given in the next sections [we found it more instructive to clarify terminology and formalism in due course, rather than solving all problems from the beginning].

Problems

P.10.5.1 Prove that in M_4 any timelike geodesic originates at ι^- and finishes at ι^+. Similarly, prove that null geodesics originate at $SCRI^-$ and end at $SCRI^+$, while spacelike geodesics originate and end at ι^0.

P.10.5.2 On suppressing two dimensions, one can represent the Einstein static universe as the cylinder $x^2 + y^2 = 1$ imbedded in three-dimensional Minkowski space M_3. Try to represent Minkowski conformal infinity as the boundary of the region in Eq. (10.5.10) in the Einstein cylinder (cf. page 122 in Hawking and Ellis 1973).

P.10.5.3 Write a monochromatic plane wave in the coordinates (t', r') defined in Eq. (10.5.8).

10.5.1.2 Penrose Diagrams for Minkowski Space-Time

Einstein's static universe and Einstein's cylinder are just a device to introduce conformal infinity for Minkowskian geometry. Since the corresponding diagram (cf. Problem 10.5.2) is (a bit) involved, it is convenient to reduce by one the number of dimensions. One then deals with diagrams in the (t', r') plane, where only $SCRI^+, SCRI^-, \iota^+, \iota^0, \iota^-$ are represented.

Penrose diagrams are the tool to represent this kind of asymptotic analysis. From problem 10.5.1, we know we can regard *any* timelike geodesic as originating at ι^- and finishing at ι^+, and any null geodesic as originating at past null infinity $SCRI^-$ and ending at future null infinity $SCRI^+$, whereas ι^0 is both past and future endpoint of spacelike geodesics. However, this is not yet the Penrose diagram of Minkowski space-time, which is instead shown in Figure (15-ii) of Hawking and Ellis 1973. We thus need only half of the rhombus of Figure (15-i) therein.

The dotted vertical line, corresponding to points on the line $r = 0$, denotes the polar-coordinates trivial singularity. Each point in Figure (15-i) of Hawking and Ellis 1973 represents a two-sphere S^2, and straight lines at $\pm\frac{\pi}{4}$ correspond to radial null geodesics. In fact we are studying a (special case of) spherically symmetric space-time. Penrose-diagrams theory deals with conformal infinity for these space-times. The basic representation rules to construct such diagrams are (Hawking and Ellis 1973, Beem and Ehrlich 1981) :

(1) infinity → single lines;

(2) origin of polar coordinates → dotted lines;

(3) irremovable singularities of the metric (and Riemann) → double lines.

Problems

P.10.5.4 Study electromagnetism on *SCRI*.

P.10.5.5 If we bring infinity down to a finite distance using conformal-infinity techniques, we might hope to improve considerably self-adjointness problems. Investigate this point for massless scalar and spin-$\frac{1}{2}$ fields.

10.5.1.3 Null Infinity and Compactified Minkowski Space-Time

A peculiar property of Minkowski space-time is that any null geodesic having a past endpoint A^- on $SCRI^-$ has the *same* future endpoint A^+ on $SCRI^+$. This may seem surprising at first sight, but it holds since the future light cone of a point on $SCRI^-$ is a null hyperplane in Minkowskian geometry. Similarly, the past light cone of any point on $SCRI^+$ is also a null hyperplane.

One thus finds that a null hyperplane acquires a vertex in the past on $SCRI^-$, say A^-, and a vertex in the future on $SCRI^+$, say A^+. By virtue of this correspondence between points on $SCRI^-$ and points on $SCRI^+$, it appears natural (although not compulsory) to identify $SCRI^-$ with $SCRI^+$, setting (Penrose 1974, Penrose and Rindler 1986) :

$$A^- = A^+ \quad , \quad SCRI^- = SCRI^+ = SCRI \quad .$$

By continuity, one then also finds that $\iota^- = \iota^0 = \iota^+ = I$. Note that the point I is nonsingular, and has a neighbourhood given by three separate portions of M_4. This implies I becomes a normal interior point of the identified manifold. The compact manifold so obtained is called compactified Minkowski space-time, and is denoted by $M^{\#}$, following Penrose and Rindler 1986.

Problems

P.10.5.6 Prove the statement in this section concerning past and future endpoints of null geodesics in (M_4, η).

P.10.5.7 What is the topology of $M^{\#}$? [Hint : any null geodesic in $M^{\#}$ has S^1 topology, since it is a null line in M_4 joined in a closed loop to a single point at infinity ...]

10.5.1.4 Conformal Rescalings, Conformal Mappings and Conformal Group

A *conformal rescaling* of the metric η of a smooth manifold is a transformation

$$\eta_{ab} \rightarrow \hat{\eta}_{ab} = \Omega^2 \eta_{ab} \quad ,$$

where $\Omega = \Omega(x)$ is a *smooth* and *positive-definite* scalar function. Hereafter, we denote by η the metric of Minkowski space-time. In general, the new metric $\hat{\eta}$ is not flat, although it is, by definition, conformally flat. However, for certain choices of Ω, $\hat{\eta}$ is flat. For example, if Ω is constant, or if $\Omega(x) = (x^a x_a)^{-1}$, then $\hat{\eta}$ is flat. In this last case Ω is singular on the set $\{x^a : x^a x_a = 0\}$, i.e. the light cone of the origin O, so that such a cone has to be removed from space-time (Ward and Wells 1990).

Given two pseudo-Riemannian manifolds (M_1, g_1) and (M_2, g_2) [thus not necessarily Lorentzian], a diffeomorphism $f : (M_1, g_1) \rightarrow (M_2, g_2)$ is called a *conformal mapping* if the metric induced on M_2 by f from the metric on M_1 is a conformal rescaling of the given metric on M_2. Thus, in particular, if f is an isometry, then f is a conformal mapping.

Example 1. Let $M_1 = M_2 = M_4$, $f : x^a \rightarrow kx^a$, where $k > 0$ and constant. The induced metric $\hat{\eta}$ is then related to the original metric η by the relation $\eta_{ab} = k^2 \hat{\eta}_{ab}$. Thus f is a conformal mapping, called *dilation*.

Example 2. Let

$$M_1 = M_4 - \{\text{light cone of } p^a\} = \{x^a : (p^a - x^a)(p_a - x_a) \neq 0\} \quad ,$$

and

$$M_2 = M_4 - \{\text{light cone of } O\} \quad .$$

We then define f as :

$$f : x^a \rightarrow \frac{(p^a - x^a)}{(p_b - x_b)(p^b - x^b)} \quad .$$

Thus f is a diffeomorphism, and the induced metric $\hat{\eta}$ on M_2 is related to the original metric η on M_2 by : $\hat{\eta}_{ab} = \Omega^2 \eta_{ab}$, where $\Omega(x) = (x^a x_a)^{-1}$. Thus f is a conformal mapping, called *inversion*. The underlying idea related to this conformal mapping is the compactification of M_4 by adjoining a light cone at infinity, I. One then obtains a compact space M containing M_4.

The Lie group of conformal mappings of M to itself is called conformal group and denoted by $C(1,3)$. It is 15-dimensional, since it is generated from Poincaré transformations (10 parameters), dilations (1 parameter), and inversions (4 parameters) [we here only study four-dimensional Minkowski space-time]. Poincaré transformations and dilations leave I invariant, whereas inversions move such a light cone I to infinity. The precise metric on M does not matter; what is relevant is the light-cone (i.e. conformal) structure.

When we study conformal transformations, there are two possible meanings for the statement that a field theory is conformally invariant. In fact :

(I) One might require it is possible to assign a *conformal weight* to any field in the theory, in such a way that field equations are invariant under *arbitrary* conformal rescalings

$$\eta_{ab} \rightarrow \Omega^2(x)\eta_{ab}$$

of the metric. A scalar, tensor or spinor field is then said to have conformal weight k if it transforms as $\varphi \rightarrow \Omega^k \varphi$ under such rescaling. Since $\Omega^2 \eta_{ab}$ is not flat, in general this leads to study field theories in curved space-time. Indeed, it is well-known that massless free fields are invariant under conformal rescalings (Ward and Wells 1990). In the case of scalar massless free fields, what is conformally invariant is the wave equation :

$$\left[\Box + \frac{1}{4}\frac{(n-2)}{(n-1)}R\right]\phi = 0 \quad .$$

One can easily see that massive theories are not conformally invariant. In fact, under the dilation $\eta_{ab} \rightarrow k^2 \eta_{ab}$, the D'Alembertian transforms as $\Box \rightarrow k^{-2}\Box$, so that the Klein-Gordon equation can be conformally invariant if and only if the mass term m is set to zero.

(II) The second possible definition of conformal invariance is to require that the field theory be invariant under the action of the conformal group $C(1,3)$. If a theory is Poincaré invariant, and also invariant under conformal rescalings (or even just under the action of those rescalings which preserve the flat metric), then it is also conformally invariant in this sense. This result may not seem *a priori* obvious. It holds since Poincaré transformations become conformal transformations according to any other conformally-rescaled flat metric. Conformal transformations obtained in this way, together with Poincaré transformations, generate the whole conformal group.

Massless free-field theories are also conformally invariant in this second sense. This implies the extension of various fields from M_4 to its compactification. One has then to be very careful : fields should be thought of as sections of some nontrivial spinor bundles on M (Ward and Wells 1990), and a conformal weight is associated to each field, according to its spin.

Problems

P.10.5.8 Try to present and prove in rigorous way the statements made in this section about the conformal group. What happens if we study two-dimensional Minkowski space-time ? Under which conditions is the conformal group still finite-dimensional ? [cf. Penrose 1974]

P.10.5.9 Prove conformal invariance of [cf. Penrose and Rindler 1986, Ward and Wells 1990] : (a) timelike, null and spacelike vectors; (b) Weyl tensor; (c) massless free-field equations; (d) Yang-Mills equations; (e) twistor equation; (f) twistor space; (g) Hodge-star operator in four dimensions.

10.5.2 The Bondi-Metzner-Sachs Group

For a generic space-time, the isometry group is simply the identity, and thus does not provide relevant information. But isometry groups play a very important role in physics.

The most important example is given by the Poincaré group, which is the group of all real transformations of Minkowski space-time :

$$x' = \Lambda x + a \quad , \tag{10.5.11}$$

which leave invariant the length $(x-y)^2$. More precisely, the Poincaré group is given by the semidirect product of the Lorentz group $O(3,1)$ and of translations T_4 in Minkowski space-time.

It is therefore very important to generalize the concept of isometry group to a suitably regular curved space-time (Penrose and Rindler 1986). The diffeomorphism group is not really useful because it is *too large* and it only preserves the differentiable structure of space-time. The concept of asymptotic symmetry group makes sense for any space-time (M,g) which tends to infinity either to Minkowski or to a Friedmann-Robertson-Walker model. The goal is achieved adding to (M,g) a boundary given by future null infinity, past null infinity or the whole of null infinity (cf. section 10.5.1). We now formulate in a precise way this idea. For this purpose let us begin by recalling that the cuts of scri are spacelike two-surfaces in scri orthogonal to the generators of scri. Each cut has S^2 topology. They can be regarded as Riemann spheres with coordinates (ζ,ζ^*), where $\zeta = x+iy$ and ζ^* is the complex conjugate of ζ, so that locally the metric is given by : $g_2 = -d\zeta d\zeta^*$. Thus, defining (Stewart 1990) :

$$\zeta \equiv e^{i\phi} \cot \frac{\theta}{2} \quad , \tag{10.5.12}$$

we find :

$$g_2 = -\frac{1}{4}(1+\zeta\zeta^*)^2 g_\Sigma \quad , \quad g_\Sigma = d\theta \otimes d\theta + (\sin\theta)^2 d\phi \otimes d\phi \quad . \tag{10.5.13}$$

Thus, if we choose a conformal factor $\Omega = \frac{2}{(1+\zeta\zeta^*)}$, each cut becomes the unit two-sphere. The choice of a chart can then be used to define an asymptotic symmetry group. Indeed, the following simple but fundamental result holds (Stewart 1990) :

Theorem 10.5.1 All holomorphic bijections f of the Riemann sphere are of the form :

$$\hat{\zeta} = f(\zeta) = \frac{(a\zeta+b)}{(c\zeta+d)} \quad , \tag{10.5.14}$$

where $ad - bc = 1$.

The transformations (10.5.14) are called fractional linear transformations (FLT). Now, if a cut has to remain a unit sphere under (10.5.14), we must perform another conformal transformation : $\hat{g}_\Sigma = K^2 g_\Sigma$, where (Stewart 1990) :

$$K = \frac{(1+\zeta\zeta^*)}{(a\zeta+b)(a^*\zeta^*+b^*)+(c\zeta+d)(c^*\zeta^*+d^*)} \quad . \tag{10.5.15}$$

Finally, for the theory to remain invariant under (10.5.15), the lengths along the generators of scri must change according to : $d\hat{u} = K du$, which implies :

$$\hat{u} = K\Big[u + \alpha(\zeta,\zeta^*)\Big] \quad . \tag{10.5.16}$$

The transformations (10.5.14-16) form the Bondi-Metzner-Sachs (BMS) asymptotic symmetry group of space-time. The subgroups of BMS are :

10.5.2.1 Supertranslations

This is the subgroup S defined by :

$$\hat{u} = u + \alpha(\zeta,\zeta^*) \quad , \quad \hat{\zeta} = \zeta \quad . \tag{10.5.17}$$

The quotient group $\frac{(BMS)}{S}$ is the orthocronous proper Lorentz group.

10.5.2.2 Translations

This four-parameter subgroup T is given by (10.5.17) plus the following relation :

$$\alpha = \frac{(A + B\zeta + B^*\zeta^* + C\zeta\zeta^*)}{(1+\zeta\zeta^*)} \quad . \tag{10.5.18}$$

The name is chosen since a translation in Minkowski space-time generates a member of T. In fact, denoting by (t,x,y,z) Cartesian coordinates in Minkowski space-time, if we define the functions :

$$u \equiv t - r \quad , \quad r^2 \equiv x^2 + y^2 + z^2 \quad , \quad Z \equiv \frac{1}{(1+\zeta\zeta^*)} \quad , \tag{10.5.19}$$

we find that (Stewart 1990) :

$$Z^2\zeta = (x+iy)\frac{(1-\frac{z}{r})}{4r} \quad , \tag{10.5.20}$$

$$x = r(\zeta+\zeta^*)Z \quad , \quad y = -ir(\zeta-\zeta^*)Z \quad , \quad z = r(\zeta\zeta^*-1)Z \ . \tag{10.5.21}$$

Thus the translation :

$$t' = t+a \quad , \quad x' = x+b \quad , \quad y' = y+c \quad , \quad z' = z+d \quad , \tag{10.5.22}$$

implies that :

$$u' = u + Z\Big(A + B\zeta + B^*\zeta^* + C\zeta\zeta^*\Big) + \mathrm{O}\left(\frac{1}{r}\right) \quad , \tag{10.5.23}$$

which agrees with (10.5.17-18).

10.5.2.3 Poincaré

A BMS transformation is obtained from a Lorentz transformation and a supertranslation. This is why there are several Poincaré groups at scri, one for each supertranslation which is not a translation, and no-one of them is preferred. This implies there is not yet agreement about how to define angular momentum in an asymptotically flat space-time (because this is related to the Lorentz group which is a part of the Poincaré group as explained before). Still, the energy-momentum tensor is well-defined, because it is only related to the translations.

The Poincaré group can be defined as the subgroup of BMS which maps good cuts into good cuts (Newman and Tod 1980). In other words, there is a four-parameter collection of cuts, called good cuts, whose asymptotic shear vanishes, and these good cuts provide the structure needed so as to reduce BMS to the Poincaré group. In fact, the asymptotic shear $\sigma^0(u,\zeta,\zeta^*)$ of the $u = constant$ null surfaces is related to the $(\sigma')^0(u',\zeta',(\zeta^*)')$ of the $u' = constant$ null surfaces by the relation :

$$(\sigma')^0(u',\zeta',(\zeta^*)') = K^{-1}\Big[\sigma^0(u,\zeta,\zeta^*) + (\mathrm{edth})^2\alpha(\zeta,\zeta^*)\Big] \quad , \tag{10.5.24}$$

where ζ' is given by the left-hand side of (10.5.14), and the operator *edth* is defined on page 8 of Newman and Tod 1980. In view of (10.5.17), for the supertranslations the relation (10.5.24) assumes the form :

$$(\sigma')^0(u,\zeta,\zeta^*) = \sigma^0(u'-\alpha,\zeta,\zeta^*) + (\text{edth})^2\alpha \quad . \tag{10.5.25}$$

For stationary space-times (which have a timelike Killing vector field), the Bondi system exists where $\sigma^0 = 0$. Therefore, a supertranslation between two Bondi systems both having $\sigma^0 = 0$, leads to the equation $(\text{edth})^2\alpha = 0$, which is solved by the translation group. This proves in turn that there is indeed a collection of good cuts as defined before. As explained in Ashtekar 1987 (see also Penrose and Rindler 1986), in geometrical terms the main ideas can be summarized as follows. The generators of scri are the integral curves of a null vector field N. A vector field X is called an (asymptotic) symmetry if it generates a diffeomorphism which leaves invariant the integral curves of N. Denoting by h the intrinsic metric on scri, one then has (Ashtekar 1987) : $L_X N = -\rho N, L_X h = 2\rho h, L_N \rho = 0$, where ρ is a smooth function. Any linear combination and any Lie bracket of symmetries is still a symmetry, so that they form a Lie algebra denoted by B, say. Given the vector field $X = \beta N$, one finds that X is a symmetry if and only if (Ashtekar 1987) : $L_N \beta = 0$. The symmetries of this form are the supertranslations $ST \subset B$. As clarified in Ashtekar 1987, ST is the Abelian infinite-dimensional ideal of B, and the quotient B/ST is found to be the Lie algebra of the Lorentz group.

As remarked in Schmidt and Stewart 1988, it should be emphasized that the basic problem in asymptotics, i.e. the existence of solutions to Einstein's equations whose asymptotic properties are described by the scri formalism, is still unsolved. We refer the reader to Schmidt and Stewart 1988 for a detailed study of this problem.

10.5.3 The Boundary of Space-Time

The singularity theorems in general relativity (Hawking and Ellis 1973) were proved using a definition of singularities based on the g-boundary. Namely, one defines a topological space, the g-boundary, whose points are equivalence classes of incomplete nonspacelike geodesics. The points of the g-boundary are then the singular points of space-time. Timelike and null geodesic incompleteness are considered minimum conditions for space-time to be singularity-free, since timelike geodesic incompleteness implies there could be freely moving observers or particles whose histories did not exist after, or before, a finite interval of proper time, and null geodesics are the histories of zero rest-mass particles (Hawking and Ellis 1973). However, as emphasized for example in Schmidt 1971, this definition has two basic drawbacks :

(1) it is based on geodesics, whereas in Geroch 1968b it was proved there are geodesically complete space-times with curves of finite length and bounded acceleration;

(2) there are several alternative ways of forming equivalence classes and defining the topology.

Schmidt's method is along the following lines :

Step 1. Connections are known to provide a parallelization of the bundle $L(M)$ of linear frames.

Step 2. This parallelization can be used to define a Riemannian metric.

Step 3. This Riemannian metric has the effect of making a connected component of $L(M)$ into a metric space. This connected component $L'(M)$ is dense in a complete metric space $L'_C(M)$.

Step 4. One defines $\overline{M}$ as the set of orbits of the transformation group on $L'_C(M)$.

Step 5. The b-boundary ∂M of M is then defined as : $\partial M \equiv \overline{M} - M$.

Step 6. Singularities of M are defined as points of the b-boundary ∂M which are contained in the b-boundary of any extension of M.

A few more details about steps 1-4 of this construction are now given.

Step 1

The parallelization of $L(M)$ is obtained by defining horizontal and vertical vector fields. For this purpose, we first introduce the mapping $\pi : L(M) \rightarrow M$ of the frame at x into x.

Definition 10.5.1 The curve γ in $L(M)$ is horizontal if the frames $Y_1(t), ..., Y_n(t)$ are parallel along $\pi(\gamma(t))$.

Definition 10.5.2 The horizontal vector fields B_i are the unique vector fields such that :

$$\pi_*((B_i)_\gamma) = Y_i \quad , \quad \pi_*((B(\xi))_\gamma) = \xi^i Y_i \quad , \tag{10.5.26}$$

if $\gamma = Y_1, ..., Y_n$.

Definition 10.5.3 Vertical vector fields are given by :

$$(E^*) = \left(\frac{d}{dt} R_{a(t)\gamma} \right)_{t=0} \quad , \tag{10.5.27}$$

where R_a is the action of the general linear group $GL(n, R)$ on $L(M)$. The parallelization of $L(M)$ is then given by $\left(E_k^{*\ i}, B_i\right)$.

Step 2

Definition 10.5.4 Denoting by $gl(n, R)$ the Lie algebra of $GL(n, R)$, a $gl(n, R)$-valued one-form ω is expressed as :

$$\omega(Y) = \omega_k{}^i(Y) E_i{}^k \quad . \tag{10.5.28}$$

Definition 10.5.5 The canonical one-forms θ^i are given by :

$$\pi_*(Y_\gamma) = \theta^i(Y_\gamma) Y_i \quad , \tag{10.5.29}$$

if $\gamma = Y_1, ..., Y_n$.

Definition 10.5.6 The Riemannian metric g is then (Schmidt 1971) :

$$g(X,Y) = \sum_i \theta^i(X)\theta^i(Y) + \sum_{i,k} \omega_k{}^i(X)\omega_k{}^i(Y) \quad . \tag{10.5.30}$$

Step 3

The Riemannian metric g defines a distance function according to (10.2.2). Thus the connected component $L'(M)$ of $L(M)$ is a metric space, and it uniquely determines a complete metric space, $L'_C(M)$. Moreover, $L'(M)$ is dense in $L'_C(M)$.

Step 4

One proves that $GL(n, R)$ is a topological transformation group on $L'_C(M)$, in that the transformations R_a are uniformly continuous and can be extended in a uniformly continuous way on the closure of $L'(M)$ in $L'_C(M)$.

However, also Schmidt's definition has some drawbacks. In fact :

(1) in a closed FRW universe the initial and final singularities form the same single point of the b-boundary (Bosshard 1976);

(2) in the FRW and Schwarzschild solutions the b-boundary points are not Hausdorff separated from the corresponding space-time (Johnson 1977).

The b-boundary has been replaced by the causal boundary (Geroch et al. 1972), which is able to distinguish between singular points and regular points at infinity. In other words, it is known that singularities cannot be regarded as points of the space-time manifold. One has thus to define suitable additional *ideal points* which, once adjoined to (M, g), give rise to a structure $\overline{M}$ incorporating both singular and nonsingular points. The construction in Geroch et al. 1972 can be applied provided no two distinct points of (M, g) have same past and same future : $\forall p, q \in M, p \neq q \Rightarrow I^-(p) \neq I^-(q)$ and $I^+(p) \neq I^+(q)$, and is conformally invariant. However, the interpretation of ideal points depends on the particular choice of conformal factor. The advantage of the method is that $\overline{M}$ can be easily found by inspection in concrete examples, and ideal points make possible a definition of

Cauchy development (cf. definition 10.4.4). However, difficulties remain connected with topological identifications which occur for causal boundaries, whose construction works smoothly only for stably causal space-times (Szabados 1988-1989).

10.6 Hamiltonian Structure

Dirac's theory of constrained Hamiltonian systems (Dirac 1964, Hanson et al. 1976) has been successfully applied to general relativity, though many unsolved problems remain on quantization (Ashtekar 1988, Ashtekar 1991, Rovelli 1991). The ADM formalism for general relativity is discussed in Misner et al. 1973, MacCallum 1975, Hanson et al. 1976, our section 2.4. The derivation of boundary terms in the action integral can be found in York 1972, Gibbons and Hawking 1977, York 1986, our section 2.4, whereas a modern treatment of the ADM phase space for general relativity in the asymptotically flat case is in Ashtekar 1988. More recently, Ashtekar's spinorial variables have given rise to a renewed interest in canonical gravity. We now analize the *new* phase space of general relativity, only paying attention to the classical theory.

The basic postulate of canonical gravity is that space-time is topologically $\Sigma \times R$, and it admits a foliation in spacelike three-manifolds S_t, which are all diffeomorphic to Σ (cf. section 2.4). Ashtekar's variables for canonical gravity are very important at least for the following reasons :

(1) they are one of the most striking applications of the spinorial formalism to general relativity;

(2) the constraint equations assume a polynomial form, which is not achieved using the old variables;

(3) they realize a formal analogy between gravity and Yang-Mills theory;

(4) they might lead to an exact solution of the constraint equations of the quantum theory.

The basic ideas of the formalism of $SU(2)$ spinors in Euclidean three-space are the following (cf. Ashtekar 1988 and section 5.8). We consider $(V, \epsilon_{AB}, G_{A'B})$, where V is a complex two-dimensional vector space with a nondegenerate symplectic form ϵ_{AB} and a positive-definite Hermitian scalar product $G_{A'B}$. Then, given a real three-manifold Σ, we consider the vector bundle B over Σ whose fibres are isomorphic to $(V, \epsilon_{AB}, G_{A'B})$. The $SU(2)$ spinor fields on Σ are thus the cross-sections of B. The isomorphism between the space of symmetric, second-rank Hermitian spinors λ^{AB} and the tangent space to Σ is realized by the soldering form $\sigma^a{}_{AB}$, and the metric h on Σ is given by :

$$h^{ab} \equiv \sigma^a{}_{AB}\,\sigma^b{}_{CD}\,\epsilon^{AC}\epsilon^{BD} = -Tr\left(\sigma^a\sigma^b\right) \quad . \tag{10.6.1}$$

The conjugation of $SU(2)$ spinors obeys the rules (5.8.60-62). We now consider a new configuration space C in the asymptotically flat case, defined as the space of all $\sigma^a{}_A{}^B$ such that (Ashtekar 1988) :

$$\sigma^a{}_A{}^B = \left[1 + \frac{P(\theta,\phi)}{r}\right]^2 (\sigma^0)^a{}_A{}^B + \mathrm{O}\left(\frac{1}{r^2}\right) \quad . \tag{10.6.2}$$

The momentum conjugate to $\sigma^a{}_A{}^B$, following Ashtekar 1988, is denoted by $M_{aA}{}^B$ and it obeys the relations :

$$Tr(M_a\sigma^a) = \mathrm{O}\left(\frac{1}{r^3}\right) \quad , \tag{10.6.3}$$

$$M_{aA}{}^B + \frac{1}{3}Tr\left(M_l\sigma^l\right)\sigma_{aA}{}^B = \mathrm{O}\left(\frac{1}{r^2}\right) \quad . \tag{10.6.4}$$

The extended phase space Γ is the space whose points are the pairs $\left(\sigma^a{}_A{}^B, M_{aA}{}^B\right)$ obeying (10.6.2-4), and the Poisson brackets among observables are defined by :

$$\{u,v\} \equiv \int_\Sigma Tr\left(\frac{\delta u}{\delta M_a}\frac{\delta v}{\delta\sigma^a} - \frac{\delta u}{\delta\sigma^a}\frac{\delta v}{\delta M_a}\right) d^3x \quad . \tag{10.6.5}$$

In going to the new phase space we have added three degrees of freedom, which lead to three new constraints :

$$C_{ab} = -Tr\left(M_{[a}\sigma_{b]}\right) = M_{[ab]} \quad , \tag{10.6.6}$$

in addition to the Hamiltonian and momentum constraints. We now consider the phase space :

$$\Gamma' \equiv \left\{ \left(\tilde{\sigma}^a{}_J{}^L, A_{aJ}{}^L \right) \right\} \quad , \tag{10.6.7}$$

where the spinorial variables $\tilde{\sigma}^a{}_J{}^L$ and $A_{aJ}{}^L$ are obtained from $\sigma^a{}_J{}^L$ and $M_{aJ}{}^L$ as follows. The variable $\tilde{\sigma}^a{}_J{}^L$ is defined by :

$$\tilde{\sigma}^a{}_J{}^L \equiv \sqrt{h}\, \sigma^a{}_J{}^L \quad . \tag{10.6.8}$$

The step leading to $A_{aJ}{}^L$ is simple but not trivial (it can be more thoroughly understood by recalling the definition of Sen connection as done in Ashtekar 1988). At first we define a new momentum variable :

$$\pi_{aJ}{}^L \equiv \frac{1}{\sqrt{h}} \left[M_{aJ}{}^L + \frac{1}{2} Tr(M_b \sigma^b) \sigma_{aJ}{}^L \right] \quad . \tag{10.6.9}$$

Now, denoting by D the connection on the real three-manifold Σ, we define a new connection $\tilde{D}$ by (Ashtekar 1988) :

$$\tilde{D}_a \lambda_M \equiv \partial_a \lambda_M + A_{aM}{}^C \lambda_C \quad . \tag{10.6.10}$$

The spinorial variable $A_{aJ}{}^L$ in (10.6.10) is obtained from (10.6.9) and the spin-connection one-form $\Gamma_{aJ}{}^L$ of D by :

$$A_{aJ}{}^L \equiv \Gamma_{aJ}{}^L + \frac{i}{\sqrt{2}} \pi_{aJ}{}^L \quad , \tag{10.6.11}$$

where the spin-connection is known to be the unique connection which annihilates the soldering form $\sigma^a{}_{JL}$, and is given by :

$$\Gamma_a{}^{JL} \equiv -\frac{1}{2} \sigma_f{}^{EL} \left[\partial_a \sigma^{fJ}{}_E + \Gamma^f_{ba} \sigma^{bJ}{}_E \right] \quad , \tag{10.6.12}$$

and $\Gamma^f_{\cdot ba}$ are the Christoffel symbols involving the three-metric h on Σ. The new variables defined in (10.6.8) and (10.6.11) obey the Poisson-bracket relations (Ashtekar 1988) :

$$\left\{ \tilde{\sigma}^a{}_J{}^L(x), \tilde{\sigma}^b{}_M{}^N(y) \right\} = 0 \quad , \tag{10.6.13}$$

$$\left\{A_a{}^{JL}(x), \widetilde{\sigma}^m{}_{MN}(y)\right\} = \frac{i}{\sqrt{2}}\delta(x,y)\delta_M{}^{(J}\delta_N{}^{L)} \quad , \tag{10.6.14}$$

$$\left\{A_{aJ}{}^L(x), A_{bM}{}^N(y)\right\} = 0 \quad . \tag{10.6.15}$$

Finally, denoting by $F_{abM}{}^N$ the curvature of $\widetilde{D}$:

$$F_{abM}{}^C\lambda_C = 2\widetilde{D}_{[a}\widetilde{D}_{b]}\lambda_M \quad , \tag{10.6.16}$$

the constraints of the theory assume the form (Ashtekar 1988) :

$$\widetilde{D}_a\left(\widetilde{\sigma}^a{}_J{}^L\right) \approx 0 \quad , \tag{10.6.17}$$

$$Tr\left(\widetilde{\sigma}^a F_{ab}\right) \approx 0 \quad , \tag{10.6.18}$$

$$Tr\left(\widetilde{\sigma}^a\widetilde{\sigma}^b F_{ab}\right) \approx 0 \quad , \tag{10.6.19}$$

where the Gauss-law constraint (10.6.17) is due to (10.6.6), and (10.6.18-19) are the momentum and Hamiltonian constraints respectively. Defining :

$$E^a \equiv \widetilde{\sigma}^a \quad , \quad B^a \equiv \frac{1}{2}\epsilon^{abc}F_{bc} \quad , \tag{10.6.20}$$

we see that the new phase space (10.6.7) can be thought as a submanifold of the constrained phase space of a complexified $SU(2)$ Yang-Mills theory (see remark (3) at the beginning of this section), and the constraints are indeed polynomial as we anticipated. One can also reverse things, and regard the $A_{aJ}{}^L$ as configuration variables, so that their momentum conjugate becomes $\widetilde{\sigma}^a{}_J{}^L$. In so doing the momentum constraints remain linear and the Hamiltonian constraint remains quadratic in the momenta (Ashtekar 1988).

Also, it should be emphasized that $\widetilde{\sigma}^a{}_J{}^L$ is real whereas $A_{aJ}{}^L$ is complex, so that they are not conjugate variables in the usual sense. We can overcome this difficulty by going to the complex regime, i.e. we consider a complex phase space Γ_C whose points are defined on a real three-manifold Σ. The real section Γ of Γ_C is then defined by (Ashtekar 1988) :

$$\left(\widetilde{\sigma}^a\right)^\dagger = -\widetilde{\sigma}^a \quad , \tag{10.6.21}$$

$$\left(A_{aJ}{}^{L} - \Gamma_{aJ}{}^{L}\right)^{\dagger} = -\left(A_{aJ}{}^{L} - \Gamma_{aJ}{}^{L}\right) \quad . \tag{10.6.22}$$

In so doing we get back to (real Lorentzian) general relativity, whereas real Euclidean general relativity is defined by the conditions where (10.6.21) is unchanged, whereas in (10.6.22) the spin-connection one-form $\Gamma_{aJ}{}^{L}$ does not appear.

10.7 The Singularity Problem for Space-Times with Torsion

The singularity theorems of Penrose, Hawking and Geroch (Geroch 1966, Hawking 1966a-b, Hawking 1967, Hawking and Penrose 1970, Hawking and Ellis 1973, Geroch and Horowitz 1979) show that Einstein's general relativity leads to the occurrence of singularities in cosmology in a rather generic way. On the other hand, much work has also been done on alternative theories of gravitation (Fuchs et al. 1988). It is by now well-known that when we describe gravity as the gauge theory of the Poincaré group, this naturally leads to theories with torsion (Hehl et al. 1976, Trautman 1980, Hehl et al. 1980, Nester 1983, Awada et al. 1986). The basic ideas can be summarized as follows (Trautman 1973, Nester 1983, Esposito 1989b). The holonomy theorems imply that torsion and curvature are related to the groups of translations and of homogeneous transformations respectively in the tangent vector spaces to a manifold. It is worth recalling how the holonomy group is defined. Given a real manifold M, we fix a point $p \in M$ and consider any closed curve γ containing p. One then makes parallel transport along γ of any basis for the tangent space at p. This leads to a new basis at p, related to the original one by an element of $GL(6, R)$. Doing so for *all* closed curves γ, one obtains a subgroup of $GL(6, R)$, independent of the original point p, and called the holonomy group of the connection. In particular, since we are interested in metric-preserving connections, our holonomy groups are subgroups of $SO(6)$.

The introduction of torsion related to spin gives rise to a strong link between gravitation and particle physics, because it extends the holonomy group to the translations. An enlightening discussion of gauge translations can be found for example in Smrz 1987, Lord and Goswami 1988. In particular, the introduction of Smrz 1987 clarifies from the beginning the main geometric role played by the translations in the gauge group : they change a principal fibre bundle having no special relationship between the points on the fibres and the base manifold into the bundle of linear frames of the base manifold. When we consider the gauge theory of the Poincaré group, we discover that the gauge fields for the translation invariance are the orthonormal frames, and the gauge field for Lorentz transformations is the part of the full connection called contorsion (Nester 1983). From the point of view of fibre-bundle theory, the possibility of defining torsion is a peculiarity of relativistic theories of gravitation. In fact, the bundle $L(M)$ of linear frames is soldered to the base $B = M$, whereas for gauge theories other than gravitation the bundle $L(M)$ is loosely connected to M (Trautman 1980). Denoting by $\theta : TL(M) \rightarrow R^4$ the soldering form and by ω a connection one-form on $L(M)$, the torsion two-form T is defined by (Trautman 1980) : $T \equiv d\theta + \omega \wedge \theta$. The Poincaré group deserves special consideration because it corresponds to an external symmetry, it yields momentum and angular-momentum conservation, and its translational part can be seen as carrying matter through space-time (Hehl et al. 1980).

At the very high densities present in the early universe, the effects of spin can no longer be neglected (Esposito 1989b). Thus it is natural to address the question : is there a rigorous theory of singularities in a space-time with torsion ? The answer can only be found discussing at first the properties of geodesics in a space-time with torsion, and trying to define what is a singularity in such a theory.

10.7.1 Space-Times with Torsion and Their Geodesics

A space-time with torsion (hereafter referred to as U_4 space-time) is defined by adding the following fourth requirement to the ones in section 10.2.1 :

(4) Given a linear C^r connection $\widetilde{\nabla}$ which obeys the metricity condition, a nonvanishing tensor :

$$S(X,Y) \equiv \widetilde{\nabla}_X Y - \widetilde{\nabla}_Y X - [X,Y] \quad , \tag{10.7.1}$$

where X and Y are arbitrary C^r vector fields and the square bracket denotes their Lie bracket. The tensor $\frac{S}{2}$ is then called the torsion tensor (compare with Hawking and Ellis 1973).

Now, it is well-known that the curve γ is defined to be a geodesic curve if its tangent vector moves by parallel transport, so that $\nabla_X X$ is parallel to $\left(\frac{\partial}{\partial t}\right)_\gamma$ (see, however, comment before definition 10.7.1). A new parameter $s(t)$, called affine parameter, can always be found such that, in local coordinates, this condition is finally expressed by the following equation :

$$\frac{d^2x^a}{ds^2} + \Gamma_{bc}{}^a \frac{dx^b}{ds}\frac{dx^c}{ds} = 0 \quad . \tag{10.7.2}$$

The geodesic equation (10.7.2) contains the effect of torsion through the symmetric part of the connection coefficients : $\Gamma_{(bc)}{}^a = \left\{ {a \atop bc} \right\} + 2S^a{}_{(bc)}$, where $S^a{}_{(bc)}$ should not be confused with the vanishing $S_{(bc)}{}^a$. It is very useful to study this equation in a case of cosmological interest. For example, in a closed FRW universe the only nonvanishing components of the torsion tensor are the ones given in Esposito 1989b : $S_{m0}{}^m = Q(t)$, $\forall m = 1,2,3$, so that (10.7.2) yields (Esposito 1990-92) :

$$\frac{d^2x^0}{ds^2} + a\frac{da}{ds}\frac{ds}{dt}c_{ii}\left(\frac{dx^i}{ds}\right)^2 = 0 \quad , \tag{10.7.3}$$

$$\frac{d^2x^m}{ds^2} + \Gamma_{ij}{}^m \frac{dx^i}{ds}\frac{dx^j}{ds} + 2\left(\frac{1}{a}\frac{da}{ds}\frac{ds}{dt} - Q\right)\frac{dx^0}{ds}\frac{dx^m}{ds} = 0 \quad . \tag{10.7.4}$$

In (10.7.3), c_{ii} are the diagonal components of the unit three-sphere metric, and we are summing over all $i = 1,2,3$. In (10.7.4), we use the result of Esposito 1989b according to which :

$$\Gamma_{0m}{}^m = \frac{\dot{a}}{a} - 2Q \quad , \quad \Gamma_{m0}{}^m = \frac{\dot{a}}{a} \quad , \quad \forall m = 1,2,3 \quad . \tag{10.7.5}$$

Of course, $\dot{a}$ denotes $\frac{da}{ds}\frac{ds}{dt}$. Now, if the field equations are such that both $\frac{1}{a}\frac{da}{ds}\frac{ds}{dt}$ and Q remain finite for all values of s, the model will be nonspacelike geodesically complete. If a torsion singularity is thought as a point where torsion is infinite, we are ruling out this possibility with our criterion, in addition to the requirement that the scale factor never shrinks to zero. Thus it seems that, whatever the physical source of torsion is (spin or theories with quadratic Lagrangians etc.), nonspacelike geodesic completeness is a concept of physical relevance even though test particles do not move along geodesics (Hehl et al. 1974).

An important comment is now in order. We have defined geodesics exactly as leading treatises do in general relativity (cf. Hawking and Ellis 1973, page 33; Beem and Ehrlich 1981, page 403) for reasons which will become even more clear studying maximal timelike geodesics in section 10.7.2. However, our definition differs from the one given in Hehl et al. 1974. In that paper, our geodesics are just called autoparallel curves, whereas the authors interpret as geodesics the curves of extremal length whose tangent vector is parallelly transported according to the Christoffel connection. Indeed, if test particles were moving along extremal curves in theories with torsion, there would be a strong reason for defining geodesics and studying singularities only as done in Hehl et al. 1974. By contrast, as explained also in Hehl et al. 1974, the trajectories of particles differ in general from extremal curves and from the curves we call geodesics. Thus, it appears important to improve our understanding by studying the mathematical properties of a singularity theory based on the definition of geodesics as autoparallel curves. This definition, involving the properties of the full connection, may be expected to have physical relevance, imposing regularity conditions on the geometry and on the torsion of the cosmological model which is studied. It is also worth recalling that in the study of classical supersymmetric particles in Galvao and Teitelboim 1980, it was proved that a supergauge exists in which a massless spinning particle in a gravitational field follows a geodesic defined as autoparallel curve (whereas the authors found no supergauge where the worldline of a massive particle is a geodesic). Moreover, since the definition of timelike, null and spacelike vectors is not affected by the presence of torsion, the whole theory of causal structure outlined in section 10.4 remains unchanged. Combining this remark, also made in Hehl et al. 1974 [see, however, Fig. 12 on page 14 of Penrose 1983], with the qualitative argument concerning the geodesic equation, we here give the following definition (Esposito 1990-92) :

Definition 10.7.1 A U_4 space-time is singularity-free if it is timelike and null geodesically complete, where geodesics are defined as curves whose tangent vector moves by parallel transport with respect to the full U_4 connection.

This definition differs from the one given in Hehl et al. 1974 because we rely on a different definition of geodesics, and it has the drawbacks already illustrated at the beginning of section 10.5.3. Moreover, the inclusion of null geodesics in definition 10.7.1 is an open problem for theories with torsion (see Garcia de Andrade 1990b and references therein). However, definition 10.7.1 is a preliminary definition which allows a direct comparison with the corresponding situation in general relativity, is generic in that it does not depend on the specific physical theory which is the source of torsion and it has physical relevance not only for a closed FRW model but also for completely arbitrary models as we said before. Hence we can now try to make the same (and eventually additional) assumptions which lead to singularity theorems in general relativity, and check whether one obtains timelike and or null geodesic incompleteness. Indeed, the extrinsic-curvature tensor and the vorticity which appears in the Raychaudhuri equation contain now explicitly the effects of torsion, and it is not *a priori* clear what is going to happen. Thus, if one adopts definition 10.7.1 as a preliminary definition of singularities in a U_4 space-time, the main issues to be studied seem to be (Esposito 1990-92) :

(1) How can we explain from first principles that a space-time which is nonspacelike geodesically incomplete may become nonspacelike geodesically complete in the presence of torsion ? And is the converse possible ?

(2) What happens in a U_4 space-time (Hehl et al. 1974) under the assumptions which lead to the theorems of Penrose, Hawking and Geroch ?

We shall partially study question (2) in the next sub-section.

10.7.2 A Singularity Theorem without Causality Assumptions for U_4 Space-Times

In this section we denote by $R(X,Y)$ the four-dimensional Ricci tensor with scalar curvature R, and by $K(X,Y)$ the extrinsic-curvature tensor of a spacelike three-surface. The energy-momentum tensor is written as $T(X,Y)$, so that the Einstein equations are :

$$R(X,Y) - \frac{1}{2}g(X,Y)R = T(X,Y) \quad . \tag{10.7.6}$$

In so doing, we are absorbing the $8\pi G$ factor into the definition of $T(X,Y)$. For the case of general relativity, it was proved in Hawking 1967 that singularities must occur under certain assumptions, even though no causality requirements are made. In fact, Hawking's result (Hawking 1967, Hawking and Ellis 1973) states that space-time cannot be timelike geodesically complete if :

(1) $R(X,X) \geq 0$ for any nonspacelike vector X (which can also be written in the form : $T(X,X) \geq g(X,X)\frac{T}{2}$);

(2) there exists a compact spacelike three-surface Σ without edge;

(3) the trace K of the extrinsic-curvature tensor $K(X,Y)$ of Σ is either everywhere positive or everywhere negative.

We now study the following problem : is there a suitable generalization of this theorem to the case of a U_4 space-time ? Indeed, a careful examination of Hawking's proof (see Hawking and Ellis 1973, page 273) shows that the arguments which should be modified or adapted in a U_4 space-time are the ones involving the Raychaudhuri equation and the results which prove the existence or the nonexistence of conjugate points. We now examine them in detail.

10.7.2.1 Raychaudhuri Equation

The generalized Raychaudhuri equation in the ECSK theory of gravity has been derived in Stewart and Hájicek 1973, Tafel 1973 (see also Raychaudhuri 1975-1979). It turns out that, denoting by $\widetilde{\omega}_{ab}$ and σ_{ab} the vorticity and the shear tensors respectively, the expansion θ for a timelike congruence of curves obeys the equation :

$$\frac{d\theta}{ds} = -\left(R(U,U) + 2\sigma^2 - 2\widetilde{\omega}^2\right) - \frac{\theta^2}{3} + \widetilde{\nabla}_a\left(\dot{U}\right)^a \quad . \tag{10.7.7}$$

In (10.7.7), U is the unit timelike tangent vector, and we have defined :

$$2\sigma^2 \equiv \sigma_{ab}\sigma^{ab} \quad , \quad 2\widetilde{\omega}^2 \equiv \left(\omega_{ab} + \frac{1}{2}\widetilde{S}_{ab}\right)\left(\omega^{ab} + \frac{1}{2}\widetilde{S}^{ab}\right) \quad , \tag{10.7.8}$$

where ω_{ab} is the vorticity tensor obtained from the Christoffel symbols, and $\widetilde{S}_{bc}$ is obtained from the spin tensor $\sigma_{bc}{}^a$ by a relation usually assumed to be of the form (Tafel 1973, Demianski et al. 1987) :

$$\sigma_{bc}{}^a = \widetilde{S}_{bc}U^a \quad . \tag{10.7.9}$$

10.7.2.2 Existence of Conjugate Points

Conjugate points are defined as in general relativity (Hawking and Ellis 1973), but bearing in mind that now the Riemann tensor is the one obtained from the connection $\widetilde{\nabla}$ appearing in (10.7.1) :

$$R(X,Y,Z,W) = \left[\widetilde{\nabla}_X\widetilde{\nabla}_Y g(W) - \widetilde{\nabla}_Y\widetilde{\nabla}_X g(W) - \widetilde{\nabla}_{[X,Y]}g(W)\right](Z) \quad . \tag{10.7.10}$$

In general relativity, if one assumes that at s_0 one has $\theta(s_0) = \theta_0 < 0$, and $R(U,U) \geq 0$ everywhere, one can then prove there is a point conjugate to q along $\gamma(s)$ between $\gamma(s_0)$ and $\gamma\left(s_0 - \frac{3}{\theta_0}\right)$, provided $\gamma(s)$ can be extended to $\gamma\left(s_0 - \frac{3}{\theta_0}\right)$. This result is then extended to prove the existence of points conjugate to a three-surface Σ along $\gamma(s)$ within a distance $\frac{3}{\theta'}$ from Σ, where θ' is the initial value of θ given by the trace K of $K(X,Y)$, provided $K < 0$ and $\gamma(s)$ can be extended to that distance (propositions 4.4.1 and 4.4.3 of Hawking and Ellis 1973). This is achieved by studying an equation of the kind (10.7.7), where $\widetilde{\omega}^2 = \omega^2$ is vanishing because ω_{ab} is constant and initially vanishing, and the last term on the right-hand side vanishes as well. However, in the ECSK theory, $\widetilde{\omega}^2$ still contributes in light of (10.7.8). Thus the inequality :

$$\frac{d\theta}{ds} \leq -\frac{\theta^2}{3} \quad ,$$

can only make sense if we assume that :

$$\left[R(U,U) - 2\widetilde{\omega}^2\right] \geq 0 \quad , \tag{10.7.11}$$

where we do not strictly need to include $2\sigma^2$ on the left-hand side of (10.7.11) because σ^2 is positive. If (10.7.11) holds, we can write (see (10.7.7) and set there $\widetilde{\nabla}_a\left(\dot{U}\right)^a = 0$) :

$$\int_{\theta_0}^{\theta} y^{-2} dy \leq -\frac{1}{3}\int_{s_0}^{s} dx \quad , \tag{10.7.12}$$

which implies :

$$\theta \leq \frac{3}{\left[s - \left(s_0 - \frac{3}{\theta_0}\right)\right]} \quad , \tag{10.7.13}$$

where $\theta_0 < 0$. Thus θ becomes infinite and there are conjugate points for some $s \in \left]s_0, s_0 - \frac{3}{\theta_0}\right]$. However, (10.7.11) can be expressed as a restriction on the torsion tensor. In fact, the equations of the ECSK theory are given by (10.7.6) plus another one more suitably written in the form used in Hehl et al. 1974 (compare with Demianski et al. 1987 and Esposito 1990) :

$$S_{bc}{}^{a} - \delta_b^a S_{dc}{}^{d} - \delta_c^a S_{bd}{}^{d} = \sigma_{bc}{}^{a} \quad . \tag{10.7.14}$$

In (10.7.14) we have absorbed the $8\pi G$ factor into the definition of $\sigma_{bc}{}^{a}$, whereas this is not done in (10.7.9). Setting $\epsilon = g(U,U) = -1$, $\rho = 8\pi G$, the insertion of (10.7.9) into (10.7.14) and the multiplication by U_a yields :

$$\widetilde{S}_{bc} = \frac{1}{\rho\epsilon}\left(U_a S_{bc}{}^{a} - U_b S_{dc}{}^{d} - U_c S_{bd}{}^{d}\right) \quad , \tag{10.7.15}$$

which implies, defining :

$$f(\omega,\omega S)\equiv \omega_{ab}\omega^{ab}+\frac{1}{2}\omega_{ab}\widetilde{S}^{ab}+\frac{1}{2}\widetilde{S}_{ab}\omega^{ab}$$

$$=\omega_{ab}\omega^{ab}+\frac{\omega_{ab}}{2\rho\epsilon}\Big(U_h S^{abh}-U^a S_h{}^{bh}-U^b S^a{}_h{}^h\Big)$$

$$+\frac{\omega^{ab}}{2\rho\epsilon}\Big(U_h S_{ab}{}^h-U_a S_{hb}{}^h-U_b S_{ah}{}^h\Big)\quad , \tag{10.7.16}$$

and using (10.7.8) and (10.7.11), that :

$$8\widetilde{\omega}^2=4f(\omega,\omega S)+\widetilde{S}_{bc}\widetilde{S}^{bc}$$

$$=4f(\omega,\omega S)+\rho^{-2}\Big(U_h S_{bc}{}^h-U_b S_{hc}{}^h-U_c S_{bh}{}^h\Big)\Big(U_f S^{bcf}-U^b S_f{}^{cf}-U^c S^b{}_f{}^f\Big)$$

$$\leq 4R(U,U)\quad . \tag{10.7.17}$$

Indeed, some cases have been studied (Stewart and Hájicek 1973) where ω_{ab} is vanishing. However, we here prefer to write the equations in general form.

Moreover, in extending (10.7.13) so as to prove the existence of conjugate points to spacelike three-surfaces, the assumption $K<0$ on the trace K of $K(X,Y)$ also implies another condition on the torsion tensor. In fact, denoting by $\chi(X,Y)$ the tensor obtained from the metric and from the lapse and shift functions as the extrinsic curvature in general relativity, in a U_4 space-time one has :

$$K(X,Y)=\chi(X,Y)+\lambda(X,Y)\quad , \tag{10.7.18}$$

where the symmetric part of $\lambda(X,Y)$ (the only one which contributes to K) is given by :

$$\lambda_{(ab)}=-2n^\mu S_{(a\mu b)}\quad . \tag{10.7.19}$$

In (10.7.19) we have changed sign with respect to Pilati 1978 because his convention for $K(X,Y)$ is opposite to Hawking's convention, and we are here following Hawking so as to avoid confusion in comparing theorems. Thus, the condition $K<0$ implies the following restriction on torsion :

$$\lambda=-2g^{ab}n^\mu S_{(a\mu b)}<-\chi\quad . \tag{10.7.20}$$

When (10.7.11) and (10.7.20) hold, one follows exactly the same technique which leads to (10.7.13) in proving there are points conjugate to a spacelike three-surface.

10.7.2.3 Maximal Timelike Geodesics

In general relativity, it is known (proposition 4.5.8 of Hawking and Ellis 1973) that a timelike geodesic curve γ from q to p is maximal if and only if there is no point conjugate to q along γ in (q,p). We now describe how this result is proved and then extended so as to rule out the existence of points conjugate to three-surfaces. This last step will then be enlightening in understanding what happens in a U_4 space-time (Esposito 1990-92).

We here follow the conventions of section 4.5 of Hawking and Ellis 1973, denoting by $L(Z_1, Z_2)$ the second derivative of the arc length defined in (10.2.3), by V the unit tangent vector $\frac{\partial}{\partial s}$ and by T_γ the vector space consisting of all continuous, piecewise C^2 vector fields along the timelike geodesic γ orthogonal to V and vanishing at q and p. We are here just interested in proving that, if the timelike geodesic γ from q to p is maximal, this implies there is no point conjugate to q. The idea is to suppose for absurd that γ is maximal but there is a point conjugate to q. One then finds that $L(Z,Z) > 0$, which in turn implies that γ is not maximal, against the hypothesis. This is achieved taking a Jacobi field W along γ vanishing at q and r, and extending it to p putting $W = 0$ in the interval $[r,p]$. Moreover, one considers a vector $M \in T_\gamma$ so that $g\left(M, \frac{D}{\partial s}W\right) = -1$ at r. In what follows, we shall just say that M is suitably chosen, in a way which will become clear later. One then defines :

$$Z \equiv \epsilon M + \epsilon^{-1} W \quad , \tag{10.7.21}$$

where ϵ is positive and constant. Thus, the general formula for $L(Z_1, Z_2)$ implies (cf. lemma 4.5.6 of Hawking and Ellis 1973) :

$$L(Z,Z) = \epsilon^2 L(M,M) + 2L(W,M) + \epsilon^{-2} L(W,W) = \epsilon^2 L(M,M) + 2 \ , \tag{10.7.22}$$

which implies that $L(Z,Z)$ is > 0, if ϵ is suitably small (as we anticipated). The same method is also used in proving there cannot be points conjugate to a three-surface Σ if the timelike geodesic γ from Σ to p is maximal. However, as proved in lemma 4.5.7 of

Hawking and Ellis 1973, in the case of a three-surface Σ, the formula for $L(Z_1, Z_2)$ is of the kind :

$$L(Z_1, Z_2) = F(Z_1, Z_2) - \chi(Z_1, Z_2) \quad , \tag{10.7.23}$$

where $\chi(X, Y)$ is the extrinsic-curvature tensor of Σ. But we know that in a U_4 space-time $\chi(X, Y)$ is replaced by the nonsymmetric tensor $K(X, Y)$ defined in (10.7.18-19), which are completed using the relation for the antisymmetric part of $\lambda(X, Y)$:

$$\lambda_{[ab]} = -n^{\mu} S_{ba\mu} \quad . \tag{10.7.24}$$

Hence the splitting (10.7.21) leads to a formula of the kind (10.7.22), where the requirement

$$L(W, M) + L(M, W) = c > 0 \quad , \tag{10.7.25}$$

involves implicitly torsion because (10.7.23) is replaced by :

$$L(Z_1, Z_2) = \widetilde{F}(Z_1, Z_2) - K(Z_1, Z_2) \quad . \tag{10.7.26}$$

In other words, the left-hand side of (10.7.25) contains $K(W, M)+K(M, W)$. The condition (10.7.25) also clarifies how to suitably choose M in a U_4 space-time. Note that only $\lambda_{(ab)}$ contributes to (10.7.25) because the contributions of $\lambda_{[ab]}$ coming from $K(M, W)$ and $K(W, M)$ add up to zero. In proving (10.7.26), the first step is the generalization of lemma 4.5.4 of Hawking and Ellis 1973 to a U_4 space-time. This is achieved by remarking that the relation :

$$\frac{\partial}{\partial u} g\left(\frac{\partial}{\partial t}, \frac{\partial}{\partial t}\right) = 2g\left(\frac{D}{\partial u}\frac{\partial}{\partial t}, \frac{\partial}{\partial t}\right) \quad , \tag{10.7.27}$$

is also valid in a U_4 space-time, where now $\frac{D}{\partial u}$ denotes the covariant derivative along the curve with respect to the full U_4 connection. In fact, denoting by X the vector $\frac{\partial}{\partial t}$ and using the definition of covariant derivative along a curve one finds (Esposito 1990-92) :

$$\frac{\partial}{\partial u} g(X, X) = 2g\left(\frac{D}{\partial u}X, X\right) + X^a X^b \frac{D}{\partial u} g_{ab} \quad , \tag{10.7.28}$$

where $\frac{D}{\partial u} g_{ab}$ is vanishing if the connection obeys the metricity condition, which is also assumed in a U_4 space-time (cf. section 10.7.1 and Hehl et al. 1976). Thus, the key role is played by the connection which obeys the metricity condition, and $\frac{\partial}{\partial u} g(X, X)$ implicitly contains the effects of torsion by virtue of the relation :

$$\frac{DX^a}{\partial u} \equiv \frac{\partial X^a}{\partial u} + \Gamma_{bc}{}^a \frac{dx^b}{du} X^c \quad . \tag{10.7.29}$$

Although this point seems to be elementary, it plays a vital role in leading to (10.7.26). This is why we chose to emphasize it (Esposito 1990-92).

10.7.2.4 The Singularity Theorem

If we now compare the results discussed or proved in sections 10.7.2.1-3 with page 273 of Hawking and Ellis 1973, we are led to state the following singularity theorem :

Theorem 10.7.1 The U_4 space-time of the ECSK theory cannot be timelike geodesically complete if :

(1) $\left[R(U, U) - 2\tilde{\omega}^2\right] \geq 0$ for any nonspacelike vector U;

(2) there exists a compact spacelike three-surface S without edge;

(3) the trace K of the extrinsic-curvature tensor $K(X, Y)$ of S is either everywhere positive or everywhere negative. This tensor also plays a role in the theory of maximal timelike geodesics (cf. (10.7.25-26)) and partial Cauchy surfaces.

Conditions (1) and (3) involve the torsion tensor defined in (10.7.1). Indeed, the second part of condition (3) can be seen as a prerequisite, but we have chosen to insert it into the statement of the theorem so as to present together all conditions which involve the extrinsic-curvature tensor $K(X, Y)$. The compatibility of (1) with the field equations of the ECSK theory is expressed by (10.7.17) if (10.7.9) makes sense. Otherwise, (10.7.17) should be replaced by a different relation. Note that if we switch off torsion, condition (1) becomes the one required in general relativity because, as explained on pp 96-97 of Hawking and Ellis 1973, the vorticity of the torsion-free connection vanishes wherever a 3×3 matrix which appears in the Jacobi fields is nonsingular. Finally, if $\tilde{\nabla}_a \left(\dot{U}\right)^a$ is not vanishing as we assumed so far (cf. (10.7.7) and comment before (10.7.12)) following Stewart and Hájicek 1973, Tafel 1973, condition (1) of our theorem should be replaced by (Esposito 1990-92) :

(1') $\left[R(U,U) - 2\tilde{\omega}^2 - \tilde{\nabla}_a\left(\dot{U}\right)^a\right] \geq 0$ for any nonspacelike vector U.

10.8 Concluding Remarks

At first we have seen our task as presenting an unified description of some aspects of the differentiable, spinor, causal, asymptotic and Hamiltonian structure of space-time. There is a very rich literature on these topics on specialized books (Hawking and Ellis 1973, Beem and Ehrlich 1981, O'Neill 1983, Penrose 1983, Penrose and Rindler 1984-1986, Gallot et al. 1987, Francaviglia 1988, Ashtekar 1988) and on the original papers (see also Lichnerowicz 1961, Carter 1971, Woodhouse 1973, Clarke 1975-1976, Hawking et al. 1976, Clarke and Schmidt 1977, Ellis and Schmidt 1977, Tipler 1977, Dodson 1978, Carter 1979, Lee 1983, Vyas and Joshi 1983, Clarke and de Felice 1984, Vyas and Akolia 1984, Newman 1984a-b, Vyas and Akolia 1986, Newman and Clarke 1987, Budinich and Trautman 1988, Dabrowski 1988, Isham et al. 1988), but we thought it was important to analyze them in a single chapter. As a partial completion of what we studied so far, the following remarks are in order.

(a) In section 10.4.2 we briefly motivated and described stable causality. Recent progress on the topology of stable causality is due to Aguirre-Dabán and Gutiérrez-López 1989. The authors give an enlightening discussion of causally convex and stably causally convex sets and of their topologies. For example, they prove that a point of space-time has arbitrarily small neighbourhoods that are stably causally convex sets if and only if stable causality holds. They also define returning sets, and analyze the structure of subsets that control the fulfillment of strong and stable causality at a point. In Dieckmann 1988 volume functions have been used to characterize strong causality, global hyperbolicity and other causality conditions, and in Rácz 1988 the causal boundary for stably causal space-times has been analyzed. In Joshi 1989, strong and stable causality have been characterized in terms of causal functions, proving two important theorems.

(b) There is a very rich literature on the asymptotic structure of space-time (Esposito and Witten 1977, Flaherty 1984). For example, the structure of the gravitational field at spatial infinity is studied through an analysis of asymptotically Euclidean and Minkowskian space-times in Persides 1979, 1980a-b. More recently, impressive progress has been made in studying the global structure of simple space-times in Newman 1989b. In that paper, the author has proved for these space-times what follows (Newman 1989b) :

(1) Future null infinity is diffeomorphic to the complement of a point in some contractible open three-manifold;

(2) The strongly causal region Σ_{SC} of future null infinity is diffeomorphic to $S^2 \times R$;

(3) Every compact, connected, spacelike two-surface in future null infinity is contained in Σ_{SC};

(4) Space-time must be globally hyperbolic.

(c) In section 10.6 we only focused on some classical aspects of Ashtekar's formalism for canonical gravity. However, its main motivation is the development of a nonperturbative approach to quantum gravity. At present, the main interest is in a representation where quantum states arise as functions of loops on a three-manifold, and in so doing a class of exact solutions to all quantum constraints has been obtained for the first time (Ashtekar et al. 1989, Ashtekar 1991, Rovelli 1991). At the classical level, other important work has been done in Dolan 1989, where the author has shown that the trace of the extrinsic-curvature tensor of the boundary of space-time is the generating function for the canonical transformation of the phase space of general relativity introduced by Ashtekar. When torsion is nonvanishing, an additional boundary term is present in the generating function, which has the effect of making the action complex.

(d) At the beginning of section 10.7, we briefly described the gauge theory of gravitation based on the Poincaré group. As explained in Gotzes and Hirshfeld 1990, other important generalizations of this local symmetry group are $SO(3,2)$ or $SO(4,1)$, the only groups reducing to the Poincaré group by Wigner-Inönü contraction. The authors choose $SO(3,2)$ because it leads to supersymmetry as a natural extension. In the $SO(3,2)$ theory, the vierbein θ and the connection ω of section 10.7 appear formally as different components of the $SO(3,2)$-valued connection. Thus the authors conclude that the $SO(3,2)$ theory is the best candidate for a gauge theory of gravitation.

(e) Very recent progress in singularity theory in cosmology is due to Newman 1989a and Kriele 1989. In Newman 1989a the author has studied an alternative interpretation according to which gravitational collapse may give rise not to singularities but to chronology violation. He has found an example of a singularity-free chronology-violating space-time with a nonachronal closed trapped surface. In Kriele 1989, a remarkable proof has been given that a nonempty chronology-violation set with compact boundary causes singularities. In other words, these papers shed new light on the problem of whether causality violations lead to the occurrence of singularities, and one can now prove that causality violations that do not extend to infinity must cause singularities (see Kriele 1989 and references therein). Even more recently, since there is no evidence for the absence of global causality violations, efforts have been made so as to generalize the singularity theorem of Hawking and Penrose to space-times with causality violations (Kriele 1990). The result obtained in Kriele 1990 preserves the original idea of the theorem, and it seems unlikely to be improved.

In Horowitz and Steif 1990a-b, Horowitz and Steif 1991, it has been shown that a large class of time-dependent solutions to Einstein's equations are classical solutions to string theory. Interestingly, these include metrics with large curvature and some with space-time singularities. This has been proved by first establishing that plane-wave solutions of general relativity are exact solutions to string theory. Later on, by studying the evolution of test strings in these backgrounds, it was found that they become infinitely excited when the amplitude of the plane wave diverges. This implies such solutions are both geodesically incomplete and singular in a sense appropriate to string theory (Horowitz and Steif 1991).

Finally, in section 10.7 we have studied other aspects of the singularity problem in cosmology. We have then taken the point of view according to which nonspacelike geodesic incompleteness can be used as a preliminary definition of singularities also in space-times with torsion. We have finally been able to show under which conditions Hawking's singularity theorem without causality assumptions can be extended to the space-time of the

ECSK theory. However, when we assume (10.7.9) and we require consistency of condition (10.7.11) with the equations of the ECSK theory, we obtain the relation (10.7.17) which explicitly involves the torsion tensor on the left-hand side (of course, the torsion tensor is also present in $R(U, U)$ through the connection coefficients, but this is an implicit appearance of torsion, and it is better not to make this splitting). Also conditions (10.7.20) and (10.7.25) involve the torsion tensor if one uses the formula (10.7.18). This is why our analysis seems to show that the presence of singularities in the ECSK theory can be less generic than in general relativity. Our result should be compared with the one in Hehl et al. 1974. The relevant differences between our work and their work are :

(1) We rely on a different definition of geodesics, as explained in section 10.7.1;

(2) We emphasize the role played by the full extrinsic-curvature tensor and by the variation formulae in U_4 theory, a remark which is absent in Hehl et al. 1974;

(3) We keep the field equations of the ECSK theory in their original form, whereas in Hehl et al. 1974 they are cast in a form analogous to general relativity, but with a modified energy-momentum tensor which contains torsion. We think this technique is not strictly needed (Esposito 1990-92). Moreover, from a Hamiltonian point of view, the splitting of the Riemann tensor into the one obtained from the Christoffel symbols plus the one explicitly related to torsion does not seem to be in agreement with the choice of the full connection as a canonical variable. In fact, if we look for example at models with quadratic Lagrangians in U_4 theory, the frame and the full connection should be regarded as independent variables (Esposito 1989b), and this choice of canonical variables has also been made for the ECSK theory (Isenberg and Nester 1980, Castellani et al. 1982, Di Stefano and Rauch 1982).

Problems to be studied for further research are the generalization to U_4 space-times of the other singularity theorems in Hawking and Ellis 1973 using our approach, and of the results in Schmidt 1971 that we outlined in section 10.5.3. Moreover, the generalization to U_4 space-times of the classification of singularities in Dodson 1978, and its relation to the preliminary definition of singularities we used in this chapter (i.e. specification of the regularity condition needed for the Riemann tensor and for the full connection) deserves careful consideration.

Recently, the singularity problem for space-times with torsion has also been studied in Pullin 1989 for the case of classical $N = 1$ supergravity. In that paper the author has used the modified weak-energy condition for theories with torsion developed in Hehl et al. 1974. Pullin has found that spin-spin contact interactions cannot avert singularities in general. Finally, he has presented a singularity-free model for a spatially homogeneous Rarita-Schwinger field in a FRW space-time. Even more recently, the generalization of singularity theorems to U_4 space-times has also been addressed in Garcia de Andrade 1990a-b. The differences between our work and his work are :

(1) Garcia de Andrade does not define geodesics as autoparallel curves;

(2) After a review of the approach in Hehl et al. 1974, he still makes a splitting so as to express $R(U,U)$ as the part formally identical to general relativity plus other contributions involving torsion;

(3) He derives the Hawking-Penrose timelike convergence condition in U_4 space-time for a shear-free and convergence-free congruence, and he obtains : $R(U,U) \geq 0$, because he separately requires (using the notation in (10.7.7-8)) :

$$\frac{1}{4}\widetilde{S}_{ab}\widetilde{S}^{ab} \geq \left[\frac{\theta^2}{3} + \frac{d\theta}{ds}\right] \quad . \tag{10.8.1}$$

However, in the generalized Raychaudhuri equation, $R(U,U)$ and $2\widetilde{\omega}^2$ occur with opposite signs, so that one has to require in general $\left[R(U,U) - 2\widetilde{\omega}^2\right] \geq 0$, for any nonspacelike vector U, if $\frac{d\theta}{ds}$ must remain less than or equal to $-\frac{\theta^2}{3}$, as we explained in section 10.7.2.2. Thus, in our work (see also Stewart and Hájicek 1973), torsion explicitly appears in (10.7.11), but (10.7.13) is as in Einstein's theory, and theorem 10.7.1 is proved under only three assumptions, as in general relativity. By contrast, in the work of Garcia de Andrade, one requires two separate conditions instead of (10.7.11), and torsion appears in (10.8.1);

(4) Garcia de Andrade does not consider the full extrinsic-curvature tensor and conditions for maximal timelike geodesics, and he does not avoid the introduction of a modified energy-momentum tensor;

(5) Garcia de Andrade studies the generalization of the singularity theorem in Hawking and Penrose 1970.

However, appendix A in Garcia de Andrade 1990a is devoted to Hawking's singularity theorem without causality assumptions in general relativity, and in the concluding section 4 of that paper it is emphasized that one could investigate singularities in space-times with torsion looking at the completeness of autoparallels (which are there called nongeodesic curves). Thus it is possible that while (or after) our monograph is completed, the derivation of our theorem 10.7.1, first appearing in Esposito 1990-92, is studied by Garcia de Andrade or even other authors. Another interesting study of the inclusion of spin in the Raychaudhuri equation can be found in Fennelly et al. 1991. This equation is applied to the behaviour of an irrotational, unaccelerated fluid, and the development of singularities in the expansion is studied for constant spin densities. The fundamental difference between our work and their work is the following. In equation (19) of Fennelly et al. 1991, a Raychaudhuri equation is written for space-times with torsion where all covariant derivatives only contain Christoffel symbols. The spin-connection has been separated from the covariant derivatives and explicitly included. However, as we already emphasized, this split is contrary to the canonical treatment of theories with torsion. Hence we can repeat the remarks made in comparing our work with the one in Hehl et al. 1974 and Garcia de Andrade 1990a-b.

Last but not least, we should clarify the following point. In theorem 10.7.1 we really deal with the covering manifold M_H of M, defined as the set of all points $(p,[\lambda])$, where $[\lambda]$ is an equivalence class of curves from S to p homotopic modulo S and p. In Hawking and Ellis 1973 (cf. pages 205 and 273 therein) it is claimed that M_H is the largest covering manifold of M such that each connected component of the image of S is homeomorphic to S. However, it was later proved in Haggman et al. 1980 that this claim is incorrect, i.e. we can only say that M_H is the largest covering manifold of M such that there exists at least one connected component of the image of S which is homeomorphic to S.

PART IV:

SUMMARY

CHAPTER ELEVEN

CONCLUSIONS

11.1 Foundational Issues

The most recurring themes of our work are the study of the eigenvalues of differential operators and the application of canonical methods to general relativity, supergravity and other gauge theories. The mathematical theories we have used are : (1) Riemannian geometry and the zeta-function technique; (2) theory of SL(2,C) spinors, SU(2) spinors and the Dirac operator; (3) twistor theory in flat space and complex manifolds; (4) self-adjointness theory; (5) constrained Hamiltonian systems and path integrals in quantum field theory; (6) spinor, causal, asymptotic and Hamiltonian structure of space-time; (7) singularity theory in cosmology.

We have thus one more example of the fundamental unity and richness of modern mathematics. However, before mentioning the unsolved issues related to our work, we think it is worth emphasizing an important, foundational problem. One of the main motivations for studying quantum cosmology is the singularity problem in cosmology. It is now well-known that, in order to avoid the singularity in the past, we should assume that a fundamental massive (Hawking 1984a-b) or massless (Esposito and Platania 1988) scalar field existed in the very early universe. Moreover, the semiclassical approximation should hold in quantum cosmology under certain conditions, so that regular matter fields can be used in solving the corresponding classical field equations. However, nothing proves that such a kind of fundamental scalar field existed (although more complicated couplings of gravity to other matter fields can be studied), nor it is sure that the Hartle-Hawking proposal is correct. Hawking has indeed greatly emphasized that his proposal cannot be deduced from first principles, but its validity can only be checked testing it against observations. This is a delicate matter, because the interpretation of data in cosmology may be uncertain, and there may be theories which agree about the present universe, but whose predictions about the singularity are different.

The problems of singularities and of typical wave functions in a quantum theory of gravity have been investigated by several authors (e.g. Gotay and Demaret 1983, Padmanabhan 1984, Smith and Bergmann 1986, Husain 1987, Louko 1987, Gibbons and Grishchuk 1989, Page 1990, Louko and Ruback 1991). In the Hartle-Hawking program, an alternative to scalar fields is given by higher-derivative theories (Hawking and Luttrell 1984b, Horowitz 1985). However, this approach has not been extended to the anisotropic case, basically in view of the presence of negative-norm states (we are indebted to Professor S. Hawking for explaining this point). In our opinion, if a theory of quantum gravity can solve the singularity problem, it should avoid the singularity in the past in a way as generic as possible. We are not aiming to speculate, but we think it is worth working on alternative viewpoints. Even a failure could be a progress in our understanding of gravitation and cosmology. In proving singularity theorems, one of the main ideas (Hawking and Ellis 1973) is to use global hyperbolicity theory so as to show there must be longest timelike curves between certain pairs of points. But, if there were no singularities, there would be conjugate points, which in turn would imply there were no longest curves between the pairs of points (cf. section 10.7.2.2). The results obtained are rather generic. We think it would be very important to get generic results also in a nonclassical theory of gravity. Two important things which seem to be lacking at present are :

(a) a detailed qualitative understanding of the solutions of a functional differential equation such as the Wheeler-De Witt on superspace (possibly without using perturbative techniques);

(b) a precise mathematical relation between the solutions of this equation and the ones of partial differential equations which occur in minisuperspace models.

It is indeed well-known that Ashtekar's new variables (section 10.6) provide a radically different approach to these issues. However, severe consistency problems remain, i.e. choice of the inner product for the quantization of the theory, suitable regularizations etc. (Ashtekar 1988, Ashtekar 1991, Rovelli 1991).

11.2 Our Results

Also our work shows that the Hartle-Hawking proposal can lead to a lot of interesting developments in fundamental physics. We have studied boundary-value problems and one-loop calculations for bosonic and fermionic fields in the case of flat Euclidean backgrounds bounded by S^3 so as to evaluate the semiclassical approximation of the Hartle-Hawking wave functional. In what follows, the $\zeta(0)$ values for gauge fields should be understood as PDF values for scale-invariant measures. The results obtained in chapters four, five, seven, eight and nine can then be summarized as follows :

Theorem 1. Using global boundary conditions where all untwiddled perturbative modes are set equal to zero on the S^3 boundary, one finds : $\zeta(0) = \frac{11}{360}$ for spin $\frac{1}{2}$, and $\zeta(0) = -\frac{289}{360}$ for spin $\frac{3}{2}$, for massless Majorana fermions.

Theorem 2. Using local boundary conditions (hereafter referred to as L. B. C.) on S^3 involving normals and field strengths for bosons, normals and undifferentiated fields for massless Majorana spin-$\frac{1}{2}$ fermions, normals and spatial components of potentials for gravitinos, one finds : $\zeta(0) = \frac{7}{45}$ for spin 0, $\zeta(0) = \frac{11}{360}$ for spin $\frac{1}{2}$, $\zeta(0) = -\frac{77}{180}$ or $\frac{13}{180}$ for spin 1, $\zeta(0) = -\frac{289}{360}$ for spin $\frac{3}{2}$, and $\zeta(0) = \frac{112}{45}$ for spin 2 when $B^{*}_{ij} = 0$ on S^3.

Corollary. $\zeta(0)$ values for fermionic fields subject to local or global boundary conditions are the same (it has not yet been possible to derive this result from first principles).

Theorem 3. Within the PDF approach, $O(N)$ supergravity theories are not one-loop finite when Euclidean backgrounds bounded by S^3 are studied in the limit of very small S^3-radius.

Theorem 4. If the linearized electric curvature is set equal to zero on S^3, there is no surface term to add to the linearized Einstein action such that the linearized Einstein equations follow from requiring the action to be stationary. Thus, the classical boundary-value problem is ill-posed.

Theorem 5. The classical boundary-value problem for the spin-$\frac{1}{2}$ field obeying the Weyl equation and subject to L. B. C. on S^3 has a unique smooth solution.

Theorem 6. Using L. B. C. the eigenvalue condition for spin $\frac{1}{2}$ is : $J_{n+1}(Ea) = \pm J_{n+2}(Ea)\ \forall n \geq 0$, and $J_{n+2}(Ea) = \pm J_{n+3}(Ea)\ \forall n \geq 0$ for spin $\frac{3}{2}$, where a is the three-sphere radius, J_n is the regular Bessel function and fermionic fields are assumed to be massless.

Theorem 7. Using L. B. C. for spin $\frac{1}{2}$, a first-order differential operator for the boundary-value problem exists which is symmetric and has self-adjoint extensions.

Theorem 8. A form of the spin-lowering operator exists which preserves L. B. C. required on S^3 for solutions to the massless free-field equations for adjacent spins s and $s+\frac{1}{2}$, where $s = 0, \frac{1}{2}, 1, \frac{3}{2}$.

Theorem 9. Let ω^L be the spin-lowering operator, and let $\left(\phi_L, \widetilde{\phi}_{L'}\right)$ be the spin-$\frac{1}{2}$ field. Then, given the eigenvalue equations (5.9.63), the scalar fields $\phi(x) \equiv \omega^L \phi_L$ and $\widetilde{\phi}(x) \equiv \widetilde{\omega}^{L'} \widetilde{\phi}_{L'}$ obey the differential equations :

$$\Box \phi(x) = \lambda^2 \phi(x) + \lambda G(x) \quad , \tag{11.2.1}$$

$$\Box \widetilde{\phi}(x) = \lambda^2 \widetilde{\phi}(x) + \lambda H(x) \quad , \tag{11.2.2}$$

where : $G(x) \equiv -2i\widetilde{\phi}^{A'} \pi^o_{A'}$ and $H(x) \equiv -2i\phi^A \widetilde{\pi}^o_A$, $\pi^o_{A'}$ and $\widetilde{\pi}^o_A$ being the constant spinor fields appearing in (5.7.9-10).

In the third part of our book (chapter ten) we have studied nonperturbative properties of classical gravity. Our main result can be described as follows.

Theorem 10. The space-time with torsion of the ECSK theory cannot be timelike geodesically complete if :

(1) $\left[R(U,U) - 2\widetilde{\omega}^2\right] \geq 0$ for any nonspacelike vector U;

(2) there exists a compact spacelike three-surface S without edge;

(3) the trace K of the full extrinsic-curvature tensor $K(X,Y)$ of S is either everywhere positive or everywhere negative. This tensor also plays a role in the theory of maximal timelike geodesics and partial Cauchy surfaces.

Corollary (from Theorem 10). The occurrence of singularities in closed cosmological models in the ECSK theory can be less generic than in general relativity.

In other words, the contributions of our work seem to be :

(1) Study of global boundary conditions for fermionic fields in one-loop quantum cosmology;

(2) Study of the relation of local boundary conditions to twistor theory in flat space;

(3) Proof of self-adjointness of boundary-value problems for fermions with local boundary conditions;

(4) Proof of lack of one-loop finiteness of supersymmetric theories in the presence of boundaries within the PDF approach;

(5) Detailed analysis of *direct* $\zeta(0)$ calculations for fields of various spins with local and global boundary conditions;

(6) Extension of Hawking's singularity theorem without causality assumptions to the space-time of the ECSK theory, when geodesics are defined as curves whose tangent vector moves by parallel transport. This definition of geodesics had not been considered previously in the literature on cosmological models with torsion in order to prove singularity theorems (see the end of section 10.8).

11.3 Our Unsolved Problems

As we already emphasized in chapter one, it is not yet clear whether any theory of quantum gravity can be consistently developed, or whether the Hartle-Hawking proposal is correct. However, we devoted our main efforts to different questions. The main problems to be addressed in the study of one-loop quantum cosmology are :

11. Conclusions

(1) Has the classical boundary-value problem for fields of spin $0, \frac{1}{2}, 1, \frac{3}{2}, 2$ an unique smooth solution, once the boundary conditions (local or nonlocal) and the background (flat or curved) have been chosen ?

(2) Do the boundary conditions respect supersymmetry ?

(3) Do the boundary conditions yield gauge-invariant amplitudes ?

(4) Are the corresponding differential operators self-adjoint ?

(5) Has the corresponding heat kernel an asymptotic expansion ?

(6) Is the corresponding $\zeta(0)$ value well-defined ?

(7) Is the measure in the path integral for gauge theories well-defined, so that no ambiguities affect the $\zeta(0)$ value for gauge fields ?

(8) What is the relation between *direct* and *indirect* techniques used in the literature for $\zeta(0)$ calculations ?

(9) Are extended supergravity theories one-loop finite in the presence of boundaries ?

Indeed, local boundary conditions are the most important in quantum field theory, because for them the answer to questions (1-6) is always affirmative. By contrast, global boundary conditions do not respect supersymmetry (question 2). However, the relation between the work in Hartle and Schleich 1987, and the work in Poletti 1990 for quantum fields inside a three-sphere, has not yet been thoroughly understood. At present it seems that the one-loop result depends on whether one first restricts the classical theory to a set of physical degrees of freedom, by choice of gauge, and then quantizes (this would yield the PDF contribution to $\zeta(0)$), or whether one quantizes the full theory in BRST-invariant fashion with gauge-averaging and ghost terms included. Further evidence of this problem has been found in Rozansky 1991, where the author has shown that, for one-loop calculations in minisuperspace models, the ghost determinant is not cancelled by the contributions of unphysical momenta. Even though the PDF method was among the original motivations of the work by Faddeev and his collaborators, who were aiming to obtain gauge-invariant amplitudes which agree with the PDF result, it is not obvious this program can be successfully implemented when boundaries are present (cf. section 6.4). The disagreement about $\zeta(0)$ values for gauge fields would be easily explained if one could show that the differential operators corresponding to gauge modes and ghost fields are the same, but obey different boundary conditions. Unfortunately, the situation is much more involved even just for Abelian gauge fields (cf. section 6.5). It therefore appears that only a very detailed and *direct* analytic calculation, together with the study of the measure in the path integral for quantized gauge fields, can shed further light on this problem.

Moreover, further discrepancies have been found in performing $\zeta(0)$ calculations for massless Majorana spin-$\frac{1}{2}$ fields with *direct* (cf. chapter eight) or *indirect* (cf. section 6.5) techniques, and severe technical difficulties exist in trying to perform gauge-invariant and *direct* $\zeta(0)$ calculations for gauge fields (cf. section 6.5). Thus the fundamental, *direct* check of $\zeta(0)$ values for supergravity theories obtained using *indirect* methods, is not yet available. It therefore appears that also the question of the one-loop finiteness of extended supergravity, in the presence of boundaries, has not yet been settled, since the main problems are as follows :

(1) The PDF method is not equivalent to the Faddeev-Popov method;

(2) *Direct* $\zeta(0)$ calculations using the Faddeev-Popov method for gauge fields seem to be too difficult at present (see, however, addendum to chapter six);

(3) So far, no $\zeta(0)$ calculation has been attempted using the Hamiltonian Batalin-Fradkin-Vilkovisky technique when boundaries are present.

Note also that we have studied the Hartle-Hawking path integral (5.8.53) for fermionic fields subject to the boundary conditions (5.8.3) on S^3. Since the Dirac operator is a first-order elliptic operator, it is not entirely obvious this is equivalent to the definition used in Poletti 1990, where the extra boundary conditions (8.5.7) are required (D'Eath and Esposito 1991a). This difference is a peculiar property of problems with boundaries. Interestingly, recent work in Barvinsky et al. 1992a, Kamenshchik and Mishakov 1992, Kamenshchik and Mishakov 1993, agrees with our flat-space $\zeta(0)$ values for spin $\frac{1}{2}$ and spin $\frac{3}{2}$ (we are grateful to Dr. P. D'Eath for bringing the first paper of this series to our attention). This seems to provide a relevant check of our *direct* $\zeta(0)$ calculations.

The unsolved problems mentioned so far seem to add evidence in favour of no end being in sight for cosmology and fundamental physics, and this can greatly stimulate our curiosity and our efforts.

PROBLEMS FOR THE READER

Problem 1.1 In section 1.2, we want to prove that the wave function obtained from the path integral satisfies the Wheeler-De Witt equation. Study first this problem in the case of minisuperspace models, following Halliwell 1988. You can then study the problem in the case of superspace. After reading Halliwell and Hartle 1991, write an essay on the invariant sums over histories they use to prove their results. Try also to give a rigorous mathematical formulation of the invariant measure and of the invariant class of paths summed over within that approach.

Problem 1.2 In section 1.2, following Esposito and Platania 1988, show how the Hartle-Hawking proposal picks out inflationary solutions in quantum cosmology.

Problem 2.1 Suppose we deal with a Hamiltonian system subject to second-class constraints. What happens if we try to define a *maximally extended effective Hamiltonian*, i.e. also including secondary second-class constraints, as $\widetilde{H}_E \equiv \widetilde{H} + v_m(q,p)\varphi_m^{(2,II)}(q,p)$? Does it make sense a comparison between the classical theories described by $\widetilde{H}_E$ and by $\widetilde{H}$ defined in Eq. (2.1.14) ?

Problem 2.2 It is now possible to quantize a constrained Hamiltonian system without eliminating second-class constraints at the classical level. Following the remarks at the end of section 2.3, Pilati 1978 and McMullan 1991, can you quantize the Einstein-Cartan-Sciama-Kibble theory of gravitation using this new approach ? What is the corresponding phase space ?

Problem 2.3 Check that Eq. (2.4.9) follows from the definition (2.4.1).

Problem 2.4 Show that for any solution of the vacuum Einstein equations, the action integral is a purely boundary term (cf. Eq. (2.4.11)). Following Gibbons and Hawking 1977, use this result to derive the entropy of a Schwarzschild black hole via path-integral methods.

Problem 2.5 What is the appropriate boundary term in the action of general relativity, if the conformal three-metric $\widetilde{h}_{ij} \equiv (det\ h_{ij})^{-\frac{1}{3}}\ h_{ij}$ and the trace K_i^i are fixed on the boundary ? [cf. York 1986]

Problem 2.6 Check Eqs. (2.4.23-24) and (2.4.26-29).

Problem 3.1 Is quantum electrodynamics Borel-summable (cf. section 3.4) ? This is indeed a very difficult problem; for this purpose the reader should first study section 8 of Feldman et al. 1988 (and possibly all this reference).

Problem 4.1 Prove the statement appearing on the three lines before Eq. (4.1.19).

Problem 5.1 Suppose that in remark (a) of section 5.7 we replace the field $\overline{\phi}$ by its analytic continuation $\widetilde{\phi}$, where $\widetilde{\phi}$ is completely independent of ϕ. How would this affect the subsequent analysis ?

Problem 5.2 Following Eq. (5.8.94), check that the operator iC has a unique self-adjoint extension.

Problem 6.1 Write an essay, where you describe in detail how classical fields can be viewed as sections of vector bundles (Ward and Wells 1990), and how quantum fields can be defined as operator-valued distributions (Streater and Wightman 1964). In particular, prove why relativity theory plays a key role in defining quantum fields as above.

Problem 6.2 Bearing in mind Eq. (6.5.9), can you find a different gauge-averaging term such that the gauge modes g_n and R_n obey decoupled eigenvalue equations ? If so, compute the various contributions to the full $\zeta(0)$ for the spin-1 problem.

Problem 7.1 Using the technique described in section 7.3, compute the other coefficients in the asymptotic heat kernel.

Problem 8.1 Using two-component spinor notation, show that the Dirac operator turns primed spinors into unprimed spinors, and viceversa. In other words, the Dirac operator can be seen as a first-order elliptic operator mapping primed spin-space S' to unprimed spin-space S, and unprimed spin-space S to primed spin-space S'. Can you now show why only half of a fermionic field can be fixed on the boundary in the classical boundary-value problem (cf. D'Eath and Halliwell 1987, D'Eath and Esposito 1991a-b) ? Finally, can you prove why, in light of the first-order nature of the Dirac operator, one has a choice of global or local boundary conditions for fermionic fields ?

Problem 9.1 Is it possible to compute $\zeta(0)$ for spin $\frac{3}{2}$, if (5.7.4) is required to hold on S^3 ? Do we get an electric and a magnetic case, as for spin 1 and spin 2 ? Are the corresponding classical boundary-value problems well-posed ?

Problem 9.2 If the effects of two-forms and three-forms are included, what are the corresponding PDF $\zeta(0)$ values (cf. Eqs. (9.3.9-10)) ?

Problem 10.1 What is the strongest causality condition on a space-time ? Do we lose global hyperbolicity if we remove a point from space-time ?

Problem 10.2 Can you find examples of space-times where closed timelike curves occur, and all nonspacelike geodesics are complete ?

Problem 10.3 It is possible to give an example of a space-time which is spacelike, timelike and null geodesically complete, but which contains a timelike curve of bounded acceleration and finite total length. For this purpose, one first takes the universal covering space of two-dimensional anti-de Sitter space-time, whose metric may be written in the form : $g = -\left(1+x^2\right) dt \otimes dt + \left(1+x^2\right)^{-1} dx \otimes dx$. Interestingly, each timelike geodesic intersecting the t axis at $t = t_0$, intersects that axis again at $t = t_0 + \pi$. Following the appendix of Geroch 1968b, try to exploit this focusing effect on timelike geodesics in order to construct the desired space-time. Can we generalize this result to four-metrics ?

Problem 10.4 Can you say why Einstein's general relativity can be seen as the gauge theory of the Lorentz group, whereas the Einstein-Cartan-Sciama-Kibble theory can be seen as the gauge theory of the Poincaré group ? What is the main difference between Yang-Mills gauge theories, and gauge theories of gravitation ? [cf. Carmeli 1982]

Problem 10.5 Find examples of U_4 space-times which are : (i) singularity-free according to Hehl et al. 1974 but not according to Esposito 1990-92; (ii) singularity-free according to Esposito 1990-92 but not according to Hehl et al. 1974; (iii) singular (or singularity-free) according to both Hehl et al. 1974 and Esposito 1990-92.

Problem 10.6 What can we say about boundary terms in the action principle for general relativity, when Ashtekar's spinorial variables are used ? Study in particular the linearized action of general relativity with these variables. What boundary conditions can be used ? Is it still impossible to fix the linearized electric curvature on the boundary (cf. section 7.2) ? [This problem has been suggested to us by Professor Chris Isham on August 1991]

As far as I know, the solution of problems 5.2, 6.2, 9.1, 10.5-6 might lead to good research papers (problems 4.1, 7.1, 8.1, 9.2 are recommended as a simpler exercise). I thus hope they will stimulate further work on these issues. I have also inserted problems 1-3 for readers who are only interested in Part I of this book. Problem 2.2, however, is an open research problem.

APPENDIX A: Two-Component Spinor Calculus and Its Applications

A complete description of two-component spinor calculus may be found in Penrose 1968, Penrose and Rindler 1984-1986 and in Stewart 1990. In this appendix we are aiming to recall at first some basic results of two-component spinor calculus following Ward 1980, D'Eath 1984 and Perjés 1989. We shall then discuss some applications which play an important role in chapters four, five, eight and nine.

Spin-space is a pair (Σ, ϵ), where Σ is a two-dimensional vector space over the complex or real numbers and ϵ a symplectic structure on Σ. This ϵ provides an isomorphism between Σ and the dual space Σ^*. One has : $\lambda^A \in \Sigma$, $\lambda_A \in \Sigma^*$. Unprimed (primed) spinor indices take the values 0 and 1 ($0'$ and $1'$). They can be raised and lowered by means of ϵ^{AB}, ϵ_{AB}, $\epsilon^{A'B'}$, $\epsilon_{A'B'}$, which are given by $\begin{pmatrix} 0 & 1 \\ -1 & 0 \end{pmatrix}$ and $\begin{pmatrix} 0' & 1' \\ -1' & 0' \end{pmatrix}$ respectively, according to the rules : $\rho^A = \epsilon^{AB}\rho_B$, $\rho_A = \rho^B \epsilon_{BA}$, $\rho^{A'} = \epsilon^{A'B'}\rho_{B'}$, $\rho_{A'} = \rho^{B'}\epsilon_{B'A'}$. Thus : $\epsilon_A^{\ \ A} = 2 = -\epsilon^A_{\ \ A}$.

An isomorphism exists between the tangent space T at a point of space-time and the tensor product of the unprimed spin-space S and the primed spin-space S' : $T \cong S \otimes S'$. The Infeld-van der Waerden symbols $\sigma^a_{\ \ AA'}$ and $\sigma_a^{\ \ AA'}$ express this isomorphism, and the correspondence between a vector v^a and a spinor $v^{AA'}$ is given by :

$$v^{AA'} = \sigma_a^{\ \ AA'} v^a \quad , \tag{A.1}$$

$$v^a = \sigma^a_{\ \ AA'} \, v^{AA'} \quad . \tag{A.2}$$

The $\sigma_a^{\ \ AA'}$ are given by :

$$\sigma_0 = -\frac{I}{\sqrt{2}} \quad , \quad \sigma_i = \frac{\Sigma_i}{\sqrt{2}} \quad , \tag{A.3}$$

where Σ_i are the Pauli matrices. Given a pseudo-orthonormal tetrad $e^a_{\ \ \mu}$, the spinor-valued one-forms used in chapter five are :

$$e^{AA'}_{\ \ \ \ \mu} = e^a_{\ \ \mu} \, \sigma_a^{\ \ AA'} \quad . \tag{A.4}$$

Denoting by n^{μ} the future-pointing unit timelike normal to a spacelike three-surface, its spinor version obeys the relations :

$$n_{AA'}e^{AA'}_{\ \ i} = 0 \quad , \tag{A.5}$$

$$n_{AA'}n^{AA'} = 1 \quad . \tag{A.6}$$

Denoting by h the induced metric on the three-surface, other useful relations are :

$$h_{ij} = -e_{AA'i}\, e^{AA'}_{\ \ j} \quad , \tag{A.7}$$

$$e^{AA'}_{\ \ 0} = Nn^{AA'} + N^{i}e^{AA'}_{\ \ i} \quad , \tag{A.8}$$

$$n_{AA'}n^{AB'} = \frac{1}{2}\epsilon_{A'}^{\ \ B'} \quad , \quad n_{AA'}n^{BA'} = \frac{1}{2}\epsilon_{A}^{\ \ B} \quad , \tag{A.9}$$

$$n_{[EB'}n_{A]A'} = \frac{\epsilon_{EA}\epsilon_{B'A'}}{4} \quad , \tag{A.10}$$

$$e_{AA'j}\, e^{AB'}_{\ \ k} = -\frac{1}{2}h_{jk}\epsilon_{A'}^{\ \ B'} - i\epsilon_{jkl}\sqrt{h}\, n_{AA'}e^{AB'l} \quad . \tag{A.11}$$

In (A.8), N and N^{i} are the lapse and shift functions respectively (Misner et al. 1973, MacCallum 1975). The first application of spinor calculus we are interested in is the construction of spinor harmonics for the expansion of the spin-$\frac{3}{2}$ field on S^{3}. For simplicity, let us describe at first the procedure for the spin-1 field strength. We consider the Euclidean four-space (D'Eath and Halliwell 1987) with metric : $g_{E} = dr \otimes dr + r^{2}\Omega_{3}$, where Ω_{3} is the metric on a unit three-sphere : $\Omega_{3} = d\chi \otimes d\chi + (\sin\chi)^{2}\Big(d\theta \otimes d\theta + (\sin\theta)^{2}d\phi \otimes d\phi\Big)$. We then consider the following homogeneous polynomials in Cartesian coordinates $x^{\mu} = -\sigma^{\mu}_{\ \ AA'}x^{AA'}$ (omitting a further degeneracy label, p, for simplicity of notation) :

$$r^{n}\rho^{n}_{AB}(\chi,\theta,\phi) = T_{ABA_{1}...A_{n}A'_{1}...A'_{n}}\, x^{A_{1}A'_{1}}...x^{A_{n}A'_{n}} \quad , \tag{A.12}$$

where $T_{ABA_{1}...A_{n}A'_{1}...A'_{n}}$ is a constant spinor of rank $2n+2$ and totally symmetric in all its indices. Thus, in this flat Euclidean four-space one has :

$$\nabla^{AA'}(r^{n}\rho^{n}_{AB}) = e^{AA'\mu}D_{\mu}(r^{n}\rho^{n}_{AB}) = 0 \quad . \tag{A.13}$$

In deriving the form of (A.13) in spherical polar coordinates, we must take into account the following relation :

$$D_\mu \rho^{AB} = \partial_\mu \rho^{AB} + \omega^{A}_{\ C\mu} \rho^{CB} + \omega^{B}_{\ C\mu} \rho^{CA} \quad , \tag{A.14}$$

which implies :

$$\left({}^{(4)}D_j \rho^{AB} - {}^{(3)}D_j \rho^{AB}\right) = \left({}^{(4)}\omega^{A}_{\ Cj} - {}^{(3)}\omega^{A}_{\ Cj}\right) \rho^{CB} + \left({}^{(4)}\omega^{B}_{\ Cj} - {}^{(3)}\omega^{B}_{\ Cj}\right) \rho^{CA} \,. \tag{A.15}$$

Moreover, it is known that :

$${}^{(4)}\omega^{AB}_{\ \ j} - {}^{(3)}\omega^{AB}_{\ \ j} = \left({}_e n^{A}_{\ B'}\right) e^{BB'i} K_{ij} \quad , \tag{A.16}$$

where, in the case of a three-sphere of radius r imbedded in flat Euclidean space, one has : $K_{ij} = -\frac{h_{ij}}{r}$. Thus, using (A.5), (A.7) and (5.1.18) we find that :

$${}^{(4)}D_j \rho^{n}_{AB} = {}^{(3)}D_j \rho^{n}_{AB} + \frac{1}{r}\left({}_e n_{A}^{\ \ B'}\right) e_{CB'j}\, \rho^{C}_{\ B} + \frac{1}{r}\left({}_e n_{B}^{\ \ B'}\right) e_{CB'j}\, \rho^{C}_{\ A} \quad . \tag{A.17}$$

Using again the identities just mentioned, setting $N = 1$ and $N^i = 0$ in (A.8) and inserting (A.17) into (A.13) we finally get :

$$\frac{e^{AA'j}}{r}\left[{}^{(3)}D_j\left(r^n \rho^{n}_{AB}\right)\right] + \left({}_e n^{AA'}\right)\left(\frac{\partial}{\partial r} + \frac{2}{r}\right)\left(r^n \rho^{n}_{AB}\right) = 0 \quad . \tag{A.18}$$

In expanding a spin-$\frac{3}{2}$ field on S^3, we have shown in chapter five that, restricting the theory to its physical degrees of freedom, it only survives the contribution of the spinor $\beta_{ABB'} = \rho^{n}_{ABC}\ n^{C}_{\ B'}$, where ρ^{n}_{ABC} is completely symmetric. In view of the relation : $D_\mu \rho^{ABC} = \partial_\mu \rho^{ABC} + \omega^{A}_{\ D\mu} \rho^{DBC} + \omega^{B}_{\ D\mu} \rho^{DCA} + \omega^{C}_{\ D\mu} \rho^{DAB}$, the same procedure now yields :

$$\frac{e^{AA'j}}{r}\left[{}^{(3)}D_j\left(r^n \rho^{n}_{ABC}\right)\right] + \left({}_e n^{AA'}\right)\left(\frac{\partial}{\partial r} + \frac{5}{2r}\right)\left(r^n \rho^{n}_{ABC}\right) = 0 \quad , \tag{A.19}$$

which is again a special case of the following general result :

$$e^{AA'\mu}D_\mu\left(r^n\rho^n_{A_1\ldots A_s}\right)=\frac{e^{AA'j}}{r}\left[{}^{(3)}D_j\left(r^n\rho^n_{A_1\ldots A_s}\right)\right]$$

$$+\left({}_e n^{AA'}\right)\left(\frac{\partial}{\partial r}+\frac{(s+2)}{2r}\right)\left(r^n\rho^n_{A_1\ldots A_s}\right)\quad . \tag{A.20}$$

In view of the relation : ${}_e n^{AA'}=-in^{AA'}$, (A.19) finally yields :

$$e^{AA'j}\left[{}^{(3)}D_j\rho^n_{ABC}\right]=i\left(n+\frac{5}{2}\right)n^{AA'}\rho^n_{ABC}\quad . \tag{A.21}$$

All orthogonality relations obeyed by these harmonics have been derived in (5.22) of Hughes 1990. Here, with our conventions (cf. section 9.1), we just recall that :

$$\int d\mu\ \rho^{npABC}n_{AA'}n_{BB'}n_{CC'}\bar{\rho}^{mqA'B'C'}=\delta^{nm}\delta^{pq}\quad . \tag{A.22}$$

Another useful application of spinor calculus is the derivation of (5.2.5-9). We may indeed recall that the α_n^{pq} appearing in the expansions of chapter five are such that (D'Eath and Halliwell 1987), for each n, they may be regarded as block-diagonal matrices α_n with blocks $\begin{pmatrix}1 & 1\\ 1 & -1\end{pmatrix}$. In addition, we know that the ρ_A harmonics obey the orthogonality relations :

$$\int d\mu\ \rho_A^{np}n^{AA'}\bar{\rho}_{A'}^{mq}=\delta^{nm}\delta^{pq}\quad , \tag{A.23}$$

$$\int d\mu\ \rho_A^{np}\epsilon^{AB}\rho_B^{mq}=\sqrt{2}\,\delta^{nm}C_n^{pq}\quad , \tag{A.24}$$

where C_n^{pq} is block-diagonal with blocks $\begin{pmatrix}0 & 1\\ -1 & 0\end{pmatrix}$, so that $C_n^{pq}=-C_n^{qp}$. Thus, in view of (5.1.5), (5.1.7-10), (5.1.17), (5.2.2) and (5.2.4) we have that :

$$k_1 I_3^{(\rho,\bar{\rho})}=i\int dt\ e^{-\alpha}\sum_{npq}\sum_{n'p'q'}\alpha_n^{pq}\alpha_{n'}^{p'q'}\bar{u}_{np}\left[m_{n'p'}^{(\alpha)}A^{\bar{u}m(\alpha)}(n,n',q,q')\right.$$

$$+m_{n'p'}^{(\beta)}A^{\bar{u}m(\beta)}(n,n',q,q')+m_{n'p'}^{(\gamma)}A^{\bar{u}m(\gamma)}(n,n',q,q')$$

$$\left.+\bar{s}_{n'p'}A^{\bar{u}\bar{s}}(n,n',q,q')\right]\quad , \tag{A.25}$$

where, using the identity (Hughes 1990) :

$$\rho^{nq}_{ABB'}n^{BB'} = i\rho^{n+1,q}_{A} \quad , \tag{A.26}$$

one finds :

$$\begin{aligned}
A^{\bar{u}m(\alpha)}(n,n',q,q') &= \int d\mu\ \bar{\rho}^{nqA'}\epsilon_{AC}n_{EA'}e^{E\ \ j}_{\ C'}\left({}^{(3)}D_j\alpha^{n'q'ACC'}\right) \\
&- \int d\mu\ \bar{\rho}^{nqA'}n_{CE'}\epsilon_{A'C'}e_{A}^{\ \ E'j}\left({}^{(3)}D_j\alpha^{n'q'ACC'}\right) \\
&= - \int d\mu\ \bar{\rho}^{nqA'}n_{CA'}\rho^{n'+1,q'\ C} \\
&+ \frac{1}{3}\left(n' + \frac{5}{2}\right)\int d\mu\ \bar{\rho}^{nqA'}n_{AA'}\rho^{n'+1,q'\ A} \\
&+ \frac{1}{3}\left(n' + \frac{5}{2}\right)\int d\mu\ \bar{\rho}^{nqA'}n_{CA'}\rho^{n'+1,q'\ C} \\
&= \left(\frac{2}{3}n' + \frac{2}{3}\right)\delta^{n'+1,n}\delta^{qq'} \quad ,
\end{aligned} \tag{A.27}$$

$$\begin{aligned}
A^{\bar{u}m(\gamma)}(n,n',q,q') &= \int d\mu\ \bar{\rho}^{nqA'}\epsilon_{AC}n_{EA'}e^{E\ \ j}_{\ C'}\left({}^{(3)}D_j\gamma^{n'q'ACC'}\right) \\
&- \int d\mu\ \bar{\rho}^{nqA'}n_{CE'}\epsilon_{A'C'}e_{A}^{\ \ E'j}\left({}^{(3)}D_j\gamma^{n'q'ACC'}\right) \\
&= i\left(\frac{n'}{2} + \frac{3}{4}\right)\delta^{nn'}\delta^{qq'} \quad ,
\end{aligned} \tag{A.28}$$

$$\begin{aligned}
A^{\bar{u}3}(n,n',q,q') &= \int d\mu\ \bar{\rho}^{nqA'}\epsilon_{AC}\epsilon^{AC}n_{EA'}e^{E\ \ j}_{\ C'}\left({}^{(3)}D_j\bar{\rho}^{n'q'C'}\right) \\
&- \int d\mu\ \bar{\rho}^{nqA'}\epsilon_{A'C'}\epsilon^{AC}n_{CE'}e_{A}^{\ \ E'j}\left({}^{(3)}D_j\bar{\rho}^{n'q'C'}\right) \\
&= i\sqrt{2}\left(n' + \frac{3}{2}\right)\delta^{nn'}C^{qq'}_{n} \quad ,
\end{aligned} \tag{A.29}$$

whereas $A^{\hat{u}m(\beta)}$ can be shown to vanish. Inserting (A.27-29) into (A.25), and taking into account (A.24), we finally get (5.2.7) and (5.2.9). In working out $\widetilde{I}_4^{(\rho,\bar{\beta})}$ as done in (5.2.8), we need the additional identity (section 5.3 of Hughes 1990) :

$$e^{DA'j}\left[{}^{(3)}D_j \rho^{nq}_{DAD'}\right] = i\left(n+\frac{3}{2}\right) n^{DA'} \rho^{nq}_{DAD'} - \epsilon_{D'}{}^{A'}\, \rho^{n+1,q}_{A} \quad . \tag{A.30}$$

Finally, an important concept which should be clarified is the one of spinor conjugation (cf. section 5.8). In so doing, we closely follow Woodhouse 1985. In spinor language we associate the 2×2 matrix :

$$\left(C^{AA'}\right) = \frac{1}{\sqrt{2}}\begin{pmatrix} t+x & y+iz \\ y-iz & t-x \end{pmatrix} \quad , \tag{A.31}$$

to the set of complex four-vector components $(C^a) = t, x, y, z$. This correspondence is such that : $g_{ab}C^aC^b = \epsilon_{AB}\epsilon_{A'B'}C^{AA'}C^{BB'}$. Therefore, given $(L^A{}_B)$ and $\left(R^{A'}{}_{B'}\right) \in SL(2,C)$, the transformation :

$$C^{AA'} \rightarrow L^A{}_B\, C^{BB'} R^{A'}{}_{B'} \quad , \tag{A.32}$$

gives a realization of the isomorphism $SO(4,C) \cong SL(2,C) \times SL(2,C)/Z_2$. In the case of Minkowski space M_4 one defines the (M) conjugation such that :

$$w^A = \begin{pmatrix} \alpha \\ \beta \end{pmatrix} \rightarrow \overline{w}^{A'} = \begin{pmatrix} \overline{\alpha} \\ \overline{\beta} \end{pmatrix} \quad , \quad z^{A'} = \begin{pmatrix} \gamma \\ \delta \end{pmatrix} \rightarrow \overline{z}^{A} = \begin{pmatrix} \overline{\gamma} \\ \overline{\delta} \end{pmatrix} \quad , \tag{A.33}$$

whereas for Euclidean space E_4 one defines the corresponding Euclidean conjugation :

$$w^A \rightarrow \left(w^A\right)^\dagger = \begin{pmatrix} \overline{\beta} \\ -\overline{\alpha} \end{pmatrix} \quad , \quad z^{A'} \rightarrow \left(z^{A'}\right)^\dagger = \begin{pmatrix} -\overline{\delta} \\ \overline{\gamma} \end{pmatrix} \quad . \tag{A.34}$$

In Woodhouse 1985 the symbol $\wedge$ is used instead of $\dagger$. However, we preserve the notation using the *dagger* $\dagger$, so as to avoid confusion. The Euclidean conjugation (A.34) is anti-linear as (A.33), but it does not interchange primed and unprimed indices. It has the important anti-involutory property :

$$\left(w^{A\dagger}\right)^\dagger = -w^A \quad , \quad \left(z^{A'\dagger}\right)^\dagger = -z^{A'} \quad . \tag{A.35}$$

However, it should be emphasized that on spinors with an even number of indices, Euclidean conjugation is involutory. Therefore one has the Hermitian inner products :

$$w^A w_A^\dagger = -\epsilon_{AB}\, w^A\, w^{B\dagger} = \alpha\overline{\alpha} + \beta\overline{\beta} \quad , \tag{A.36}$$

$$z_{A'}\left(z^{A'}\right)^\dagger = \epsilon_{A'B'}\, z^{A'}\left(z^{B'}\right)^\dagger = \gamma\overline{\gamma} + \delta\overline{\delta} \quad . \tag{A.37}$$

The transformation (A.32) preserves Euclidean conjugation and determines an element of $SO(4,R)$ when $\left(L^A{}_B\right)$ and $\left(R^{A'}{}_{B'}\right) \in SU(2)$. In other words, the Euclidean form of the isomorphism is $SO(4,R) \cong SU(2) \times SU(2)/Z_2$. If we require the preservation of both conjugations, the transformation of (t,x,y,z) will be a rotation in the spacelike hyperplane $\Sigma \equiv (t=0; x,y,z \in R) = M_4 \cap E_4$. This implies that the isomorphism is further reduced to $SO(3) \cong SU(2)/Z_2$, so that we have :

$$\left(w^A\right)^\dagger = \sqrt{2}\, t^A{}_{A'}\, \overline{w}^{A'} \quad , \tag{A.38}$$

$$\left(z^{A'}\right)^\dagger = -\sqrt{2}\, \overline{z}^A\, t_A{}^{A'} \quad , \tag{A.39}$$

where $t^a = (1,0,0,0)$ is the unit timelike normal to Σ in M_4. In section 5.8, we have used the notation $\delta^A{}_{A'}$ for an identity matrix in the case of Euclidean four-space, so as to avoid confusion with the spinorial form of the normal to the S^3 boundary. The reader should also bear in mind that in (5.8.65) our convention for $\left(\widetilde{\psi}^{A'}\right)^\dagger$ is opposite to the one of Woodhouse appearing in (A.34), whereas our convention for $\left(\psi^A\right)^\dagger$ is in agreement with Woodhouse 1985.

APPENDIX B: The Generalized Zeta-Function

The study of the eigenvalues of differential operators is a very important and far-reaching problem in mathematics. A related mathematical tool is the generalized Riemann zeta-function (hereafter referred to as zeta-function). Let A be an elliptic, second-order, self-adjoint differential operator with positive-definite spectrum, and let $\lambda_{n,m}$ be its eigenvalues having multiplicity $d_m(n)$. The zeta-function is then defined by :

$$\zeta(s) \equiv \sum_{n,m} d_m(n) \lambda_{n,m}^{-s} \quad . \tag{B.1}$$

We need a double sum because, for each value of the integer n, there may be infinitely many eigenvalues (this happens for example in studying the eigenvalue condition $J_n(\epsilon) = 0$). In four dimensions, $\zeta(s)$ converges for $Re(s) > 2$ (Hawking 1977, Allen 1983a), and can be analytically extended to a meromorphic function of s which only has poles at $s = 1, 2$ (Seeley 1967, Hawking 1977). For problems with boundaries involving fermionic fields, poles may also occur at $s = \frac{1}{2}, \frac{3}{2}$.

Other remarkable results are the regularity of ζ at $s = 0$, and the relation : $det\ A = e^{-\zeta'(0)}$. It may happen quite often that the eigenvalues of A are not known. However, in such a case one can study the heat equation : $\frac{\partial}{\partial \tau} F(x, y, \tau) + AF(x, y, \tau) = 0$, where A acts on x and the initial condition is : $F(x, y, 0) = \delta(x, y)$. Let : $A\phi_n = \lambda_n \phi_n$ be the eigenvalue equation. If the eigenfunctions ϕ_n are a complete set, so that a scalar field ϕ can be expanded according to the relation : $\phi = \sum_n a_n \phi_n$, one finds (Hawking 1977) :

$$F(x, y, \tau) = \sum_n e^{-\lambda_n \tau} \phi_n(x) \otimes \phi_n(y) \quad , \tag{B.2}$$

$$G(\tau) = \int_M Tr\ F(x, x, \tau) \sqrt{g}\, d^4x = \sum_n e^{-\lambda_n \tau} \quad , \tag{B.3}$$

$$\zeta(s) = \frac{1}{\Gamma(s)} \int_0^\infty \tau^{s-1} G(\tau) d\tau \quad . \tag{B.4}$$

These relations can be generalized to the case of tensor or spinor fields. $F(x,y,\tau)$ is called Green's function and $G(\tau)$ is called (integrated) heat kernel of F. A good treatment of kernel functions of elliptic operators on manifolds may be found in Gilkey 1975. More recent work on heat kernels and their asymptotic expansions appears in Davies 1989, Gusynin 1989 and Cognola et al. 1990.

In section 4.1, we are interested in the heat equation for the Laplacian operator acting on transverse-traceless tensors :

$$\left(\frac{\partial}{\partial\tau} - \nabla_f\nabla^f\right) G_{ab,cd}(x,x',\tau) = \delta_{ab,cd}(x,x')\delta(\tau) \quad ,$$

and in (4.1.7) we take its Laplace transform. The basic underlying result used in that section is the following. As $\tau \to 0^+$, one has the asymptotic expansion (Greiner 1971, Schleich 1985) :

$$G(\tau) \sim \sum_i g_{\frac{i}{2}} \tau^{\left(\frac{i}{2}-2\right)} \quad , \tag{B.5}$$

where $g_{\frac{i}{2}}$ depends on a. Hence one finds (cf. (B.4)) :

$$\zeta(s) \sim \frac{1}{\Gamma(s)}\left[\sum_i \frac{g_{\frac{i}{2}}}{(\frac{i}{2}+s-2)} + \int_1^\infty \tau^{s-1}G(\tau)d\tau\right] \quad . \tag{B.6}$$

The integral in (B.6) is finite in view of (B.3) and of the positivity ($\lambda_n > 0$) of the spectrum of the Laplacian satisfying Dirichlet boundary conditions (Chavel 1984, page 9). This latter result can also be understood by studying the corresponding eigenvalue condition : $J_n(\sqrt{E}a) = 0$, where the E's denote the eigenvalues of the Laplacian. It is well-known that, if $\nu > 1$, $J_\nu(z)$ has no zeros which are not real (Watson 1966, page 482). Thus $\sqrt{E}$ is real and $E = \lambda_n$ is positive, $\forall n$. It is then easy to show that $\zeta(0)$ is equal to the constant coefficient g_2 appearing in the asymptotic expansion (B.5), because $\frac{1}{\Gamma(s)}$ has a simple zero at $s = 0$.

APPENDIX C: Euler-Maclaurin Formula and Free Part of the Heat Kernel for the Spin-$\frac{3}{2}$ Field

If f is a function which obeys suitable differentiability and integrability conditions, the Euler-Maclaurin formula is a very useful tool which can be used to estimate the sum $\sum_{i=0}^{n} f(i)$. Denoting by $\widetilde{B}_s$ the Bernoulli numbers, defined by the expansion (Jeffreys and Jeffreys 1946, Wong 1989) :

$$\frac{t}{(e^t-1)} = \sum_{s=0}^{\infty} \widetilde{B}_s \frac{t^s}{s!} \quad , \quad |\, t \,| < 2\pi \quad , \tag{C.1}$$

the following theorem holds (Wong 1989).

Theorem C.1 Let f be a real- or complex-valued function defined on $0 \leq t < \infty$. If $f^{(2m)}(t)$ is absolutely integrable on $(0,\infty)$ then, for $n = 1, 2, \ldots$:

$$\sum_{i=0}^{n} f(i) - \int_0^n f(x)\, dx = \frac{1}{2}\Big[f(0) + f(n)\Big]$$
$$+ \sum_{s=1}^{m-1} \frac{\widetilde{B}_{2s}}{(2s)!}\Big[f^{(2s-1)}(n) - f^{(2s-1)}(0)\Big] + R_m(n) \, , \tag{C.2}$$

where the remainder $R_m(n)$ satisfies :

$$|\, R_m(n) \,| \leq \Big(2 - 2^{1-2m}\Big) \frac{|\, \widetilde{B}_{2m} \,|}{(2m)!} \int_0^n |\, f^{(2m)}(x) \,|\, dx \quad . \tag{C.3}$$

Equation (C.2) can be used to evaluate infinite sums, setting $n = \infty$, if the corresponding derivatives, and the integrals in (C.2-3) are well-defined (e.g. Itzykson and Zuber 1985, page 140). Typically one considers $\widetilde{n}$ terms in the sum involving Bernoulli numbers in (C.2), and neglects terms starting with some value $s = \widetilde{n} + 1 \leq (m-1)$. The inequality (C.3) can be used to show that the remainder $R_m(n)$ is bounded in absolute value by a constant times the first neglected term in the sum involving Bernoulli numbers in (C.2), provided $f^{(2m)}$ obeys suitable conditions specified on page 38 of Wong 1989.

The free part $G^F(\widetilde{\tau})$ of the heat kernel for a single set of modes for the spin-$\frac{3}{2}$ field is given in section 5.5 by :

$$G^F(\widetilde{\tau}) = \int_0^{\frac{a^2}{2\widetilde{\tau}}} \sum_{n=2}^{\infty} \Big[(n+2)(n-1)\Big] \frac{I_n(y)}{2} e^{-y} dy$$

$$= \frac{1}{2}\sum_{n=1}^{\infty} \int_0^{\frac{a^2}{2\widetilde{\tau}}} n^2 I_n(y) e^{-y} dy + \frac{1}{2}\sum_{n=1}^{\infty} \int_0^{\frac{a^2}{2\widetilde{\tau}}} n I_n(y) e^{-y} dy$$

$$- \sum_{n=1}^{\infty} \int_0^{\frac{a^2}{2\widetilde{\tau}}} I_n(y) e^{-y} dy \quad . \qquad (C.4)$$

Using the relations (4.4.2-7) we realize that, in computing (C.4), the only calculation not yet performed is the one of the third term on the right-hand side of (C.4). Indeed, one has

$$\sum_{n=1}^{\infty} I_n(y) = -\frac{1}{2} I_0(y) + \frac{e^y}{2} \quad , \qquad (C.5)$$

which implies :

$$\sum_{n=1}^{\infty} \int_0^{\frac{a^2}{2\widetilde{\tau}}} I_n(y) e^{-y} dy = \frac{a^2}{4\widetilde{\tau}} - \frac{1}{2} \int_0^{\frac{a^2}{2\widetilde{\tau}}} I_0(y) e^{-y} dy$$

$$= \frac{a^2}{4\widetilde{\tau}} - \frac{a^2}{4\widetilde{\tau}} e^{-\frac{a^2}{2\widetilde{\tau}}} \left[I_0\left(\frac{a^2}{2\widetilde{\tau}}\right) + I_1\left(\frac{a^2}{2\widetilde{\tau}}\right) \right] \quad . \qquad (C.6)$$

Hence the relations (C.4-6) imply :

$$G^F(\widetilde{\tau}) = \frac{a^4}{32\widetilde{\tau}^2} + \frac{a^2}{8\widetilde{\tau}} e^{-\frac{a^2}{2\widetilde{\tau}}} I_0\left(\frac{a^2}{2\widetilde{\tau}}\right) + \frac{1}{6}\Big[y^2 I_0(y) e^{-y} + (y^2+y) I_1(y) e^{-y}\Big]_0^{\frac{a^2}{2\widetilde{\tau}}}$$

$$- \frac{1}{4}\Big[y e^{-y}(I_0(y) + I_1(y))\Big]_0^{\frac{a^2}{2\widetilde{\tau}}} - \frac{a^2}{4\widetilde{\tau}} + \frac{a^2}{4\widetilde{\tau}} e^{-\frac{a^2}{2\widetilde{\tau}}} \left[I_0\left(\frac{a^2}{2\widetilde{\tau}}\right) + I_1\left(\frac{a^2}{2\widetilde{\tau}}\right) \right] . \qquad (C.7)$$

Defining $z \equiv \frac{a^2}{2\widetilde{\tau}}$, and using the asymptotic expansions (Abramowitz and Stegun 1964) :

$$I_0(z) \sim \frac{e^z}{\sqrt{2\pi z}} \left[1 + \frac{1}{8z} + \frac{9}{128 z^2} + \ldots \right] \quad , \qquad (C.8)$$

$$I_1(z) \sim \frac{e^z}{\sqrt{2\pi z}}\left[1 - \frac{3}{8z} - \frac{15}{128z^2} - \dots\right] \quad , \tag{C.9}$$

valid as $z \to \infty$, we find from (C.7-9) that, as $\widetilde{\tau} \to 0^+$, the free part of the heat kernel has the asymptotic expansion :

$$G^F(\widetilde{\tau}) \sim \frac{a^4}{32}\widetilde{\tau}^{-2} + \frac{a^3}{12\sqrt{\pi}}\widetilde{\tau}^{-\frac{3}{2}} - \frac{a^2}{4}\widetilde{\tau}^{-1} + \frac{7}{16}\frac{a}{\sqrt{\pi}}\widetilde{\tau}^{-\frac{1}{2}} + O(\sqrt{\widetilde{\tau}}) \quad . \tag{C.10}$$

The relation (C.10) clearly proves that $G^F(\widetilde{\tau})$ does not contribute to the PDF value of $\zeta(0)$. However, it is worth remarking a difference with respect to the spin-$\frac{1}{2}$ case, i.e. that in our case the asymptotic form of $G^F(\widetilde{\tau})$ also contains a term proportional to $\widetilde{\tau}^{-1}$.

APPENDIX D: Complex Manifolds

The statement of theorem 5.7.1 needs some further clarification. This is why we here briefly describe complex manifolds and their real sections.

A complex manifold (Chern 1979) is a paracompact Hausdorff space covered by neighbourhoods U_k each homeomorphic to an open set in C^m such that, if $(z^1, ..., z^m)$ are local coordinates in U_i and if $(w^1, ..., w^m)$ are local coordinates in U_j, then where they are both defined we have $w^i = w^i(z^1, ..., z^m)$, each w^i being a holomorphic function of the z's. Moreover, the functional determinant $\partial(w^1, ..., w^m)/\partial(z^1, ..., z^m)$ is nonvanishing. Examples of complex manifolds are (Greene 1987) : C^n, the unit ball B^n in C^n, variations of B^n, complex projective space CP^n, submanifolds of CP^n.

Let us now assume (Gibbons 1978) that real space-time $(M, g_{\alpha\beta})$ with contravariant metric $g^{\alpha\beta}\frac{\partial}{\partial x^\alpha}\frac{\partial}{\partial x^\beta}$ is imbedded as a real four-dimensional submanifold in a complex four-manifold M_c, called the complexification of M. We also assume that in M_c a complex contravariant tensor field $g_c^{\alpha\beta}$ is given of rank 2 and type $(2,0)$, such that the restriction of $g_c^{\alpha\beta}$ to the cotangent space of the submanifold is real. M is then said to be a real section of M_c. Local coordinates for M_c are (x, x^*), and M is defined by : $x = x^*$. In general a real Lorentzian metric has not a section of the complexified space-time where the metric is real and positive-definite (also called Euclidean), and viceversa. However, this is possible for the Schwarzschild solution and for Friedmann-Robertson-Walker metrics.

In section 5.7, the situation is simple but not trivial. In fact, the basic mathematical objects of our theorem are :

(1) A solution ω^L of the twistor equation in flat Euclidean four-space. Therefore we can define its independent primed counterpart $\widetilde{\omega}^{L'}$;

(2) Linearized independent field strengths ϕ_{ABCD} and $\widetilde{\phi}_{A'B'C'D'}$ (as well as ϕ_{ABC} and $\widetilde{\phi}_{A'B'C'}$). These field strengths do not live in flat space, but they still satisfy a massless free-field equation. This is why ω^L and $\widetilde{\omega}^{L'}$ can play a role (our background is flat). Note that, for pure gravity, the massless free-field equation $\nabla^{AA'}\phi_{ABCD} = 0$ holds by virtue of the Bianchi identity (cf. Ward and Wells 1990, page 454), and that in section 5.7 and

chapter 7 we denote by ϕ_{ABCD} the spinor written instead as ψ_{ABCD} in section 6.8 of Penrose and Rindler 1986 (see also Ward and Wells 1990). More precisely, the situation can be described as explained in section 5.7 of Penrose and Rindler 1984. If u is the perturbative parameter for the expansion of the metric, the curvature tensor K_{abcd} to first order in u can be expressed in terms of the spinor :

$$\phi_{ABCD} \equiv \lim_{u \rightarrow 0} \Big(u^{-1} \psi_{ABCD}(u) \Big)$$

as in (5.7.1). If ϕ_{ABCD} is regarded as a massless field, its massless free-field equation corresponds to the Bianchi identity of K_{abcd}.

Of course, when the field strengths are no longer linearized (the fully curved case), our theorem cannot be generalized because there is no solution of the twistor equation in a generic curved space-time, as explained in Penrose and Rindler 1986 (see, however, Lewandowski 1991).

From a mathematical point of view, the significance of theorem 5.7.1 is that we should consider a flat complexified space-time and its Euclidean section if we want to find a spin-lowering operator which preserves the local boundary conditions (5.7.3-4). This is explained in detail in the comment (c) after theorem 5.7.1. In physical language, we can say that rigid supersymmetry transformations exist in Euclidean quantum gravity which realize a mapping between solutions of the linearized massless free-field equations, subject to (5.7.3-4) on the three-sphere boundary, whose spins differ by $\frac{1}{2}$ (see also section 1.4).

APPENDIX E: Lorentzian ADM Formulae for the Curvature

We are here interested in the ADM formulae for the curvature in a torsion-free space-time manifold, whose metric may be cast in the form (1.1.1). This problem is studied by a number of authors in the literature (e.g. Hawking and Ellis 1973, Misner et al. 1973, Lightman et al. 1975, Hanson et al. 1976, Boulware 1984, Wald 1984, York 1986, Ashtekar 1988). To our knowledge, the most explicit formulae including all possible contributions from lapse and shift functions are the ones in Boulware 1984. However, we define the extrinsic-curvature tensor as (cf. (2.4.1) and (2.4.9)) :

$$K_{ij} \equiv -\frac{1}{2}(L_n h)_{ij} = \frac{1}{2N}\left(-\frac{\partial h_{ij}}{\partial t} + 2N_{(i|j)}\right) \quad , \tag{E.1}$$

whereas, in Boulware 1984, K_{ij} is defined as $\frac{1}{2}(L_n h)_{ij}$. In so doing we agree with the convention used in Misner et al. 1973, Lightman et al. 1975, Hanson et al. 1976, York 1986 and in most papers on quantum cosmology. The curvature tensor in component language (Boulware 1984) is defined by :

$${}^{(4)}R^{\mu}{}_{\nu\lambda\sigma} \equiv \Gamma^{\mu}_{\nu\sigma,\lambda} - \Gamma^{\mu}_{\nu\lambda,\sigma} + \Gamma^{\tau}_{\nu\sigma}\Gamma^{\mu}_{\tau\lambda} - \Gamma^{\tau}_{\nu\lambda}\Gamma^{\mu}_{\tau\sigma} \quad . \tag{E.2}$$

We also have :

$$n^{\mu} = \frac{1}{N}(1, -N^{m}) \quad , \quad n_{\mu} = (-N, 0) \quad , \tag{E.3}$$

so that $n_{\mu}n^{\mu} = -1$. The curvature calculation is performed as in Boulware 1984 using the relations :

$${}^{(4)}R^{0}{}_{l0m} = -n_{\mu}\left[{}^{(4)}R^{\mu}{}_{l\sigma m}\right]n^{\sigma} \quad , \tag{E.4}$$

$${}^{(4)}R^{0}{}_{lmn} = -n_{\mu}\left[{}^{(4)}R^{\mu}{}_{lmn}\right] \quad , \tag{E.5}$$

$${}^{(4)}R_{jlmn} = {}^{(3)}R_{jlmn} + K_{jm}K_{ln} - K_{jn}K_{lm} \quad , \tag{E.6}$$

$${}^{(4)}R^{00} = -g^{mn}\left[{}^{(4)}R^{0}{}_{m0n}\right] \quad , \tag{E.7}$$

$$^{(4)}R = g^{kl}\left[{}^{(4)}R_{kl}\right] - {}^{(4)}R^{00} = g^{kl}\left[{}^{(4)}R_{kl}\right] + {}^{(4)}R^0{}_0 \quad , \tag{E.8}$$

which yield (cf. (7.2.22-25)) :

$$\begin{aligned} {}^{(4)}R_{0i0j} &= -{}^{(4)}R^0{}_{i0j} \\ &= \frac{1}{N}\frac{\partial K_{ij}}{\partial t} + K_i{}^l K_{lj} \\ &- \frac{1}{N}\left(N^m K_{ij|m} + N^m{}_{|i}\, K_{jm} + N^m{}_{|j}\, K_{im}\right) + \frac{N_{|ij}}{N} \quad , \end{aligned} \tag{E.9}$$

$$^{(4)}R^0{}_{ijl} = -(K_{il|j} - K_{ij|l}) \quad , \tag{E.10}$$

$$^{(4)}R^0{}_i = -\left(K_{|i} - K^l{}_{i|l}\right) \quad , \tag{E.11}$$

$$^{(4)}R^0{}_0 = -\left[\frac{1}{N}\frac{\partial K}{\partial t} - K_{ij}K^{ij} - \frac{N^l}{N}K_{|l} + \frac{\nabla^2 N}{N}\right] \quad , \tag{E.12}$$

$$\begin{aligned} {}^{(4)}R_{ij} &= {}^{(3)}R_{ij} + \left[K_{ij}K - \frac{1}{N}\frac{\partial K_{ij}}{\partial t} - 2K_{im}K_j{}^m\right] \\ &+ \frac{1}{N}\left(N^m K_{ij|m} + N^m{}_{|i}\, K_{jm} + N^m{}_{|j}\, K_{im}\right) - \frac{N_{|ij}}{N} \quad , \end{aligned} \tag{E.13}$$

$$^{(4)}R = {}^{(3)}R + K_{ij}K^{ij} + K^2 - \frac{2}{N}\frac{\partial K}{\partial t} - \frac{2}{N}\left(\nabla^2 N - N^p K_{|p}\right) \quad . \tag{E.14}$$

In deriving (E.9-14) it is useful to recall (appendix A of Boulware 1984) that, given a covariant second-rank tensor T_{kl} orthogonal to the normal to the three-surface, its Lie derivative along the normal is :

$$\begin{aligned} (L_n T)_{kl} &= n^\mu T_{kl,\mu} - \frac{1}{N}\left[\left(N^m{}_{,k}\right)T_{ml} + \left(N^m{}_{,l}\right)T_{km}\right] \\ &= \frac{1}{N}\left[\frac{\partial T_{kl}}{\partial t} - \left(N^m\right)T_{kl|m} - \left(N^m{}_{|k}\right)T_{ml} - \left(N^m{}_{|l}\right)T_{km}\right] \quad . \end{aligned} \tag{E.15}$$

APPENDIX F: $\zeta(0)$ Calculations

In chapters four and five we have used the Watson-transform technique, which is preferred in the literature. However, this is not strictly needed, because also when one deals with infinite sums whose terms are even functions of n, one obtains the same PDF result for $\zeta(0)$ using the Euler-Maclaurin formula. This is why our $\zeta(0)$ values for fermionic fields contain no ambiguities, in that the two techniques used are equivalent. We now clarify this point by performing the PDF $\zeta(0)$ calculation when B_i vanishes on S^3 for the spin-1 field. In that case the interacting heat kernel $G^{int}(t)$, as $t \to 0^+$, has the asymptotic expansion :

$$G^{int}(t) \sim -\sum_{n=2}^{\infty}\left(n^2-1\right)\sum_{i=1}^{4} f_i(n,t) + O(\sqrt{t}) \quad , \tag{F.1}$$

where :

$$f_1(n,t) = \frac{1}{2}e^{-n^2 t} \quad , \tag{F.2}$$

$$f_2(n,t) = -\frac{2}{3\sqrt{\pi}}t^{\frac{3}{2}}n^2 e^{-n^2 t} \quad , \tag{F.3}$$

$$f_3(n,t) = \frac{1}{8}\left[t - 6t^2 n^2 + \frac{5}{2}t^3 n^4\right] e^{-n^2 t} \quad , \tag{F.4}$$

$$f_4(n,t) = \frac{1}{\sqrt{\pi}}\left[\frac{t^{\frac{3}{2}}}{6} - \frac{53}{30}t^{\frac{5}{2}}n^2 + \frac{168}{105}t^{\frac{7}{2}}n^4 - \frac{50}{189}t^{\frac{9}{2}}n^6\right] e^{-n^2 t} \quad . \tag{F.5}$$

Thus, using the following relations obtained from the Euler-Maclaurin formula :

$$\sum_{n=1}^{\infty} e^{-n^2 t} = -\frac{1}{2} + \frac{\sqrt{\pi}}{2}t^{-\frac{1}{2}} \quad , \quad \sum_{n=1}^{\infty} n^2 e^{-n^2 t} = \frac{\sqrt{\pi}}{4}t^{-\frac{3}{2}} \quad , \tag{F.6}$$

$$\sum_{n=1}^{\infty} n^4 e^{-n^2 t} = \frac{3}{8}\sqrt{\pi}\, t^{-\frac{5}{2}} \quad , \quad \sum_{n=1}^{\infty} n^6 e^{-n^2 t} = \frac{15}{16}\sqrt{\pi}\, t^{-\frac{7}{2}} \quad , \tag{F.7}$$

$$\sum_{n=1}^{\infty} n^8 e^{-n^2 t} = \frac{105}{32}\sqrt{\pi}\, t^{-\frac{9}{2}} \quad , \tag{F.8}$$

we find the PDF value :

$$\zeta_B(0) = -\frac{1}{6} - \frac{1}{90} - \frac{1}{4} = -\frac{77}{180} \quad , \tag{F.9}$$

where $-\frac{1}{4}$ is due to $f_1(n,t)$ (cf. (F.2)) and (F.6), whereas $-\frac{1}{6}$ and $-\frac{1}{90}$ are due to (F.3) and (F.5) respectively. In other words, if we use the asymptotic formula (F.1), $f_1(n,t)$ leads to the same contribution due to the poles appearing in the Watson-transform technique. In section 5.9, we also find the PDF value :

$$\zeta_E(0) = -\frac{1}{6} - \frac{1}{90} + \frac{1}{4} = \frac{13}{180} \quad , \tag{F.10}$$

since $\widetilde{f}_1(n,t)$ has opposite sign with respect to $f_1(n,t)$ (cf. (5.9.47)). The reader can now understand why the PDF $\zeta(0)$ values in the cases (F.9) and (F.10) are different. Also if one checks the PDF $\zeta(0)$ calculation of Schleich 1985 using the Euler-Maclaurin formula, one finds of course the same result. Again, the contribution $-\frac{5}{2}$ due to the poles is reproduced by the contribution of $f_1(n,t)$ to the asymptotic expansion :

$$G^{int}(t) \sim -\sum_{n=3}^{\infty}\left(n^2-4\right)\sum_{i=1}^{4} f_i(n,t) + O(\sqrt{t}) \quad . \tag{F.11}$$

In fact the contribution of $f_1(n,t)$ in (F.11) to the asymptotic form of $G^{int}(t)$ is found to be :

$$-\frac{\sqrt{\pi}}{8}t^{-\frac{3}{2}} + 2\left(-\frac{1}{2} + \frac{\sqrt{\pi}}{2}t^{-\frac{1}{2}}\right) - \frac{3}{2}e^{-t} \quad ,$$

which finally leads to $-\frac{5}{2}$ because, as $t \to 0^+$, one has : $e^{-t} = 1 + O(t)$.

The formula (7.3.16) is proved by using the relations 9.3.7, 9.3.9, 9.3.11 and 9.3.13 on page 366 of Abramowitz and Stegun 1964. In those formulae, the argument of J_n and J'_n is $\frac{n}{\cosh\gamma}$, where γ is fixed and positive and n is large and positive. If we put $\frac{n}{\cosh\gamma} \equiv x$, we find that :

$$e^{\gamma} = \frac{n}{x} \pm \sqrt{\frac{n^2}{x^2} - 1} \quad , \quad \tanh\gamma = \pm\frac{1}{n}\sqrt{n^2 - x^2} \quad . \tag{F.12}$$

Thus, making the analytic continuation $x \rightarrow ix$ and then defining $\alpha_n \equiv \sqrt{n^2+x^2}$, we get the following asymptotic expansions which are uniformly valid in the order as $|x| \rightarrow \infty$:

$$J_n(ix) \sim \frac{(ix)^n}{\sqrt{2\pi}} \alpha_n{}^{-\frac{1}{2}} e^{\alpha_n} e^{-n\log(n+\alpha_n)} \Sigma_1(n, \alpha_n(x)) \quad , \tag{F.13}$$

$$J'_n(ix) \sim \frac{(ix)^{n-1}}{\sqrt{2\pi}} \alpha_n{}^{\frac{1}{2}} e^{\alpha_n} e^{-n\log(n+\alpha_n)} \Sigma_2(n, \alpha_n(x)) \quad , \tag{F.14}$$

where $\Sigma_1(n,\alpha_n(x)) \sim \sum_{k=0}^{\infty} \frac{u_k(\frac{n}{\alpha_n})}{n^k}$, $\Sigma_2(n,\alpha_n(x)) \sim \sum_{k=0}^{\infty} \frac{v_k(\frac{n}{\alpha_n})}{n^k}$. This is why, in the case of Dirichlet boundary conditions ($J_n(z)=0$, $\forall n \geq 3$), $-\frac{1}{2}$ is multiplied by $\log(\alpha_n)$ in the asymptotic formula for $\log\Bigl((ix)^{-n} J_n(ix)\Bigr)$, whereas for boundary conditions of the type studied in section 7.3 one finds :

$$\log\Bigl[(ix)^{-n}\left(J_n(ix) + ixJ'_n(ix)\right)\Bigr] \sim \log\left(\alpha_n{}^{-\frac{1}{2}} F(\alpha_n)\Sigma_1 + \alpha_n{}^{\frac{1}{2}} F(\alpha_n)\Sigma_2\right)$$

$$= \log(F(\alpha_n)) + \frac{1}{2}\log(\alpha_n) + \log\left(\frac{\Sigma_1}{\alpha_n} + \Sigma_2\right) \quad ,$$

which is cast in the form (7.3.16) expanding in inverse powers of α_n the $\log\left(\frac{\Sigma_1}{\alpha_n} + \Sigma_2\right)$. We are very much indebted to Dr. I. Moss for helping us in clarifying this point by sending us his notes. A further important check of the signs in our asymptotic expansions is given by section 7.4, where we find agreement with the PDF $\zeta(0)$ values obtained using the Laplace transform of the heat kernel for bosonic fields.

We now write the infinite sums used for the $\zeta(0)$ calculation of chapter eight. In (8.2.2) we have :

$$\rho_1 \equiv \sum_{m=0}^{\infty} m^3 \quad , \quad \rho_2 \equiv \sum_{m=0}^{\infty} m^4 \alpha_m^{-1} \quad , \quad \rho_3 \equiv \sum_{m=0}^{\infty} m^4 \alpha_m^{-3} \quad , \tag{F.15}$$

$$\rho_4 \equiv \sum_{m=0}^{\infty} m^4 \alpha_m^{-5} \quad , \quad \rho_5 \equiv \sum_{m=0}^{\infty} m^2 \quad , \quad \rho_6 \equiv \sum_{m=0}^{\infty} m^3 \alpha_m^{-1} \quad , \tag{F.16}$$

$$\rho_7 \equiv \sum_{m=0}^{\infty} m^3 \alpha_m^{-3} \quad , \quad \rho_8 \equiv \sum_{m=0}^{\infty} m^3 \alpha_m^{-5} \quad , \quad \rho_9 \equiv \sum_{m=0}^{\infty} m \quad , \tag{F.17}$$

$$\rho_{10} \equiv \sum_{m=0}^{\infty} m^2 \alpha_m^{-1} \quad , \quad \rho_{11} \equiv \sum_{m=0}^{\infty} m^2 \alpha_m^{-3} \quad , \quad \rho_{12} \equiv \sum_{m=0}^{\infty} m^2 \alpha_m^{-5} \quad . \tag{F.18}$$

Of course, the sums (F.15-18) do not by themselves make sense as they are written (D'Eath and Esposito 1991a). However, as discussed in section 8.2, the sums appear in convergent linear combinations and their values are regularized as explained following the asymptotic formula (7.3.43). Note also that the occurrence of sums such as for example ρ_1 in (F.15) is not a peculiarity of the spin-$\frac{1}{2}$ problem, since ρ_1 also appears in all bosonic one-loop calculations (cf. sections 7.3-4), including scalar fields.

Moreover, the integrals used in section 8.4 are given by :

$$I_1^{(np)} \equiv \int_0^\infty y^2 \left(y+\sqrt{x^2+y^2}\right)^{-p} \left(x^2+y^2\right)^{\frac{p}{2}-\frac{n}{2}-3} dy \quad , \tag{F.19}$$

$$I_2^{(np)} \equiv \int_0^\infty y \left(y+\sqrt{x^2+y^2}\right)^{-p} \left(x^2+y^2\right)^{\frac{p}{2}-\frac{n}{2}-3} dy \quad , \tag{F.20}$$

$$I_3^{(np)} \equiv \int_0^\infty y^2 \left(y+\sqrt{x^2+y^2}\right)^{-p-1} \left(x^2+y^2\right)^{\frac{p}{2}-\frac{n}{2}-\frac{5}{2}} dy \quad , \tag{F.21}$$

$$I_4^{(np)} \equiv \int_0^\infty y \left(y+\sqrt{x^2+y^2}\right)^{-p-1} \left(x^2+y^2\right)^{\frac{p}{2}-\frac{n}{2}-\frac{5}{2}} dy \quad , \tag{F.22}$$

$$I_5^{(np)} \equiv \int_0^\infty y^2 \left(y+\sqrt{x^2+y^2}\right)^{-p-2} \left(x^2+y^2\right)^{\frac{p}{2}-\frac{n}{2}-2} dy \quad , \tag{F.23}$$

$$I_6^{(np)} \equiv \int_0^\infty y \left(y+\sqrt{x^2+y^2}\right)^{-p-2} \left(x^2+y^2\right)^{\frac{p}{2}-\frac{n}{2}-2} dy \quad , \tag{F.24}$$

$$I_7^{(np)} \equiv \int_0^\infty y^2 \left(y+\sqrt{x^2+y^2}\right)^{-p-3} \left(x^2+y^2\right)^{\frac{p}{2}-\frac{n}{2}-\frac{3}{2}} dy \quad , \tag{F.25}$$

$$I_8^{(np)} \equiv \int_0^\infty y \left(y+\sqrt{x^2+y^2}\right)^{-p-3} \left(x^2+y^2\right)^{\frac{p}{2}-\frac{n}{2}-\frac{3}{2}} dy \quad , \tag{F.26}$$

and the integrals used in section 9.2 for $\widetilde{H}_{\infty}^{n,A}$ are :

$$\widetilde{I}_1^{(np)} \equiv \int_0^\infty \left(y+\sqrt{x^2+y^2}\right)^{-p} \left(x^2+y^2\right)^{\frac{p}{2}-\frac{n}{2}-3} dy \quad , \tag{F.27}$$

$$\widetilde{I}_2^{(np)} \equiv \int_0^\infty \left(y+\sqrt{x^2+y^2}\right)^{-p-1} \left(x^2+y^2\right)^{\frac{p}{2}-\frac{n}{2}-\frac{5}{2}} \, dy \quad , \tag{F.28}$$

$$\widetilde{I}_3^{(np)} \equiv \int_0^\infty \left(y+\sqrt{x^2+y^2}\right)^{-p-2} \left(x^2+y^2\right)^{\frac{p}{2}-\frac{n}{2}-2} \, dy \quad , \tag{F.29}$$

$$\widetilde{I}_4^{(np)} \equiv \int_0^\infty \left(y+\sqrt{x^2+y^2}\right)^{-p-3} \left(x^2+y^2\right)^{\frac{p}{2}-\frac{n}{2}-\frac{3}{2}} \, dy \quad . \tag{F.30}$$

We think it can now be useful to list the techniques used in our book and in the existing literature on the asymptotic heat kernel for manifolds with boundary :

(1) *Direct* technique based on the Laplace transform of the heat equation, and later use of the Watson transform and of the Euler-Maclaurin formula (Stewartson and Waechter 1971, Schleich 1985, Louko 1988a-b, D'Eath and Esposito 1991b-c, Esposito 1991, our chapters four and five);

(2) *Direct* technique based on the properties of the function appearing in the eigenvalue condition (Moss 1989, D'Eath and Esposito 1991a-1991c, Esposito 1991, our chapters seven, eight and nine);

(3) *Direct* technique described in Barvinsky et al. 1992a-b. This seems to generalize the *direct* techniques appearing in our monograph;

(4) Use of charge layers or dipole distributions on the plane tangent to the boundary (Kennedy 1978-1979);

(5) *Indirect* geometrical technique expressing the result in terms of the geometrical objects of the problem (Melmed 1988, Moss 1989, Moss and Dowker 1989, Moss and Poletti 1990a-b, Poletti 1990). This is very important in understanding the structure of the one-loop counterterms, and the values of their coefficients (D'Eath 1986);

(6) General technique presented in Branson and Gilkey 1990;

(7) Generalization of De Witt's ansatz for the heat kernel for manifolds with boundary (McAvity and Osborne 1991a-b);

(8) Numerical method in Gusynin 1989;

(9) Numerical method in Cognola et al. 1990.

REFERENCES

Abramowitz M. and Stegun I. A. (1964) *Handbook of Mathematical Functions* (New York: Dover).

Aguirre-Dabán E. and Gutiérrez-López M. (1989) *Gen. Rel. Grav.* **21**, 45.

Ahlfors L. V. (1966) *Complex Analysis* (New York: Mc Graw-Hill).

Allen B. (1983a) *Vacuum Energy and General Relativity*, Ph.D. Thesis, University of Cambridge.

Allen B. (1983b) *Nucl. Phys.* **B 226**, 228.

Allen B. and Davis S. (1983) *Phys. Lett.* **B 124**, 353.

Amsterdamski P. (1985) *Phys. Rev.* **D 31**, 3073.

Ashtekar A. (1987) *Asymptotic Quantization* (Napoli: Bibliopolis).

Ashtekar A. (1988) *New Perspectives in Canonical Gravity* (Napoli: Bibliopolis).

Ashtekar A. , Husain V. , Rovelli C. , Samuel J. and Smolin L. (1989) *Class. Quantum Grav.* **6**, L185.

Ashtekar A. (1991) *Lectures on Nonperturbative Canonical Gravity* (Singapore: World Scientific).

Atiyah M. F. , Patodi V. K. and Singer I. M. (1975) *Math. Proc. Camb. Phil. Soc.* **77**, 43.

Atiyah M. F. (1988a) *Collected Works*, Vol. **4** (Oxford: Clarendon Press).

Atiyah M. F. (1988b) *Collected Works*, Vol. **5** (Oxford: Clarendon Press).

Awada M. A. , Gibbons G. W. and Shaw W. T. (1986) *Ann. Phys.* **171**, 52.

Bailey T. N. and Baston R. J. (1990) *Twistors in Mathematics and Physics* (Cambridge: Cambridge University Press).

Barth N. H. and Christensen S. M. (1983) *Phys. Rev.* **D 28**, 1876.

Barth W. , Peters C. and van de Ven A. (1984) *Compact Complex Surfaces* (Berlin: Springer-Verlag).

Barvinsky A. O. , Kamenshchik A. Y. , Karmazin I. P. and Mishakov I. V. (1992a) *Class. Quantum Grav.* **9**, L27.

Barvinsky A. O. , Kamenshchik A. Y. and Karmazin I. P. (1992b) *Ann. Phys.* **219**, 201.

Barvinsky A. O. (1993) *Phys. Rep.* **230**, 237.

Beem J. K. and Ehrlich P. E. (1981) *Global Lorentzian Geometry* (New York: Dekker).

Berry M. V. and Mondragon R. J. (1987) *Proc. Roy. Soc. London* **A 412**, 53.

Bimonte G. (1988) *BRST Quantization Schemes and Gauging of the BRST Symmetry*, Degree Thesis, University of Napoli.

Birrell N. D. and Davies P. C. W. (1982) *Quantum Fields in Curved Space* (Cambridge: Cambridge University Press).

Bonora L. and Cotta-Ramusino P. (1983) *Commun. Math. Phys.* **87**, 589.

Bosshard B. (1976) *Commun. Math. Phys.* **46**, 263.

Boulware D. (1984) in *Quantum Theory of Gravity*, ed. S. M. Christensen (Bristol: Adam Hilger) p 267.

Brandenberger R. (1985) *Rev. Mod. Phys.* **57**, 1.

Branson T. P. and Gilkey P. B. (1990) *Commun. in Partial Differential Equations* **15**, 245.

Breitenlohner P. and Freedman D. Z. (1982) *Ann. Phys.* **144**, 249.

Budinich P. and Trautman A. (1987) *J. Geom. Phys.* **4**, 361.

Budinich P. and Trautman A. (1988) *The Spinorial Chessboard* (Berlin: Springer-Verlag).

Burges C. J. C. , Freedman D. Z. , Davis S. and Gibbons G. W. (1986) *Ann. Phys.* **167**, 285.

Calabi E. (1954) *Proc. Intern. Congr. Math.* **2**, 206.

Carmeli M. (1982) *Classical Fields: General Relativity and Gauge Theory* (New York: John Wiley and Sons).

Carter B. (1971) *Gen. Rel. Grav.* **1**, 349.

Carter B. (1979) in *Recent Developments in Gravitation, Cargèse 1978*, eds. M. Lévy and S. Deser (New York: Plenum Press) p 41.

Castellani L. , van Nieuwenhuizen P. and Pilati M. (1982) *Phys. Rev.* **D 26**, 352.

Chavel I. (1984) *Eigenvalues in Riemannian Geometry* (New York: Academic Press).

Chern S. S. (1979) *Complex Manifolds without Potential Theory* (Berlin: Springer-Verlag).

Clarke C. J. S. (1975) *Commun. Math. Phys.* **41**, 65.

Clarke C. J. S. (1976) *Commun. Math. Phys.* **49**, 17.

Clarke C. J. S. and Schmidt B. G. (1977) *Gen. Rel. Grav.* **8**, 129.

Clarke C. J. S. and de Felice F. (1984) *Gen. Rel. Grav.* **16**, 139.

Cognola G. , Vanzo L. and Zerbini S. (1990) *Phys. Lett.* **B 241**, 381.

Collins J. C. (1984) *Renormalization* (Cambridge: Cambridge University Press).

Corichi A. (1992) *J. Math. Phys.* **33**, 4066.

Costa M. E. V. , Girotti H. D. and Simoes T. J. M. (1985) *Phys. Rev.* **D 32**, 405.

Dabrowski L. (1988) *Group Actions on Spinors* (Napoli: Bibliopolis).

Davies E. B. (1989) *Heat Kernels and Spectral Theory* (Cambridge: Cambridge University Press).

Davis S. (1985) *Supersymmetry in anti-de Sitter Space*, Ph.D. Thesis, University of Cambridge.

Dayi O. F. (1989) *Phys. Lett.* **B 228**, 435.

D'Eath P. D. (1984) *Phys. Rev.* **D 29**, 2199.

D'Eath P. D. (1986) *Nucl. Phys.* **B 269**, 665.

D'Eath P. D. and Halliwell J. J. (1987) *Phys. Rev.* **D 35**, 1100.

D'Eath P. D. and Esposito G. (1991a) *Phys. Rev.* **D 43**, 3234.

D'Eath P. D. and Esposito G. (1991b) *Phys. Rev.* **D 44**, 1713.

D'Eath P. D. and Esposito G. (1991c) in *Proceedings of the 9th Italian Conference on General Relativity and Gravitational Physics*, eds. R. Cianci, R. de Ritis, M. Francaviglia, G. Marmo, C. Rubano and P. Scudellaro (Singapore: World Scientific) p 644.

Demianski M. , de Ritis R. , Platania G. , Scudellaro P. and Stornaiolo C. (1987) *Phys. Rev.* **D 35**, 1181.

De Witt B. S. (1964) *Phys. Rev. Lett.* **13**, 114.

De Witt B. S. (1967a) *Phys. Rev.* **160**, 1113.

De Witt B. S. (1967b) *Phys. Rev.* **162**, 1195.

De Witt B. S. (1967c) *Phys. Rev.* **162**, 1239.

De Witt B. S. and Anderson A. (1986) *Found. Phys.* **16**, 91.

De Witt-Morette C. (1987) in *Functional Integration with Emphasis on the Feynman Integral, Supp. Rend. Circ. Mat. Palermo*, Serie II, Vol. **17**, 211.

Dieckmann J. (1988) *Gen. Rel. Grav.* **20**, 859.

Dirac P. A. M. (1964) *Lectures on Quantum Mechanics*, Belfer Graduate School of Science (New York: Yeshiva University).

Di Stefano R. and Rauch R. T. (1982) *Phys. Rev.* **D 26**, 1242.

Dodson C. T. J. (1978) *Intern. J. Theor. Phys.* **17**, 389.

Dolan B. P. (1989) *Phys. Lett.* **B 233**, 89.

Duff M. J. (1982) in *Supergravity 1981*, eds. S. Ferrara and J. G. Taylor (Cambridge: Cambridge University Press) p 197.

Dunne G. V. , Jackiw R. and Trugenberger C. A. (1989) *Ann. Phys.* **194**, 197.

Dyson F. J. (1990) *Am. J. Phys.* **58**, 209.

Ellicott P. , Kunstatter G. and Toms D. J. (1989) *Mod. Phys. Lett.* **A 4**, 2397.

Ellis G. F. R. (1971) in *General Relativity and Cosmology, Enrico Fermi School, Course XLVII*, ed. B. K. Sachs (New York: Academic Press) p 104.

Ellis G. F. R. and Schmidt B. G. (1977) *Gen. Rel. Grav.* **8**, 915.

Esposito F. P. and Witten L. (1977) *Asymptotic Structure of Space-Time* (New York: Plenum Press).

Esposito G. (1988) in *Gauge Theory and the Early Universe*, eds. P. Galeotti and D. N. Schramm, NATO ASI Series C, Vol. **248** (Dordrecht: Kluwer Academic Publishers) p 327.

Esposito G. and Platania G. (1988) *Class. Quantum Grav.* **5**, 937.

Esposito G. (1989a) *Boundary-Value Problems in Quantum Cosmology*, Knight Prize Essay, University of Cambridge.

Esposito G. (1989b) *Nuovo Cimento* **B 104**, 199; *Nuovo Cimento* **B 106**, 1315.

Esposito G. (1990) *Nuovo Cimento* **B 105**, 75; *Nuovo Cimento* **B 106**, 1315.

Esposito G. (1991) *Perturbative Properties of Quantum Cosmology*, Ph.D. Thesis, University of Cambridge.

Esposito G. (1992) *Fortschr. Phys.* **40**, 1.

Feldman J. S. , Hurd T. R. , Rosen L. and Wright J. D. (1988) *QED: A Proof of Renormalizability* (Berlin: Springer-Verlag).

Fennelly A. J. , Krisch J. P. , Ray J. R. and Smalley L. L. (1991) *J. Math. Phys.* **32**, 485.

Fisher A. E. (1970) in *Relativity*, eds. M. Carmeli, S. I. Fickler and L. Witten (New York: Plenum Press) p 303.

Flaherty F. J. (1984) *Asymptotic Behaviour of Mass and Space-Time Geometry* (Berlin: Springer-Verlag).

Frampton P. H. (1987) *Gauge Field Theories* (Reading, Massachusetts: Benjamin Cummings).

Francaviglia M. (1975) *Global Methods of Differential Geometry in General Relativity*, Degree Thesis, University of Torino.

Francaviglia M. (1988) *Elements of Differential and Riemannian Geometry* (Napoli: Bibliopolis).

Fuchs H. , Kasper V. , Liebscher D. E. , Muller V. and Schmidt H. J. (1988) *Fortschr. Phys.* **36**, 427.

Gallot S. , Hulin D. and Lafontaine J. (1987) *Riemannian Geometry* (Berlin: Springer-Verlag).

Galvao C. A. P. and Teitelboim C. (1980) *J. Math. Phys.* **21**, 1863.

Garcia de Andrade L. C. (1990a) *Found. Phys.* **20**, 403.

Garcia de Andrade L. C. (1990b) *Intern. J. Theor. Phys.* **29**, 997.

Gelfand I. M. and Shilov G. E. (1964) *Generalized Functions*, Vol. **1** (New York: Academic Press).

Gerlach U. H. and Sengupta U. K. (1978) *Phys. Rev.* **D 18**, 1773.

Geroch R. P. (1966) *Phys. Rev. Lett.* **17**, 445.

Geroch R. P. (1967) *Singularities in the Space-Time of General Relativity*, Ph.D. Thesis, University of Princeton.

Geroch R. P. (1968a) *J. Math. Phys.* **9**, 1739.

Geroch R. P. (1968b) *Ann. Phys.* **48**, 526.

Geroch R. P. (1970) *J. Math. Phys.* **11, 437**

Geroch R. P. , Kronheimer E. H. and Penrose R. (1972) *Proc. Roy. Soc. London* **A 327**, 545.

Geroch R. P. and Horowitz G. T. (1979) in *General Relativity, an Einstein Centenary Survey*, eds. S. W. Hawking and W. Israel (Cambridge: Cambridge University Press) p 212.

Gibbons G. W. and Hawking S. W. (1977) *Phys. Rev.* **D 15**, 2752.

Gibbons G. W. , Hawking S. W. and Perry M. J. (1978) *Nucl. Phys.* **B 138**, 141.

Gibbons G. W. (1978) in *Differential Geometrical Methods in Mathematical Physics II*, eds. K. Bleuler, H. Petry and A. Reetz (Berlin: Springer-Verlag) p 513.

Gibbons G. W. and Hawking S. W. (1979) *Commun. Math. Phys.* **66**, 291.

Gibbons G. W. and Pope C. N. (1979) *Commun. Math. Phys.* **66**, 267.

Gibbons G. W. (1983) in *Proceedings of the Ninth International Conference on General Relativity and Gravitation*, ed. E. Schmutzer (Cambridge: Cambridge University Press) p 165.

Gibbons G. W. and Grishchuk L. P. (1989) *Nucl. Phys.* **B 313**, 736.

Gibbons G. W. and Hawking S. W. (1991) *Euclidean Quantum Gravity* (Singapore: World Scientific).

Gilkey P. B. (1975) *The Index Theorem and the Heat Equation* (Boston: Publish or Perish).

Gitman D. M. and Tyutin I. V. (1990) *Quantization of Fields with Constraints* (Berlin: Springer-Verlag).

Glimme J. and Jaffe A. (1987) *Quantum Physics, A Functional Integral Point of View* (Berlin: Springer-Verlag).

Gotay M. J. and Demaret J. (1983) *Phys. Rev.* **D 28**, 2402.

Gotzes S. and Hirshfeld A. C. (1990) *Ann. Phys.* **203**, 410.

Govaerts J. and Troost W. (1991) *Class. Quantum Grav.* **8**, 1723.

Govaerts J. (1991) *Hamiltonian Quantization and Constrained Dynamics*, Leuven Notes in Mathematical and Theoretical Physics, Series B, Volume 4 (Leuven: Leuven University Press).

Gradshteyn I. S. and Ryzhik I. M. (1965) *Table of Integrals, Series and Products* (New York: Academic Press).

Green M. B. , Schwarz J. H. and Witten E. (1987) *Superstring Theory* (Cambridge: Cambridge University Press).

Greene R. (1987) in *Differential Geometry*, ed. V. L. Hansen (Berlin: Springer-Verlag) p 228.

Greiner P. (1971) *Archs. Ration. Mech. Analysis* **41**, 163.

Gusynin V. P. (1989) *Phys. Lett.* **B 225**, 233.

Guven J. and Ryan Jr. M. P. (1992) *Phys. Rev.* **D 45**, 3559.

Haggman B. C. , Horndeski G. W. and Mess G. (1980) *J. Math. Phys.* **21**, 2412.

Halliwell J. J. and Hawking S. W. (1985) *Phys. Rev.* **D 31**, 1777.

Halliwell J. J. (1987) *Phys. Rev.* **D 36**, 3626.

Halliwell J. J. (1988) *Phys. Rev.* **D 38**, 2468.

Halliwell J. J. and Hartle J. B. (1991) *Phys. Rev.* **D 43**, 1170.

Hanson A. , Regge T. and Teitelboim C. (1976) *Constrained Hamiltonian Systems*, Contributi del Centro Linceo Interdisciplinare di Scienze Matematiche e loro Applicazioni, n. 22 (Roma: Accademia Nazionale dei Lincei).

Hartle J. B. and Hawking S. W. (1983) *Phys. Rev.* **D 28**, 2960.

Hartle J. B. and Schleich K. (1987) in *Quantum Field Theory and Quantum Statistics: Essays in Honour of the Sixtieth Birthday of E. S. Fradkin*, Vol. 2, eds. I. A. Batalin, G. A. Vilkovisky and C. J. Isham (Bristol: Adam Hilger) p 67.

Hawking S. W. (1966a) *Proc. Roy. Soc. London* **A 294**, 511.

Hawking S. W. (1966b) *Proc. Roy. Soc. London* **A 295**, 490.

Hawking S. W. (1967) *Proc. Roy. Soc. London* **A 300**, 187.

Hawking S. W. and Penrose R. (1970) *Proc. Roy. Soc. London* **A 314**, 529.

Hawking S. W. (1971) *Gen. Rel. Grav.* **1**, 393.

Hawking S. W. and Ellis G. F. R. (1973) *The Large-Scale Structure of Space-Time* (Cambridge: Cambridge University Press).

Hawking S. W. , King A. R. and Mc Carthy P. J. (1976) *J. Math. Phys.* **17**, 174.

Hawking S. W. (1977) *Commun. Math. Phys.* **55**, 133.

Hawking S. W. and Pope C. N. (1978) *Nucl. Phys.* **B 146**, 381.

Hawking S. W. (1979a) in *Recent Developments in Gravitation, Cargèse 1978*, eds. M. Lévy and S. Deser (New York: Plenum Press) p 145.

Hawking S. W. (1979b) in *General Relativity, an Einstein Centenary Survey*, eds. S. W. Hawking and W. Israel (Cambridge: Cambridge University Press) p 746.

Hawking S. W. (1982) in *Pontificiae Academiae Scientiarum Scripta Varia* **48**, 563.

Hawking S. W. (1983) *Phys. Lett.* **B 126**, 175.

Hawking S. W. (1984a) in *Relativity, Groups and Topology II*, Les Houches 1983, Session XL, eds. B. S. De Witt and R. Stora (Amsterdam: North Holland) p 333.

Hawking S. W. (1984b) *Nucl. Phys.* **B 239**, 257.

Hawking S. W. and Luttrell J. C. (1984a) *Phys. Lett.* **B 143**, 83.

Hawking S. W. and Luttrell J. C. (1984b) *Nucl. Phys.* **B 247**, 250.

Hawking S. W. (1985) *Phys. Rev.* **D 32**, 2489.

Hawking S. W. (1987) in *300 Years of Gravitation*, eds. S. W. Hawking and W. Israel (Cambridge: Cambridge University Press) p 631.

Hehl F. W. , von der Heyde P. and Kerlick G. D. (1974) *Phys. Rev.* **D 10**, 1066.

Hehl F. W. , von der Heyde P. , Kerlick G. D. and Nester J. M. (1976) *Rev. Mod. Phys.* **48**, 393.

Hehl F. W. , Nitsch J. and von der Heyde P. (1980) in *General Relativity and Gravitation*, Vol. 1, ed. A. Held (New York: Plenum Press) p 329.

Henneaux M. (1985) *Phys. Rep.* **126**, 1.

Henneaux M. and Teitelboim C. (1992) *Quantization of Gauge Systems* (Princeton: Princeton University Press).

Higgs P. W. (1958) *Phys. Rev. Lett.* **1**, 373.

Higgs P. W. (1959) *Phys. Rev. Lett.* **3**, 66.

Hitchin N. J. (1984) in *Global Riemannian Geometry*, eds. T. J. Willmore and N. J. Hitchin (Chicester: Ellis Horwood) p 115.

Hojman S. A. and Shepley L. C. (1991) *J. Math. Phys.* **32**, 142.

Horowitz G. T. (1985) *Phys. Rev.* **D 31**, 1169.

Horowitz G. T. and Steif A. R. (1990a) *Phys. Rev. Lett.* **64**, 260.

Horowitz G. T. and Steif A. R. (1990b) *Phys. Rev.* **D 42**, 1950.

Horowitz G. T. and Steif A. R. (1991) *Phys. Lett.* **B 258**, 91.

Huggett S. A. and Tod K. P. (1985) *An Introduction to Twistor Theory* (Cambridge: Cambridge University Press).

Hughes D. I. (1990) *Supersymmetric Quantum Cosmology*, Ph.D. Thesis, University of Cambridge.

Husain V. (1987) *Class. Quantum Grav.* **4**, 1587.

Isenberg J. and Nester J. M. (1980) in *General Relativity and Gravitation*, Vol. 1, ed. A. Held (New York: Plenum Press) p 23.

Isham C. J. (1984) in *Relativity, Groups and Topology II*, Les Houches 1983, Session XL, eds. B. S. De Witt and R. Stora (Amsterdam: North Holland) p 1059.

Isham C. J. and Kakas A. C. (1984a) *Class. Quantum Grav.* **1**, 621.

Isham C. J. and Kakas A. C. (1984b) *Class. Quantum Grav.* **1**, 633.

Isham C. J. and Kuchar K. V. (1985a) *Ann. Phys.* **164**, 288.

Isham C. J. and Kuchar K. V. (1985b) *Ann. Phys.* **164**, 316.

Isham C. J. , Pope C. N. and Warner N. P. (1988) *Class. Quantum Grav.* **5**, 1297.

Isham C. J. (1989) *Class. Quantum Grav.* **6**, 1509.

Isham C. J. , Kubyshin Y. and Renteln P. (1990) *Class. Quantum Grav.* **7**, 1053.

Itzykson C. and Zuber J. B. (1985) *Quantum Field Theory* (Singapore: McGraw-Hill Book Co).

Ivić A. (1985) *The Riemann Zeta-Function* (New York: John Wiley and Sons).

Jacobson T. (1988) *Class. Quantum Grav.* **5**, 923.

Jeffreys H. and Jeffreys B. S. (1946) *Methods of Mathematical Physics* (Cambridge: Cambridge University Press).

Johnson R. A. (1977) *J. Math. Phys.* **18**, 898.

Joshi P. S. (1989) *Gen. Rel. Grav.* **21**, 1227.

Kamenshchik A. Y. and Mishakov I. V. (1992) *Int. J. Mod. Phys.* **A 7**, 3713.

Kamenshchik A. Y. and Mishakov I. V. (1993) *Phys. Rev.* **D 47**, 1380.

Kennedy G. (1978) *J. Phys.* **A 11**, L173.

Kennedy G. (1979) *Some Finite-Temperature Quantum-Field-Theory Calculations in Manifolds with Boundaries*, Ph.D. Thesis, University of Manchester.

Kheyfets A. , La Fave N. J. and Miller W. A. (1989) *Class. Quantum Grav.* **6**, 659.

Kheyfets A. , La Fave N. J. and Miller W. A. (1990a) *Phys. Rev.* **D 41**, 3628.

Kheyfets A. , La Fave N. J. and Miller W. A. (1990b) *Phys. Rev.* **D 41**, 3637.

Kodaira K. (1986) *Complex Manifolds and Deformation of Complex Structures* (Berlin: Springer-Verlag).

Kriele M. (1989) *Class. Quantum Grav.* **6**, 1607.

Kriele M. (1990) *Proc. Roy. Soc. London* **A 431**, 451.

Kulshreshtha U. (1992) *J. Math. Phys.* **33**, 633.

Kunstatter G. (1992) *Class. Quantum Grav.* **9**, 1469.

Kupsch J. and Thacker W. D. (1990) *Fortschr. Phys.* **38**, 35.

Laenen E. and van Nieuwenhuizen P. (1991) *Ann. Phys.* **207**, 77.

Laflamme R. and Shellard E. P. S. (1987) *Phys. Rev.* **D 35**, 2315.

Laflamme R. (1988) *Time and Quantum Cosmology*, Ph.D. Thesis, University of Cambridge.

Le Brun C. R. (1988) *Commun. Math. Phys.* **118**, 591.

Lee C. W. (1983) *Gen. Rel. Grav.* **15**, 21.

Leray J. (1952) *Hyperbolic Differential Equations*, lecture notes, Institute for Advanced Study (Princeton, N.J.).

Levichev A. (1989) *Gen. Rel. Grav.* **21**, 1027.

Lewandowski J. (1991) *Class. Quantum Grav.* **8**, L11.

Lichnerowicz A. (1961) *Collège de France, Cours 1960-61.*

Lightman P. A. , Press W. H. , Price R. H. and Teukolsky S. A. (1975) *Problem Book in Relativity and Gravitation* (Princeton: Princeton University Press).

Linde A. (1984) *Rep. Prog. Phys.* **47**, 925.

Lord E. A. and Goswami P. (1988) *J. Math. Phys.* **29**, 258.

Louko J. (1987) *Phys. Rev.* **D 35**, 3760.

Louko J. (1988a) *Ann. Phys.* **181**, 318.

Louko J. (1988b) *Phys. Rev.* **D 38**, 478.

Louko J. (1991a) *Class. Quantum Grav.* **8**, L37.

Louko J. (1991b) *Class. Quantum Grav.* **8**, 1947.

Louko J. and Ruback P. J. (1991) *Class. Quantum Grav.* **8**, 91.

Luckock H. C. and Moss I. G. (1989) *Class. Quantum Grav.* **6**, 1993.

Luckock H. C. (1991) *J. Math. Phys.* **32**, 1755.

Luke Y. L. (1962) *Integrals of Bessel Functions* (New York: McGraw-Hill Book Co).

MacCallum M. A. H. (1975) in *Quantum Gravity: an Oxford Symposium*, eds. C. J. Isham, R. Penrose and D. W. Sciama (Oxford: Clarendon Press) p 174.

Manin Y. I. (1988) *Gauge Field Theory and Complex Geometry* (Berlin: Springer-Verlag).

Marmo G. , Mukunda N. and Samuel J. (1983) *Riv. Nuovo Cimento* **6**, 1.

McAvity D. M. and Osborne H. (1991a) *Class. Quantum Grav.* **8**, 603.

McAvity D. M. and Osborne H. (1991b) *Class. Quantum Grav.* **8**, 1445.

McMullan D. (1991) *Nucl. Phys.* **B 363**, 451.

Mehta M. R. (1991) *Mod. Phys. Lett.* **A6**, 2811.

Melmed J. (1988) *J. Phys.* **A 21**, L1131.

Menotti P. and Pelissetto A. (1987) *Phys. Rev.* **D 35**, 1194.

Milnor J. W. (1962) *Morse Theory* (Princeton: Princeton University Press).

Milnor J. W. and Stasheff J. D. (1974) *Characteristic Classes* (Princeton: Princeton University Press).

Misner C. W. , Thorne K. S. and Wheeler J. A. (1973) *Gravitation* (S. Francisco: Freeman).

Moss I. G. (1989) *Class. Quantum Grav.* **6**, 759.

Moss I. G. and Dowker J. S. (1989) *Phys. Lett.* **B 229**, 261.

Moss I. G. and Poletti S. (1990a) *Nucl. Phys.* **B 341**, 155.

Moss I. G. and Poletti S. (1990b) *Phys. Lett.* **B 245**, 355.

Nash C. and Sen S. (1983) *Topology and Geometry for Physicists* (London: Academic Press).

Nester J. M. and Isenberg J. (1977) *Phys. Rev.* **D 15**, 2078.

Nester J. M. (1983) in *An Introduction to Kaluza-Klein Theories*, ed. H. C. Lee (Singapore: World Scientific) p 83.

Newman E. T. and Tod K. P. (1980) in *General Relativity and Gravitation*, Vol. 2, ed. A. Held (New York: Plenum Press) p 1.

Newman R. P. A. C. (1984a) *Gen. Rel. Grav.* **16**, 1163.

Newman R. P. A. C. (1984b) *Gen. Rel. Grav.* **16**, 1177.

Newman R. P. A. C. and Clarke C. J. S. (1987) *Class. Quantum Grav.* **4**, 53.

Newman R. P. A. C. (1989a) *Gen. Rel. Grav.* **21**, 981.

Newman R. P. A. C. (1989b) *Commun. Math. Phys.* **123**, 17.

Olver F. W. J. (1954) *Philos. Trans. Roy. Soc. London*, Ser. **A 247**, 328.

O'Neill B. (1983) *Semi-Riemannian Geometry with Applications to Relativity* (New York: Academic Press).

Padmanabhan T. (1984) *Class. Quantum Grav.* **1**, 149.

Page D. (1985) *Phys. Rev.* **D 32**, 2496.

Page D. (1990) *Class. Quantum Grav.* **7**, 1841.

Pascual P. , Taron J. and Tarrach R. (1989) *Phys. Rev.* **D 39**, 2993.

Pekonen O. (1987) *J. Geom. Phys.* **4**, 493.

Penrose R. (1968) in *Battelle Rencontres*, eds. C. M. De Witt and J. A. Wheeler (New York: Benjamin) p 121.

Penrose R. (1974) in *Group Theory in Nonlinear Problems*, ed. A. O. Barut (Dordrecht: Reidel Publishing Company) p 1.

Penrose R. (1975) in *Quantum Gravity: an Oxford Symposium*, eds. C. J. Isham, R. Penrose and D. W. Sciama (Oxford: Clarendon Press) p 268.

Penrose R. (1976) *Gen. Rel. Grav.* **7**, 31-52; 171-176.

Penrose R. (1983) *Techniques of Differential Topology in Relativity* (Bristol: Society for Industrial and Applied Mathematics).

Penrose R. (1986) in *General Relativity and Gravitation*, ed. M. A. H. MacCallum (Cambridge: Cambridge University Press) p 158.

Penrose R. and Rindler W. (1984, 1986) *Spinors and Space-Time*, Vol. **1** and **2** (Cambridge: Cambridge University Press).

Penrose R. (1987) in *Gravitation and Geometry*, eds. W. Rindler and A. Trautman (Napoli: Bibliopolis) p 341.

Perjés Z. (1989) *Proc. Centre Math. Anal. Austral. Nat. Univ.* **19**, 207.

Persides S. (1979) *J. Math. Phys.* **20**, 1731.

Persides S. (1980a) *J. Math. Phys.* **21**, 135.

Persides S. (1980b) *J. Math. Phys.* **21**, 142.

Pilati M. (1978) *Nucl. Phys.* **B 132**, 138.

Poénaru V. and Tanasi C. (1986) *Supp. Rend. Circ. Mat. Palermo*, Serie II, n. **13**, 7.

Poletti S. (1990) *Phys. Lett.* **B 249**, 249.

Pope C. N. (1980) *Instantons in Quantum Gravity*, Ph.D. Thesis, University of Cambridge.

Pope C. N. (1981) in *Quantum Gravity, a second Oxford Symposium*, eds. C. J. Isham, R. Penrose and D. W. Sciama (Oxford: Clarendon Press) p 377.

Pullin J. (1989) *Annalen der Physik* **46**, 167.

Rácz I. (1988) *Gen. Rel. Grav.* **20**, 893.

Raychaudhuri A. K. (1975) *Phys. Rev.* **D 12**, 952.

Raychaudhuri A. K. (1979) *Theoretical Cosmology* (Oxford: Clarendon Press).

Reed M. and Simon B. (1972) *Methods of Modern Mathematical Physics*, Vol. **1** (New York: Academic Press).

Reed M. and Simon B. (1975) *Methods of Modern Mathematical Physics*, Vol. **2** (New York: Academic Press).

Rocek M. and Williams R. (1982) in *Quantum Structure of Space and Time*, eds. M. J. Duff and C. J. Isham (Cambridge: Cambridge University Press) p 105.

Romano J. D. and Tate R. S. (1989) *Class. Quantum Grav.* **6**, 1487.

Rovelli C. (1991) *Class. Quantum Grav.* **8**, 1613.

Rozansky L. (1991) *Phys. Lett.* **B 254**, 75.

Schleich K. (1985) *Phys. Rev.* **D 32**, 1889.

Schleich K. (1987) *Phys. Rev.* **D 36**, 2342.

Schleich K. (1990) *Class. Quantum Grav.* **7**, 1529.

Schmidt B. G. (1971) *Gen. Rel. Grav.* **1**, 269.

Schmidt B. G. and Stewart J. M. (1988) *Proc. Roy. Soc. London* **A 420**, 355.

Schröder M. (1989) *Rep. Math. Phys.* **27**, 259.

Schulman L. S. (1981) *Techniques and Applications of Path Integration* (New York: John Wiley and Sons).

Seeley R. T. (1967) *Amer. Math. Soc. Proc. Symp. Pure Math.* **10**, 288.

Smith G. J. and Bergmann P. G. (1986) *Phys. Rev.* **D 33**, 3570.

Smrz P. K. (1987) *J. Math. Phys.* **28**, 2824.

Stelle K. S. (1977) *Phys. Rev.* **D 16**, 953.

Stewart J. M. and Hájicek P. (1973) *Nature Physical Science* **244**, 96.

Stewart J. M. (1990) *Advanced General Relativity* (Cambridge: Cambridge University Press).

Stewartson K. and Waechter R. T. (1971) *Proc. Camb. Philos. Soc.* **69**, 353.

Streater R. F. and Wightman A. S. (1964) *P. C. T. , Spin and Statistics, and All That* (New York: Benjamin).

Suen W. M. and Young K. (1989) *Phys. Rev.* **D 39**, 2201.

Sundermeyer K. (1982) *Constrained Dynamics* (Berlin: Springer-Verlag).

Szabados L. B. (1988) *Class. Quantum Grav.* **5**, 121.

Szabados L. B. (1989) *Class. Quantum Grav.* **6**, 77.

Tafel J. (1973) *Phys. Lett.* **45 A**, 341.

Tarski J. (1980) in *Functional Integration*, eds. J. P. Antoine and E. Tirapegui (New York: Plenum Press) p 143.

Teitelboim C. (1983) *Phys. Rev.* **D 28**, 297.

Tipler F. (1977) *Ann. Phys.* **108**, 1.

Titchmarsh E. C. (1939) *The Theory of Functions* (London: Oxford University Press).

Topiwala P. N. (1985) *A Twistor Approach to the Einstein Metric on $K3$*, Ph.D. Thesis, University of Michigan.

Topiwala P. N. (1987a) *Invent. Math.* **89**, 425.

Topiwala P. N. (1987b) *Invent. Math.* **89**, 449.

Trautman A. (1973) *Symp. Math.* **12**, 139.

Trautman A. (1980) in *General Relativity and Gravitation*, Vol. 1, ed. A. Held (New York: Plenum Press) p 287.

van Nieuwenhuizen P. (1981) *Phys. Rep.* **68**, 189.

Vilenkin A. (1986) *Phys. Rev.* **D 33**, 3560.

Vilenkin A. (1988) *Phys. Rev.* **D 37**, 888.

Vyas U. D. and Joshi P. S. (1983) *Gen. Rel. Grav.* **15**, 553.

Vyas U. D. and Akolia G. M. (1984) *Gen. Rel. Grav.* **16**, 1045.

Vyas U. D. and Akolia G. M. (1986) *Gen. Rel. Grav.* **18**, 309.

Wald R. M. (1984) *General Relativity* (Chicago: The University of Chicago Press).

Ward R. S. (1980) *Commun. Math. Phys.* **78**, 1.

Ward R. S. and Wells R. O. (1990) *Twistor Geometry and Field Theory* (Cambridge: Cambridge University Press).

Warner N. P. (1982) *Nonperturbative Quantum Gravity*, Ph.D. Thesis, University of Cambridge.

Warr B. J. (1988) *Ann. Phys.* **183**, 1.

Watson G. N. (1966) *A Treatise on the Theory of Bessel Functions* (Cambridge: Cambridge University Press).

Wightman A. S. (1979) in *The Whys of Subnuclear Physics*, ed. A. Zichichi (New York: Plenum Press) p 983.

Wong R. (1989) *Asymptotic Approximations of Integrals* (New York: Academic Press).

Woodhouse N. M. J. (1973) *J. Math. Phys.* **14**, 495.

Woodhouse N. M. J. (1985) *Class. Quantum Grav.* **2**, 257.

Yau S. T. (1978) *Commun. Pure Appl. Math.* **31**, 339.

York J. W. (1972) *Phys. Rev. Lett.* **28**, 1082.

York J. W. (1986) *Found. Phys.* **16**, 249.

Zhi Q. (1992) *Intern. J. Theor. Phys.* **31**, 115.

Springer-Verlag and the Environment

We at Springer-Verlag firmly believe that an international science publisher has a special obligation to the environment, and our corporate policies consistently reflect this conviction.

We also expect our business partners – paper mills, printers, packaging manufacturers, etc. – to commit themselves to using environmentally friendly materials and production processes.

The paper in this book is made from low- or no-chlorine pulp and is acid free, in conformance with international standards for paper permanency.

Lecture Notes in Physics

For information about Vols. 1–384
please contact your bookseller or Springer-Verlag

Vol. 385: A. Treves, G. C. Perola, L. Stella (Eds.), Iron Line Diagnostics in X-Ray Sources. Proceedings, Como, Italy 1990. IX, 312 pages. 1991.

Vol. 386: G. Pétré, A. Sanfeld (Eds.), Capillarity Today. Proceedings, Belgium 1990. XI, 384 pages. 1991.

Vol. 387: Y. Uchida, R. C. Canfield, T. Watanabe, E. Hiei (Eds.), Flare Physics in Solar Activity Maximum 22. Proceedings, 1990. X, 360 pages. 1991.

Vol. 388: D. Gough, J. Toomre (Eds.), Challenges to Theories of the Structure of Moderate-Mass Stars. Proceedings, 1990. VII, 414 pages. 1991.

Vol. 389: J. C. Miller, R. F. Haglund (Eds.), Laser Ablation-Mechanisms and Applications. Proceedings. IX, 362 pages, 1991.

Vol. 390: J. Heidmann, M. J. Klein (Eds.), Bioastronomy - The Search for Extraterrestrial Life. Proceedings, 1990. XVII, 413 pages. 1991.

Vol. 391: A. Zdziarski, M. Sikora (Eds.), Ralativistic Hadrons in Cosmic Compact Objects. Proceedings, 1990. XII, 182 pages. 1991.

Vol. 392: J.-D. Fournier, P.-L. Sulem (Eds.), Large-Scale Structures in Nonlinear Physics. Proceedings. VIII, 353 pages. 1991.

Vol. 393: M. Remoissenet, M.Peyrard (Eds.), Nonlinear Coherent Structures in Physics and Biology. Proceedings. XII, 398 pages. 1991.

Vol. 394: M. R. J. Hoch, R. H. Lemmer (Eds.), Low Temperature Physics. Proceedings. X,374 pages. 1991.

Vol. 395: H. E. Trease, M. J. Fritts, W. P. Crowley (Eds.), Advances in the Free-Lagrange Method. Proceedings, 1990. XI, 327 pages. 1991.

Vol. 396: H. Mitter, H. Gausterer (Eds.), Recent Aspects of Quantum Fields. Proceedings. XIII, 332 pages. 1991.

Vol. 398: T. M. M. Verheggen (Ed.), Numerical Methods for the Simulation of Multi-Phase and Complex Flow. Proceedings, 1990. VI, 153 pages. 1992.

Vol. 399: Z. Švestka, B. V. Jackson, M. E. Machedo (Eds.), Eruptive Solar Flares. Proceedings, 1991. XIV, 409 pages. 1992.

Vol. 400: M. Dienes, M. Month, S. Turner (Eds.), Frontiers of Particle Beams: Intensity Limitations. Proceedings, 1990. IX, 610 pages. 1992.

Vol. 401: U. Heber, C. S. Jeffery (Eds.), The Atmospheres of Early-Type Stars. Proceedings, 1991. XIX, 450 pages. 1992.

Vol. 402: L. Boi, D. Flament, J.-M. Salanskis (Eds.), 1830-1930: A Century of Geometry. VIII, 304 pages. 1992.

Vol. 403: E. Balslev (Ed.), Schrödinger Operators. Proceedings, 1991. VIII, 264 pages. 1992.

Vol. 404: R. Schmidt, H. O. Lutz, R. Dreizler (Eds.), Nuclear Physics Concepts in the Study of Atomic Cluster Physics. Proceedings, 1991. XVIII, 363 pages. 1992.

Vol. 405: W. Hollik, R. Rückl, J. Wess (Eds.), Phenomenological Aspects of Supersymmetry. VII, 329 pages. 1992.

Vol. 406: R. Kayser, T. Schramm, L. Nieser (Eds.), Gravitational Lenses. Proceedings, 1991. XXII, 399 pages. 1992.

Vol. 407: P. L. Smith, W. L. Wiese (Eds.), Atomic and Molecular Data for Space Astronomy. VII, 158 pages. 1992.

Vol. 408: V. J. Martínez, M. Portilla, D. Sàez (Eds.), New Insights into the Universe. Proceedings, 1991. XI, 298 pages. 1992.

Vol. 409: H. Gausterer, C. B. Lang (Eds.), Computational Methods in Field Theory. Proceedings, 1992. XII, 274 pages. 1992.

Vol. 410: J. Ehlers, G. Schäfer (Eds.), Relativistic Gravity Research. Proceedings, VIII, 409 pages. 1992.

Vol. 411: W. Dieter Heiss (Ed.), Chaos and Quantum Chaos. Proceedings, XIV, 330 pages. 1992.

Vol. 412: A. W. Clegg, G. E. Nedoluha (Eds.), Astrophysical Masers. Proceedings, 1992. XX, 480 pages. 1993.

Vol. 413: Aa. Sandqvist, T. P. Ray (Eds.); Central Activity in Galaxies. From Observational Data to Astrophysical Diagnostics. XIII, 235 pages. 1993.

Vol. 414: M. Napolitano, F. Sabetta (Eds.), Thirteenth International Conference on Numerical Methods in Fluid Dynamics. Proceedings, 1992. XIV, 541 pages. 1993.

Vol. 415: L. Garrido (Ed.), Complex Fluids. Proceedings, 1992. XIII, 413 pages. 1993.

Vol. 416: B. Baschek, G. Klare, J. Lequeux (Eds.), New Aspects of Magellanic Cloud Research. Proceedings, 1992. XIII, 494 pages. 1993.

Vol. 417: K. Goeke P. Kroll, H.-R. Petry (Eds.), Quark Cluster Dynamics. Proceedings, 1992. XI, 297 pages. 1993.

Vol. 418: J. van Paradijs, H. M. Maitzen (Eds.), Galactic High-Energy Astrophysics. XIII, 293 pages. 1993.

Vol. 419: K. H. Ploog, L. Tapfer (Eds.), Physics and Technology of Semiconductor Quantum Devices. Proceedings, 1992. VIII, 212 pages. 1993.

Vol. 420: F. Ehlotzky (Ed.), Fundamentals of Quantum Optics III. Proceedings, 1993. XII, 346 pages. 1993.

Vol. 421: H.-J. Röser, K. Meisenheimer (Eds.), Jets in Extragalactic Radio Sources. XX, 301 pages. 1993.

Vol. 422: L. Päivärinta, E. Somersalo (Eds.), Inverse Problems in Mathematical Physics. Proceedings, 1992. XVIII, 256 pages. 1993.

Vol. 423: F. J. Chinea, L. M. González-Romero (Eds.), Rotating Objects and Relativistic Physics. Proceedings, 1992. XII, 304 pages. 1993.

Vol. 424: G. F. Helminck (Ed.), Geometric and Quantum Aspects of Integrable Systems. Proceedings, 1992. IX, 224 pages. 1993.

New Series m: Monographs

Vol. m 1: H. Hora, Plasmas at High Temperature and Density. VIII, 442 pages. 1991.

Vol. m 2: P. Busch, P. J. Lahti, P. Mittelstaedt, The Quantum Theory of Measurement. XIII, 165 pages. 1991.

Vol. m 3: A. Heck, J. M. Perdang (Eds.), Applying Fractals in Astronomy. IX, 210 pages. 1991.

Vol. m 4: R. K. Zeytounian, Mécanique des fluides fondamentale. XV, 615 pages, 1991.

Vol. m 5: R. K. Zeytounian, Meteorological Fluid Dynamics. XI, 346 pages. 1991.

Vol. m 6: N. M. J. Woodhouse, Special Relativity. VIII, 86 pages. 1992.

Vol. m 7: G. Morandi, The Role of Topology in Classical and Quantum Physics. XIII, 239 pages. 1992.

Vol. m 8: D. Funaro, Polynomial Approximation of Differential Equations. X, 305 pages. 1992.

Vol. m 9: M. Namiki, Stochastic Quantization. X, 217 pages. 1992.

Vol. m 10: J. Hoppe, Lectures on Integrable Systems. VII, 111 pages. 1992.

Vol. m 11: A. D. Yaghjian, Relativistic Dynamics of a Charged Sphere. XII, 115 pages. 1992.

Vol. m 12: G. Esposito, Quantum Gravity, Quantum Cosmology and Lorentzian Geometries. Second Corrected and Enlarged Edition. XVIII, 349 pages. 1994.

Vol. m 13: M. Klein, A. Knauf, Classical Planar Scattering by Coulombic Potentials. V, 142 pages. 1992.

Vol. m 14: A. Lerda, Anyons. XI, 138 pages. 1992.

Vol. m 15: N. Peters, B. Rogg (Eds.), Reduced Kinetic Mechanisms for Applications in Combustion Systems. X, 360 pages. 1993.

Vol. m 16: P. Christe, M. Henkel, Introduction to Conformal Invariance and Its Applications to Critical Phenomena. XV, 260 pages. 1993.

Vol. m 17: M. Schoen, Computer Simulation of Condensed Phases in Complex Geometries. X, 136 pages. 1993.